TRANSMISSION LINES FOR DIGITAL AND COMMUNICATION NETWORKS

Universal Curve for Wavelength vs. Frequency

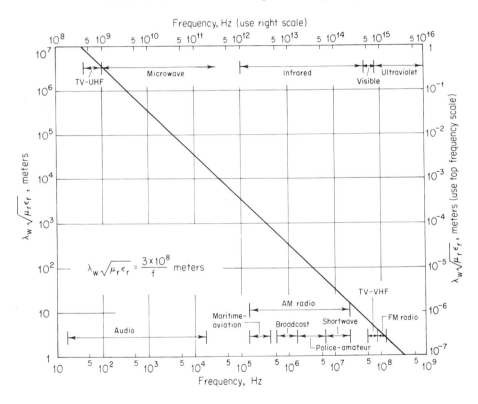

TRANSMISSION LINES FOR DIGITAL AND COMMUNICATION NETWORKS

An Introduction to Transmission Lines, High-frequency
and High-speed Pulse Characteristics and Applications

Richard E. Matick

Member, Technical Staff of Director of Research

IBM Corporation

McGRAW-HILL BOOK COMPANY

New York   *San Francisco*   *Toronto*   *London*   *Sydney*

PREFACE

The rapid advances being made in various fields such as high-speed computers, communications, pulse instrumentation, and the electronics industry in general have resulted in a continued increase in speed and frequency response of electronic equipment. The innovation of the sampling oscilloscope coupled with the increasing emphasis on pulse excitation, pulse-code modulation, and the digital computer as a universal tool has resulted in greater need for an understanding of, as well as for applications of, transmission lines by those who previously did not require such knowledge. Whereas in the past, most circuit and device design employed classical lumped parameter analysis, the present emphasis on speed has required a consideration of the transmission-line properties of various circuits and systems. In addition to requiring a broader understanding of transmission lines, the technological advances have required investigations of either relatively new phenomena and devices such as superconducting transmission lines, anomalous skin effect, and pulse transformers, or have required additional knowledge or insights about well-known phenomena such as propagation velocity, skin effect, and directional couplers.

This book attempts to satisfy these two needs by means of a coherent study of transmission lines, always starting from elementary first principles and proceeding to more complex behavior. Thus, the purpose of this book is two-fold. The first is to provide a means by which those not acquainted with transmission lines can very easily acquire the basic concepts using nothing more than simple ac circuit analysis; it is also intended as a review for those who have some past knowledge, or for use as a reference.

The second purpose of this book is to present various subjects of interest to the novice as well as the expert which are fundamental in nature and are at the same time becoming increasingly important. Many of the subjects cannot be found in books but rather only scattered throughout the literature; in many cases this is quite inadequate. Particular emphasis has been placed on high-frequency and pulse behavior although the analyses are quite general, with the low-frequency cases usually being only a special limiting case.

The content and organization of this book, according to chapters, is as follows:

Chapter 1 derives the basic concepts of transmission lines and wave propagation using only elementary ac theory. The important concepts are treated in separate sections with physical insights into their meaning. A separate section on the more general network equations is also presented.

Chapter 2 is an extension of Chap. 1, using elementary circuit fundamentals to develop the classical differential equations of a transmission line. Also included are all the basic concepts such as traveling and standing waves, reflection coefficients, impedance as a function of wavelength, etc. These first two chapters are written in such a fashion that anyone familiar with ac theory and/or differential equations can master the basic concepts and gain physical insights without much difficulty.

Chapter 3 deals with the general theory of velocity of propagation of waves, a subject which is very old but which is often all too confusing to physicists and engineers. The true meaning and relationships between group and phase velocity as well as the actual velocity of propagation of a wave when dispersion is present has not been adequately discussed in the past, especially from the practical point of view. Most students leave college without this understanding, and since this subject matter is becoming increasingly important, this chapter is intended to fill this void.

Chapter 4 covers the classical theory of skin effect but, in addition, covers a subject usually completely neglected in college, namely, that of anomalous skin effect. The latter phenomena occur generally at low temperatures. With the increasing emphasis on cryogenics, low temperature phenomena, and high speed, this information needs to be put into a coherent form useful to engineers and scientists.

Chapter 5 presents the important characteristics and behavior of transmission lines with pulse excitation. The waveforms obtained with wide and narrow pulses for various unmatched conditions are presented. The relationship between transmission line and circuit analysis is described in detail in addition to other important topics such as skin effect distortion of pulses.

Chapter 6 presents the "nearly ideal" transmission line made of superconductors with an attempt to show its fundamental limitations as well as its improved behavior. First, the important properties of superconductors which must be grasped in order to pave the way for an understanding of superconducting transmission lines are presented and then an analysis of such lines is undertaken. The surface impedance for general and specific cases is derived as a function of frequency and temperature, and the analogy and relationship to normal and anomalous skin effect is included. Pulse characteristics of some typical superconducting lines show the tremendous improvement in behavior over ordinary lines. A brief discussion of the fundamental concepts prerequisite to an understanding of power transmission lines and the extent of work in this field is included.

Chapter 7 undertakes the study of transmission lines that are coupled to one another in various ways. The fundamental concepts and properties of coupled lines are analyzed, and the phenomena of directional coupling are considered in detail for various conditions. This chapter should be of particular interest to those working with high speed circuits or memory arrays where coupling of adjacent lines is of particular concern.

Chapter 8 brings together the concepts, analyses, and techniques for determining transmission-line parameters. Various cases not usually considered in textbooks are presented, and discussions concerning accuracy and high-frequency limitations are included. Useful graphs are given which serve as a valuable reference for the determination of commonly needed parameters.

The author has tried throughout the writing of this book to focus on "usefulness," both from a concept point of view by making analyses start from or relate to first principles, and by showing results and comparisons so as to have a greater impact on the reader. Needless to say, many topics are left untouched or only partially covered. In a book which covers such a wide variety of topics, this is unavoidable. However, the attempt has always been to show how the complex behavior is easily related to the basic properties of transmission lines. References and bibliographies have been given where possible to provide further details or supplementary information.

Richard E. Matick

Peekskill, New York
1968

ACKNOWLEDGMENTS

This book represents knowledge and experience acquired over many years of practice at the IBM Research Division. A significant portion was acquired before any plans for writing a book were formulated and therefore many persons whose identities are lost have contributed in various ways. Those to whom I am indebted whose identities are not lost include D. Eastman, J. Matisoo, and A. Toxen for valuable discussions, J. Griffith for his encouragement and reading of Chap. 3, and E. W. Pugh for his encouragement and his suggestion of an appropriate title for the book. A debt of gratitude is due to the IBM Corporation for the opportunity to acquire the background experience upon which much of this book is based, and also for permission to publish this book.

I am especially indebted to my wife Doris for her patience over the past two and a half years and for her unceasing efforts in typing two drafts in addition to nearly all of the final manuscript. Without this assistance, the writing of this book would have been quite difficult and unduly prolonged.

Richard E. Matick

CONTENTS

SYMBOLS AND CONSTANTS

The RMKS system of units will be used throughout this book unless otherwise specified. Similarly, constants and symbols will be as defined below unless it is specifically stated otherwise. When one symbol has more than one meaning, the correct one will be obvious from the context.

Å	Angstrom = 10^{-10} meters
B	Magnetic flux density (webers/meter^2)
c	Velocity of light in free space $\approx 3 \times 10^8$ meters/second (more accurately, 299792.5 ± 0.1 kilometers/second)
C	Capacitance per unit length (farads/meter)
$\mathbf{D} = \epsilon \mathbf{E}$	Displacement (coulombs/meter^2)
e	2.71828183 - base of natural logarithm
E	Electric field (volts/meter)
f	Frequency (hertz)
$G = 1/R$	Conductance per unit length (mhos/meter)
H	Magnetic field (ampere turns/meter)
i_x or $i(x, t)$	Time and distance varying current (amperes)
I	Constant value (amplitude) of current (amperes)
I_n	Amplitude of current traveling in $+x$ direction on line n
I'_n	Amplitude of current traveling in $-x$ direction on line n
$j = \sqrt{-1}$	Complex number
J	Current density (amperes/meter^2)
J_0	Bessel function
k_L	Inductive coupling coefficient

k_C	Capacitive coupling coefficient
ℓ	Length of transmission line or portion thereof (meters)
ln	Natural logarithm (base e)
log	Logarithm to base 10
L	Inductance per unit length (henries/meter)
L_{ex}	External or circuit inductance
L_{in}	Internal or surface conductor inductance
M	Mutual inductance (henries/meter)
m, n, p, q	Integer constants
$n = c/v$	Index of refraction
ns	Nanoseconds = 10^{-9} second
r	Radial distance (meters)
R	Resistance per unit length (ohms/meter)
$s = j\omega$	Complex operator
S	Conductor separation (meters)
t	Time (seconds)
T	Temperature (degrees Kelvin) or conductor thickness (meters)
T_c	Critical (transition) temperature of a superconductor
T_0	Delay time (one way) of a transmission line
T_R	Rise time of an applied pulse
u	Group velocity
v_x or $v(x, t)$	Time and distance varying voltage
v	Phase velocity
V_n	Amplitude of voltage traveling in positive x direction on line n
V'_n	Amplitude of voltage traveling in negative x direction on line n
W	Conductor width (meters)
x	Distance along a line from origin (meters)
y, z	Distances from origin in cartesian coordinates (meters)
Z_0	Characteristic impedance of transmission line (ohms)

GREEK SYMBOLS

α	Attenuation constant (radians/meter)
β	Phase constant (radians/meter)
$\gamma = \alpha + j\beta$	Propagation constant in main direction of propagation
Γ	Propagation constant transverse to main direction of propagation
δ	Penetration (skin) depth (meters); also used as differential operator
Δ	Incremental operator
ϵ_0	8.85×10^{-12} farads/meter = dielectric constant (permittivity) of free space
$\epsilon = \epsilon' - j\epsilon''$	Complex dielectric constant: ϵ' represents the reactive and ϵ'' the loss component; the latter is usually considered negligible unless otherwise specified
$\epsilon_r = \epsilon'_r - j\epsilon''_r$	Relative dielectric constant
θ	Angular variation (degrees)
λ	Superconducting penetration depth (meters)
λ_w	Wavelength of a single frequency (meters)
Λ	Electron mean free path length (meters or centimeters)
μ_0	$4\pi \times 10^{-7}$ henries/meter = permeability of free space
$\mu = \mu' - j\mu''$	Complex permeability; μ' represents the reactive and μ'' the loss component (latter is negligible unless otherwise specified)
$\mu_r = \mu'_r - j\mu''_r$	Relative permeability
$\mu\mu F$	Micromicrofarads = 10^{-12} farads
μs	Microseconds = 10^{-6} second
ξ	Superconducting coherence length (meters)
π	3.14159265
ρ	Dc resistivity (ohm-meters or ohm-centimeters); also damping factor
ρ_a	Dc resistivity at onset of anomalous effects
ρ_v, ρ_i	Voltage and current reflection coefficients

$\sigma = 1/\rho$	Dc conductivity (mhos/meter or mhos/centimeter)
Σ	Summation notation
τ	Time constant; also transmission coefficient $1 + \rho_v$
ϕ	Magnetic flux (webers)
$\omega = 2\pi f$	Angular velocity (radians/second)

TRANSMISSION LINES FOR DIGITAL AND COMMUNICATION NETWORKS

1 FUNDAMENTAL CONCEPTS OF TRANSMISSION LINES DERIVED FROM AC THEORY

1.1 INTRODUCTION

Almost everyone is familiar with radio and television waves which travel through air, but the concept of waves on an electrical line is not as commonly appreciated. Lights, television sets, radios, and numerous other appliances are used daily without the slightest thought being given to the manner in which the energy is transmitted to these devices through conductors, even though the underlying phenomenon is of fundamental importance. Consequently, the student who is being exposed for the first time to the concept of a transmission line usually finds the idea difficult to comprehend until certain questions have been answered in his own mind, the major question usually being just what is a transmission line and how does it differ from other lines or circuits. The answer is that any two conductors between which voltage is applied can be considered to be a transmission line. In general, there is no clear-cut distinction between the usual concept of a circuit and a transmission line except for extreme cases with usual circuits on one end and transmission lines on the other. Fortunately, many applications fall into one of these extreme cases, but this is occurring less and less as advances are being made in high-speed circuits and systems.

In order to understand the significance of this, let us consider a very simple, but nevertheless quite profound example, namely, the turning on of a light in a home which is supplied with 60 Hz power. When the switch is activated, the light appears instantaneously, for all practical purposes. If we wished to analyze the electrical system within the house itself, we would use simple ac circuit theory and would never consider whether or not the lines might be transmission lines. However, the power company which supplies the electricity is not so fortunate because the generating station may be located a long distance from the house, as is shown in Fig. 1-1. When a switch is turned on at station A, as in Fig. 1-1, for example, we know that the power does not appear instantaneously at station B, located x miles

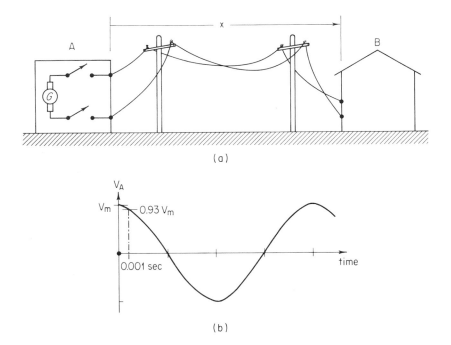

Fig. 1-1. Simplified power distribution system; (b) voltage at power generator.

away. If the cables are suspended in air, then the power will travel at the velocity of light in air, which is about 186,000 miles per second or 3×10^8 meters per second. If A is located 186 miles from B, the power will reach B at a time of 10^{-3} seconds after the switch is closed. Suppose that the generator at station A is supplying 60 Hz voltage and we close the switching at the exact instant that this voltage has reached peak value, as in Fig. 1-1(b) at time $t = 0$. This voltage of V_m travels toward station B at the speed of light and arrives 0.001 second later. In the meantime, the generator at A has continued alternating the voltage so that at the time that V_m arrives at station B, the voltage at A has decreased to 0.93 V_m. Thus, there is now a potential (voltage) difference along the wire itself as a result of the finite time required for the voltage to reach station B. Because of this voltage difference along the line, it is necessary to consider this case as a transmission line. The reason for this is that besides numerous other problems which must be considered, it is necessary to maintain the proper phase relations between various generators that are connected into a power distribution system. For instance, if there were another generator at a third station located a considerable distance away and connected by a similar line in series or parallel with generator A, it would be necessary that the two generators operate in phase; otherwise power would be transferred between the generators, a very undesirable situation.

Returning to the light switch in our homes, we see that the distances involved are so small that we need not worry about phase differences, since the voltage at one end of a given line is essentially the same as that at any other point on the line, exclusive, of course, of any series resistive, inductive, or capacitive voltage drop. If the frequency of the ac power were to be increased, for example, to 100 megacycles per second (typical FM radio frequency), then a voltage difference would exist along the lines in our homes and it would be necessary to consider them as transmission lines. Thus, from this very simple illustration, we can see that whether or not a line is to be considered as a transmission line depends on both the length of the line and the frequency of the applied voltage or, more specifically, it depends on the ratio of the length of line to the wavelength of the applied frequency. If the wavelength is very long compared to the line length, simple circuit analyses are applicable. The wavelength of 60 Hz in air is (from chart on frontispiece) 5×10^6 meters or over 3,100 miles, but it reduces to 3 meters at 100 megacycles per second. Thus, it is apparent that distances within the home are quite small compared to the wavelength at usual power frequencies, but distances between power stations are not. Furthermore, the wavelength of the FM signals is not large but instead is comparable to distances within the home, so that antenna and lead-in wire, for example, must be considered as transmission lines. This is a fundamental idea which will be encountered at various times and will be extended in Chap. 5 to cases where the applied voltage is a single pulse, thereby containing all frequencies.

In this chapter, we will usually be concerned with lines that are either infinitely long or that appear to be so. Thus, a voltage difference will appear on the conductors themselves because of the finite velocity of propagation of energy, as was shown previously, and it is necessary to treat them as transmission lines. In order to consider such lines from the more familiar circuit point of view, all the fundamental concepts of waves on transmission lines will be developed from simple ac theory as applied to circuits.* But in order to use ac circuit analysis, it is first necessary to obtain the equivalent circuit representation of any two-conductor line. This will be done in the next section. Once this is obtained, the transmission-line properties such as characteristic impedance, phase shift, phase velocity, and attenuation are easily obtained in terms of these circuit parameters which can be measured or calculated (see Chap. 9).

---

*For a review of ac theory, see [2] or any source which treats ac circuits.

1.2 EQUIVALENT CIRCUIT OF A SIMPLE TRANSMISSION LINE

In order to analyze a transmission line in terms of ac circuit theory, it is necessary first to obtain the equivalent circuit of the line. Since any passive network can be composed only of combinations of resistive, capacitive, and inductive elements, the final circuit must contain only combinations of these. Let us consider two very long parallel wires which are suspended in air, as in Fig. 1-2, and let us examine a small subsection of the line included between the dotted lines in the figure. One restriction placed on our equivalent circuit is that the length ℓ of each subsection must be much smaller than the wavelength of the applied frequency, so that each subsection can be considered a circuit and the elements within the subsection can be precisely defined. When many subsections are tied together to form an infinitely long line, ac analysis can still be applied to each individual subsection, as will be seen. Another restriction is that the frequency must be sufficiently low and/or the conductor conductivity sufficiently high so that the series losses are small; otherwise, the resistive and inductive elements cannot be evaluated in any simple manner. This is discussed in greater detail in Sec. 4.6, with the

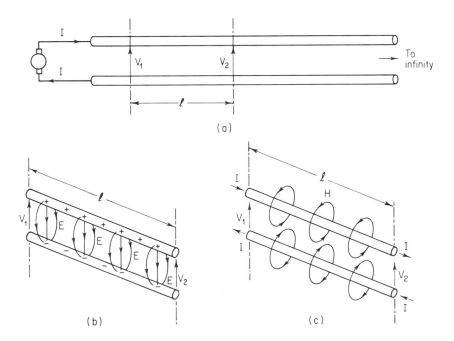

Fig. 1-2. Infinite parallel wire transmission line. (a) Subsection and notation used; (b) electric field and charge within subsection for an applied voltage; (c) magnetic field within subsection for an applied current.

restrictions given by Eqs. (4-90) and (4-91). These restrictions are usually of little concern is most applications and we will assume that all necessary line parameters can easily be evaluated.*

A voltage is applied across the wires such that a current I flows in the top conductor and an equal and opposite current flows in the bottom conductor. This voltage can be supplied from a battery, i.e., dc source, an ac generator, or any other source. We want to determine the equivalent circuit for this small subsection of line. Since the wires have series resistance, the voltage V_1 at the input end of the small subsection will be larger than the voltage V_2 at the output end, as a result of the IR drop through the wire. Thus, the small subsection must have a series resistance component in the equivalent circuit.

If the voltage across the line is not changing with time, then it is apparent that the voltage can be supported only by a static electric field since

$$V = \int E \cdot dl \qquad (1\text{-}1)$$

The presence of an electric field requires that there be free charges of opposite polarity on the two conductors, as in Fig. 1-2(b), since static electric fields can arise only from such free charges as are described by Coulomb's law

$$E = \frac{q}{4\pi\epsilon r^2} \qquad (1\text{-}2)$$

with E in volts per meter, r in meters, q in coulombs, and $\epsilon = 8.85 \times 10^{-12}$ farads per meter (in vacuum).

The free (stored) charge, accompanied by a voltage, represents a capacitor, since $C = q/V$. Thus, the equivalent circuit for the small subsection must contain a capacitive component.

In addition to the static electric field present between the conductors, there will also be a magnetic field or flux as a result of the current flow as given by either the Biot-Savart law or Ampere's law; these laws are, respectively

$$dB = \frac{\mu I dl \times r}{4\pi r^3} \qquad \oint H \cdot dl = I \qquad (1\text{-}3)$$

*A further restriction occurs when the wavelength is comparable to or smaller than the cross-sectional dimensions at which time other modes of propagation can be excited or radiation can occur in open structures. These effects are briefly considered in Sec. 8.8. The restrictions imposed by series losses are usually more significant; thus these effects are ignored here.

with B in webers per square meter, dl and r in meters, I in amperes, H in amperes per meter, and $\mu = 4\pi \times 10^{-7}$ henries per meter (in vacuum). The magnetic field associated with the current in the two parallel wires is illustraded in Fig. 1-2(c). If this magnetic flux linking the two wires is changing with time, then voltages V_1 and V_2 at the ends of the small subsection will differ not only by the resistive drop as described above, but also by the induced voltage as given by Faraday's law

$$V_1 - V_2 = \frac{d\phi}{dt} \tag{1-4}$$

where ϕ is the total flux within the subsection loop. Such an induced voltage, or voltage drop resulting from the time changing flux, is identified as inductance and is related by

$$e = L\frac{di}{dt} = V_1 - V_2 \tag{1-5}$$

Thus, the equivalent circuit must contain an inductive component for the subsection of line.

One further component in the equivalent circuit remains to be identified, namely, that associated with any current flow across the insulator between the conductors. Such a current flow can result from ordinary conduction through the insulator or can result from losses associated with time changing electric and magnetic fields, e.g., dielectric or magnetic hysteresis losses. Generally speaking, the electronic conduction for common insulators used in transmission lines is very small and can be neglected, but as this is not always true, in order to be completely general, this effect should be included. Likewise the hysteresis losses associated with ordinary insulators are usually negligible, especially at low frequencies, but for very high frequencies, all insulators generally become more lossy; again, in order to be general, such terms must be included. Since both insulator conduction and other loss terms merely represent a current flow between the conductors which is in phase with the voltage, such mechanisms can be represented by a shunt resistor between the conductors.

Thus, we see that the equivalent circuit contains a series resistance, series inductance, shunt capacitance, and shunt resistance. In the interest of clarity and consistency, these parameters will always be taken as the per-unit-length values, that is, R in ohms per unit length, L in henries per unit length, etc. Thus, it is necessary to multiply these by ℓ, the length of the subsection, to get the total R, L, C, and G of each subsection. We must now determine how to connect these terms together to form the equivalent circuit, since various configurations are possible. Two such possibilities are shown in Fig. 1-3. Since, by definition, an equivalent circuit must give an exact representation regardless of what form it takes, then it does not matter which form is used. The two circuits of Fig. 1-3(a) and (b) are equivalent and we shall arbitrarily use that of (a) for the remainder of this chapter.

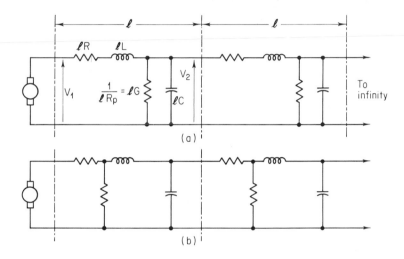

Fig. 1-3. Two possible equivalent circuits for a transmission line with subsections of length ℓ.

1.3 IMPEDANCE OF AN IDEAL TRANSMISSION LINE

If a transmission line has no losses, it can be considered to be a repeated array of small inductors and capacitors in a ladder network. This could be obtained from Fig. 1-3 by letting $R = G = 0$. If the line is uniform such that all the incremental inductors are equal and all the capacitors are likewise equal, it is of interest to determine what the ac impedance looking into the terminals a-b in Fig. 1-4 is.

In order to determine this input impedance, which we call Z_1, a number of techniques are possible; the one customarily used is to derive and solve the transmission-line equations, from which this impedance is automatically obtained. However, this method obscures the most essential idea associated with a transmission-line, namely, that the input impedance of a ladder network of reactive (nonresistive) elements looks like a pure resistance. In order to demonstrate this fundamental concept, we shall invoke nothing more than simple ac theory.

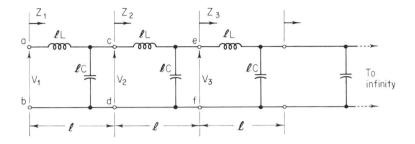

Fig. 1-4. Equivalent circuit of an ideal transmission line.

The line is assumed to extend to infinity toward the right. Let the impedance looking into the terminals a-b be Z_1, that into c-d be Z_2, that into e-f be Z_3, etc. From simple ac analysis, the input impedance is the impedance of the first inductor $L\ell$ in series with the parallel combination of Z_2 and the impedance of capacitor $C\ell$; thus

$$Z_1 = j\omega L\ell + \frac{Z_2(1/j\omega C\ell)}{Z_2 + (1/j\omega C\ell)} \tag{1-6}$$

Now if the line is really infinitely long, then the same impedance should be seen looking into terminals a-b or c-d or e-f, etc.; thus $Z_1 = Z_2$. Substituting this into Eq. (1-6) and collecting terms

$$Z_1{}^2 - Z_1 j\omega L\ell - \frac{j\omega L}{j\omega C} = 0$$

or

$$Z_1{}^2 - j\omega L\ell Z_1 = \frac{L}{C} \tag{1-7}$$

This equation contains a complex component which is frequency-dependent. This term can easily be eliminated by allowing ℓ, the length of our subsection of line, to become very small and by recognizing that the ratio L/C remains constant regardless of line length. The complex term can therefore be neglected and the impedance becomes

$$Z_1 = \sqrt{\frac{L}{C}} \tag{1-8}$$

Thus, the ac input impedance of this line has no reactive component but rather looks like a pure resistance of value given by Eq. (1-8). If this infinite line is broken at any point and is terminated in a resistor of this value, it would not be possible to distinguish this difference by any impedance measurement. We shall derive this same equation in Chap. 2 by a more involved technique.

Incidentally, the same result would have been obtained if we had chosen a different equivalent circuit, such as the T-network in Fig. 1-3(b) for the line. Since the circuits all represent the same structure, it is obvious that the results must by definition be the same.

It is interesting to note that this line, which is composed only of reactive components L and C, looks like a pure resistor to the external world for ac or dc excitation. In other words, during any transient or at steady state, the line presents its characteristic impedance to the buildup of current; we shall consider such cases more fully in Chap. 5 since they are of fundamental importance in the pulse behavior of such lines.

1.4 IMPEDANCE OF A LINE WITH SERIES LOSSES

When series resistance is present in the conductors of the line, the equivalent circuit is modified to that of Fig. 1-5. The input impedance is now different from that of the previous section but can be obtained in a similar manner. If the line is assumed to be infinitely long, the impedance looking in at any point along the line must be the same, or, as in Fig. 1-5(b), Z_{in} must equal Z_0, the characteristic impedance of the line. Proceeding as in Sec. 1.3

$$Z_{in} = R\ell + j\omega L\ell + \frac{Z_0/j\omega C\ell}{Z_0 + (1/j\omega C\ell)} = Z_0 \qquad (1\text{-}9)$$

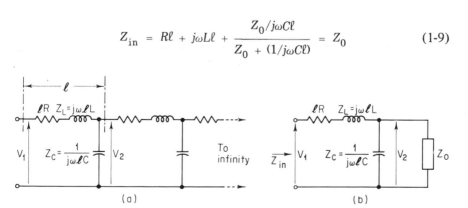

Fig. 1-5. Equivalent circuit of a line with series losses.

Multiplying through and collecting terms

$$Z_0^2 - Z_0 (R + j\omega L)\ell = \frac{R + j\omega L}{j\omega C} \qquad (1\text{-}10)$$

Once again, allowing the subsection length ℓ to become very small, the characteristic impedance Z_0 must remain constant but the total series impedance of each subsection $\ell(R + j\omega L)$ will become very small in comparison. Thus, neglecting this term, Eq. (1-10) becomes

$$Z_0 = \sqrt{\frac{R + j\omega L}{j\omega C}} \qquad (1\text{-}11a)$$

This equation is general and is valid for small or large values of R. If the series losses are small but not negligible, they will have a small effect on the characteristic impedance. This effect can be obtained by writing Eq. (1-11a) as

$$Z_0 = \left[\frac{L}{C}\left(1 + \frac{R}{j\omega L}\right)\right]^{1/2} \qquad (1\text{-}11b)$$

The square root can be expanded by using the identify* $(1 \pm \nu)^{1/2} \approx 1 \pm \nu/2$ for small ν. If the losses are small, then $R/\omega L$ will be small and Eq. (1-11b) can be approximated by

$$Z_0 \approx \sqrt{\frac{L}{C}}\left(1 - j\frac{R}{2\omega L}\right) \qquad (1\text{-}11c)$$

Since $R/2\omega L$ will be much smaller than unity, it is apparent that small series losses have little effect on the characteristic impedance of the line. In subsequent sections, it will be seen that small series losses, and losses in general, do have other more significant effects on the line behavior which must be taken into account.

---

*See [1, p.2], [3, p. 98], or Eq. (1-25).

1.5   PHASE SHIFT AND PROPAGATION CONSTANT OF AN IDEAL LINE

The phase constant, or as it is more commonly known, the propagation constant, of a lossless line can be determined from simple theory as was done in the previous section. This phase constant is merely a measure of the small shift in phase between V_1 and V_2 in Fig. 1-5 or, in other words, the amount of phase shift introduced by each small subsection of the line. Included in this is also the determination of the change in amplitude, if any, introduced by the subsection in question.

It is desirable to determine this factor for an ideal line, i.e., assuming that there are no losses in the line. Also, we arbitrarily specify that we wish to determine the angle by which V_1 leads V_2, or, in other words, we wish to determine V_1/V_2.

Referring to Fig. 1-5, if $R = 0$, then $Z_0 = \sqrt{L/C}$, and it is apparent that

$$V_2 = V_1 \frac{Z_C Z_0}{Z_C + Z_0} \frac{1}{Z_L + [Z_C Z_0/(Z_C + Z_0)]} \qquad (1\text{-}12)$$

or

$$\frac{V_1}{V_2} = \frac{Z_L(Z_C + Z_0 + Z_C Z_0)}{Z_C Z_0}$$

This simplifies to

$$\frac{V_1}{V_2} = 1 + Z_L \left(\frac{1}{Z_0} + \frac{1}{Z_C} \right) \qquad (1\text{-}13)$$

Substituting the values for Z_L, Z_C, and Z_0

$$\frac{V_1}{V_2} = 1 - \omega^2 LC\ell^2 + j\omega\ell\sqrt{LC} \qquad (1\text{-}14a)$$

$$\text{Amplitude} \equiv \left| \frac{V_1}{V_2} \right| = [(1 - \omega^2 \ell^2 LC)^2 + \omega^2 LC\ell^2]^{1/2} \qquad (1\text{-}14b)$$

$$\text{Phase angle} \equiv \tan \beta_\ell = \frac{\omega\ell\sqrt{LC}}{1 - \omega^2 LC\ell^2} \qquad (1\text{-}14c)$$

In order to evaluate the phase angle for this expression, it is desirable to make one further assumption, namely, that the subsection length ℓ is small enough so that the applied frequency is well below the resonant frequency $1/\ell\sqrt{LC}$ of each subsection; in other words

$$\omega^2 \ell^2 LC \ll 1 \qquad (1\text{-}15)$$

Thus, this term can be neglected in Eq. (1-14c) and the phase angle between V_1 and V_2 is therefore given by

$$\tan \beta_\ell = \omega \ell \sqrt{LC}$$

From the assumption of Eq. (1-15), it is obvious that $\tan \beta_\ell$ must be very small and can be replaced by β_ℓ (in radians); in other words, $\tan \beta_\ell = \beta_\ell$ for small angles. Thus, we arrive at the fundamental relationship that the phase shift of each small subsection with no losses is

$$\beta_\ell = \omega \ell \sqrt{LC} \qquad (1\text{-}16)$$

or the phase shift per unit length is

$$\beta = \frac{\beta_\ell}{\ell} = \omega \sqrt{LC} \qquad (1\text{-}17)$$

The angle β_ℓ represents the amount by which the input voltage V_1 leads the output voltage V_2, or, conversely, the amount by which V_2 lags behind V_1. The amplitude ratio, given by Eq. (1-14b), can be simplified by using the above condition of small β_ℓ

$$\left| \frac{V_1}{V_2} \right| = 1 \qquad (1\text{-}18)$$

Thus, there is no decrease in amplitude of the voltage along the line, but only a shift in phase, an important fundamental feature, as we shall see later.

Since β_ℓ merely represents the phase shift of the subsection, it is apparent that we could write

$$\frac{V_1}{V_2} = e^{j\beta_\ell} \qquad (1\text{-}19)$$

From the definition of a natural logarithm, it is clear that taking the ln of both sides of Eq. (1-19) gives

$$j\beta_\ell = \ln \frac{V_1}{V_2} \tag{1-20}$$

A number of fundamental concepts concerning the nature of transmission lines can be deduced from these simple expressions for the amplitude and phase angle of Eqs. (1-16) and (1-18). One important idea concerns the frequency limit of this transmission line. In order to derive Eq. (1-16), it was assumed that the applied frequency was well below the resonant frequency of the subsection as given by Eq. (1-15). If the applied frequency is increased such that this is no longer true, then not only will the amount of phase shift change, but there will be a change in amplitude of the voltage along the line as given by Eq. (1-14b). Fortunately, it is a simple matter to further divide the subsection into small subsections, i.e., make ℓ smaller, for which the applied frequency is once again well below the resonant frequency. In principle, it is possible to subdivide the line until the subsections are vanishingly small so that any frequency can be applied and the voltage at points along the line will have the same amplitude but will differ only in phase. Unfortunately, as the frequency is increased, losses and other difficulties appear which cannot be avoided,* so that practical transmission lines have this ideal behavior only at the lower frequencies where losses are negligible. Nevertheless, this concept of constant amplitude and shift in phase along the line is often true to a large extent and is important in understanding traveling waves along a transmission line. We shall see in Chap. 2 that this progressive phase shift along the line really represents a wave traveling down the line with a velocity determined by the inverse of the phase shift per section. In particular, the velocity is

$$\nu = \frac{\omega}{\beta} = \frac{1}{\sqrt{LC}} \tag{1-21}$$

and is independent of the applied frequency, as might be expected since all the line parameters were assumed to be independent of frequency. It should be noted that the larger the phase-shifting ability of the line, i.e., the larger L and C, the smaller will be this velocity of propagation.

*See Sec. 4.6.

1.6   PROPAGATION CONSTANT FOR A LINE
WITH SMALL SERIES LOSSES

In the previous section, we saw that for a lossless transmission line
$(R = G = 0)$, the voltages at various points along the line have the same
amplitude but differ only in phase with respect to each other, progressively
lagging the applied voltage as one proceeds down the line. When small
losses are present due to series wire resistance, for instance, a similar phenomenon exists in that the voltage experiences phase shift along the line.
However, in addition, the amplitude of the voltage no longer remains constant but decreases in value as one proceeds down the line. This is a result
of the small voltage drop across the series resistors, as can easily be understood if one considers the equivalent circuit of Fig. 1-5.

We wish to derive both the phase shift and the decrease in amplitude
introduced by each small subsection of the line. The phase angle can be
determined in a manner analogous to that used in Sec. 1.5 by evaluation of
V_1/V_2. For the decrease in amplitude, or as it is more commonly labeled,
attenuation, it is desirable to determine the voltage at a given point on the
line as a fraction of the applied voltage. This would be given by the amplitude of V_2/V_1, which can be determined from an ac analysis of Fig. 1-5 in a
manner analogous to that of Sec. 1.5. The voltage ratio is obtained simply
by adding $R\ell$ to Z_L in Eq. (1-13)

$$\frac{V_1}{V_2} = 1 + (R\ell + Z_L)\left(\frac{1}{Z_0} + \frac{1}{Z_C}\right) \tag{1-22}$$

If the subsection of line is again allowed to become very small, then $1/Z_C$
will approach zero while Z_0 remains constant. Thus, the last term in Eq.
(1-22) can be neglected

$$\frac{V_1}{V_2} = 1 + \frac{R\ell + Z_L}{Z_0} \tag{1-23}$$

In Sec. 1.4, we derived the characteristic impedance of a line with losses to
be that of Eq. (1-11). Even though we are presently considering a line with
small losses, it is necessary to retain the loss term in the characteristic impedance of Eq. (1-11) in order to obtain the correct answer. This results
from the fact that Z_0 is multiplied by $R + j\omega L$, and neglecting the loss
term in Z_0 would neglect important cross-product terms.

Substitution of Eq. (1-11) into Eq. (1-23) yields

$$\frac{V_1}{V_2} = 1 + \frac{R\ell + Z_L}{[Z_C(R\ell + Z_L)]^{1/2}} = 1 + [(R\ell + j\omega L\ell)j\omega C\ell]^{1/2} \qquad (1\text{-}24)$$

In order to evaluate this, it is necessary to express the term under the radical in a binomial expansion. This can best be done by using the series expansion

$$(1 \pm \nu)^{1/2} = 1 \pm \frac{\nu}{2} - \frac{\nu^2}{2 \cdot 4} \pm \frac{3\nu^3}{2 \cdot 4 \cdot 6} - \text{etc.}$$

$$\approx 1 \pm \frac{\nu}{2} \quad \text{for} \quad \nu \ll 1 \qquad (1\text{-}25)$$

The term under the radical can be rewritten

$$\ell[-(\omega^2 LC - j\omega RC)]^{1/2} = j\ell\omega\sqrt{LC}\left(1 - \frac{jR}{\omega L}\right)^{1/2} \qquad (1\text{-}26)$$

Since $R/\omega L$ was assumed to be much smaller than 1, then Eq. (1-24) becomes, using the first two terms of the series expansion

$$\frac{V_1}{V_2} = 1 + j\ell\omega\sqrt{LC} - j^2\frac{R}{2\omega L}\sqrt{LC}$$

$$= 1 + \frac{\ell R}{2\sqrt{L/C}} + j\ell\omega\sqrt{LC} \qquad (1\text{-}27)$$

The phase shift or phase angle between V_1 and V_2 is obtained by dividing the imaginary by the real part

$$\tan\beta_\ell = \frac{\ell\omega\sqrt{LC}}{1 + (\ell R/2\sqrt{L/C})} \qquad (1\text{-}28)$$

If the subsection length ℓ is chosen to be small enough, then the denominator equals unity, since the second term can be neglected. Also, the phase shift for the subsection will be small

$$\tan\beta_\ell \approx \beta_\ell = \omega\ell\sqrt{LC} \qquad (1\text{-}29)$$

or the phase shift per unit length is

$$\beta = \frac{\beta_\ell}{\ell} = \omega\sqrt{LC} \tag{1-30}$$

This is the same expression as was obtained for a lossless line; it may thus be concluded that small series losses have no effect on the phase constant to a first-order approximation.

The attenuation is determined by the peak amplitude ratio, which is given by Eq. (1-14b) as the square root of the sum of the squares of the real and imaginary parts of Eq. (1-14a). However, since we have assumed that the losses are small, the imaginary part of Eq. (1-27) is very small compared to the real part (as evidenced by the small value of phase shift). Thus, the magnitude of the voltage ratio is approximately

$$\frac{V_1}{V_2} = 1 + \frac{1}{2}\frac{R\ell}{\sqrt{L/C}} \tag{1-31}$$

It was initially assumed that ℓR was very small compared with $\sqrt{L/C}$. In order to get this into exponential form, we can make use of the identity

$$e^u = 1 + u + \frac{u^2}{2!} + \frac{u^3}{3!} + \text{etc.}$$

or for $u \ll 1$

$$e^u \approx 1 + u \tag{1-32}$$

Then it follows that

$$\frac{V_1}{V_2} = e^{\alpha\ell} \quad \text{or} \quad \frac{V_2}{V_1} = e^{-\alpha\ell} \tag{1-33}$$

where

$$\alpha_\ell = \frac{R\ell}{2\sqrt{L/C}}$$

The attenuation constant per unit length is then

$$\alpha = \frac{\alpha_\ell}{\ell} = \frac{R}{2\sqrt{L/C}} \tag{1-34}$$

In addition to the attenuation, we could also have obtained the phase shift directly from the identity of Eq. (1-32). For instance, from Eq. (1-27)

$$\frac{V_1}{V_2} = 1 + \left(\frac{R\ell}{2\sqrt{L/C}} + j\omega\ell\sqrt{LC} \right) \tag{1-35}$$

Using Eq. (1-32)

$$\frac{V_1}{V_2} = \exp\left[\frac{R\ell}{2\sqrt{L/C}} + j\omega\ell\sqrt{LC} \right] \tag{1-36a}$$

or

$$\frac{V_2}{V_1} = \exp\left[-\frac{R\ell}{2\sqrt{L/C}} - j\omega\ell\sqrt{LC} \right] \tag{1-36b}$$

The imaginary exponent represents the phase shift and the real exponent represents the attenuation for a line with small series losses.

It becomes apparent from the above analysis that the phase and attenuation constants could both have been obtained directly from the natural log of V_1/V_2. This results from the identity

$$\ln(1 + v) = v - \frac{v^2}{2} + \frac{v^3}{3} - \text{etc.} \tag{1-37a}$$

$$\approx v \quad \text{for small } v \tag{1-37b}$$

The voltage ratio of Eq. (1-35) is of the form $1 + v$, so that for small v (or in this case small ℓ)

$$\ln \frac{V_1}{V_2} = \ln e^{\alpha_\ell + j\beta_\ell} = \alpha_\ell + j\beta_\ell \tag{1-38}$$

In other words, the total propagation constant is

$$\gamma_\ell = \alpha_\ell + j\beta_\ell = \ln \frac{V_1}{V_2} \tag{1-39}$$

where γ_ℓ can be expressed as Eq. (1-19)

$$V_2 = V_1 e^{-\gamma_\ell} \tag{1-40}$$

Of course, the same result could have been obtained by taking the natural log of both sides of Eq. (1-36). These two methods are, in fact, equivalent.

It should be understood that α_ℓ and β_ℓ are the total attenuation constant and the phase constant, respectively, per subsection. If the length of the subsection is changed, these parameters will also change, since a longer section will obviously have more series resistance and therefore more attenuation than a shorter section. Similarly, a longer section will have more phase shift. However, α and β, the per-unit-length parameters, are constant, independent of ℓ.

Even though in all the above analyses, we have been concerned with cases where the losses are small, it is generally true that even for large losses, where higher-order terms cannot be neglected, the propagation constant is given by Eq. (1-39). Proof of this is left as an exercise to the reader.

1.7 TOTAL PROPAGATION CONSTANT FOR MANY IDENTICAL SUBSECTIONS

For cascaded subsections, numbered 1 through n as in Fig. 1-6, it is apparent that the ratio of input voltage to the voltage at the nth section is

$$\frac{V_0}{V_n} = \frac{V_0}{V_1} \frac{V_1}{V_2} \frac{V_2}{V_3} \cdots \frac{V_{n-1}}{V_n} \tag{1-41}$$

The total propagation constant of n sections is the natural logarithm of the ratio of input to output voltage or the natural logarithm of Eq. (1-41). Taking the logarithm of both sides

$$\ln \frac{V_0}{V_n} = \ln \frac{V_0}{V_1} + \ln \frac{V_1}{V_2} + \cdots + \ln \frac{V_{n-1}}{V_n}$$

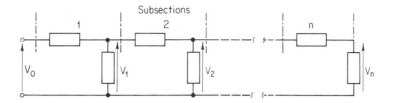

Fig. 1-6. Identical cascaded subsections.

or

$$\gamma_T = \gamma_1 + \gamma_2 + \gamma_3 + \cdots + \gamma_n = n\gamma_1 \qquad (1\text{-}42)$$

Thus, the phase shift of n sections in series is just n times the phase shift of each individual subsection of line. Similarly, the attenuation constant is the attenuation of one subsection multiplied by the number of subsections.

1.8 GENERAL NETWORK EQUATIONS FOR A UNIFORM LINE

In all the preceding sections, we have considered cases for which the conductance G always equaled zero and the series resistance was either small (but finite) or zero. This is usually the case for practical lines, since otherwise large losses cause serious distortion and attenuation and are therefore avoided in the design. As can also be seen from the preceding analyses, the assumption of small or negligible losses greatly facilitates the evaluation of line characteristics in terms of simple, closed-form expressions. If these simplifying assumptions are not made, it is still possible to analyze the network as before, but the expressions become very involved. We shall now derive some general expressions for the general line and then attempt to derive the various parameters without making simplifying assumptions.

Consider the general case of the uniform, repeated array of Fig. 1-7(a) with series impedance elements of

$$Z_s = (R + j\omega L)\ell \qquad (1\text{-}43)$$

and parallel admittance of

$$Y_p = (G + j\omega C)\ell = \frac{1}{Z_p} \qquad (1\text{-}44)$$

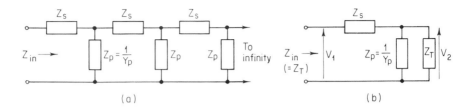

Fig. 1-7. General cascaded network.

Since the line is assumed to be infinitely long and uniform, the impedance seen looking into any subsection must be the same as that of all other subsections, just as in Sec. 1.4. Thus, Fig. 1-7(a) can be simplified to the form of (b) where Z_T is the characteristic impedance of the line and Z_{in} must equal Z_T.

It is easily seen that

$$Z_{in} = Z_s + \frac{Z_p Z_T}{Z_p + Z_T} = Z_T$$

Multiplying through by $Z_p + Z_T$ and collecting terms

$$Z_T^2 - Z_s Z_T - Z_s Z_p = 0 \tag{1-45}$$

The solution to this equation can easily be obtained with the aid of the quadratic formula

$$Z_T = \frac{Z_s \pm \sqrt{Z_s^2 + 4 Z_s Z_p}}{2} \tag{1-46}$$

Using the definition of parameters given by Eqs. (1-43) and (1-44), the above becomes

$$Z_T = \frac{1}{2}\ell(R + j\omega L) \pm \frac{1}{2}\sqrt{\ell^2(R + j\omega L)^2 + 4\frac{R + j\omega L}{G + j\omega C}} \tag{1-47}$$

In order to obtain a simplified, closed-form solution for Z_T, it is necessary to recognize that as the size of the subsection is reduced, all the parameters

ℓR, ℓL, ℓG, and ℓC decrease in the same proportion since R, L, G, and C are constant. Because of this, the ratio $(R + j\omega L)/(G + j\omega C)$, that is, the second term under the radical in Eq. (1-47), will remain constant as the subsection length ℓ is reduced. Thus, the subsection size can be reduced sufficiently so that both terms involving $(R + j\omega L)\ell$ are negligible compared to the constant ratio, and then Eq. (1-47) becomes

$$Z_T = \sqrt{\frac{R + j\omega L}{G + j\omega C}} = \sqrt{Z_s Z_p} = \sqrt{\frac{Z_s}{Y_p}} \qquad (1\text{-}48)$$

This equation represents a simple expression for the characteristic impedance of the general line where the losses are not small. It can be seen that this impedance will vary with the applied frequency, having real and imaginary components for the general case.

In the derivation of Eq. (1-48), it was necessary to choose a subsection small enough to neglect certain terms. In the derivation of the differential equations of a transmission line in Chap. 2, we will let the size of the subsection approach 0 and therefore expect the same result. This will be found to be the case. As a matter of fact, Eq. (1-48) will naturally fall out of the equations without any need to neglect certain terms, as might be expected.

The propagation constant can also be determined for this general case with the aid of the relationship given by Eq. (1-39). The ratio of input to output voltage for the circuit of Fig. 1-7(b) can be found from

$$V_2 = V_1 \frac{Z_T Z_p}{Z_T + Z_p} \frac{1}{Z_s + [Z_T Z_p/(Z_T + Z_p)]}$$

Multiply through and collect terms

$$\frac{V_1}{V_2} = 1 + \frac{Z_s}{Z_p} + \frac{Z_s}{Z_T} \qquad (1\text{-}49)$$

Z_T is the characteristic impedance given by Eq. (1-48). This equation is identical in form to Eq. (1-13), except that now we have included R and G in the impedance terms.

Making use of Eq. (1-39)

$$\gamma_\ell = \alpha_\ell + j\beta_\ell = \ln\left[1 + Z_s\left(\frac{1}{Z_p} + \frac{1}{Z_T}\right)\right] \qquad (1\text{-}50)$$

Substituting Eq. (1-48) for Z_T and Y_p for $1/Z_p$

$$\gamma_\ell = \ln\left[1 + Z_s\left(Y_p + \sqrt{\frac{Y_p}{Z_s}}\right)\right] \qquad (1\text{-}51)$$

Once again we will allow the subsection to become small enough so that Y_p becomes negligible compared to $\sqrt{Y_p/Z_s}$, the latter remaining constant as the subsection is decreased. Thus, Eq. (1-51) reduces to

$$\gamma_\ell = \ln\left(1 + Z_s\sqrt{\frac{Y_p}{Z_s}}\right)$$
$$= \ln\left(1 + \sqrt{Z_s Y_p}\right) \qquad (1\text{-}52)$$

Making use of Eq. (1-37a) for the series expansion of the natural log

$$\gamma_\ell = \sqrt{Z_s Y_p} - \frac{\left(\sqrt{Z_s Y_p}\right)^2}{2} + \cdots \text{ etc.}$$

Since we have already specified that the subsection is allowed to become very small, then $\sqrt{Z_s Y_p}$ will be much less than 1, so that the higher-order terms in the above equation can be neglected to yield

$$\gamma_\ell = \sqrt{Z_s Y_p} = \ell\sqrt{(R + j\omega L)(G + j\omega C)} \qquad (1\text{-}53)$$

for each subsection. The propagation constant per unit length is thus

$$\gamma = \frac{\gamma_\ell}{\ell} = \sqrt{(R + j\omega L)(G + j\omega C)} \qquad (1\text{-}54)$$

with all parameters taken per unit length. If R and G are allowed to approach zero, the attenuation terms must be zero with the result that Eq. (1-53) becomes

$$\gamma_\ell = j\beta_\ell = j\omega l \sqrt{LC} \tag{1-55}$$

which is identical to Eq. (1-16), as was expected.

This propagation constant will be derived in Chap. 2 from the differential equation. It will be seen there that no simplifying assumptions are necessary, primarily because the subsection length must approach zero in order to arrive at the differential equations.

Thus far, we have only been concerned with the derivation of some fundamental parameters of transmission lines and have not used the results to examine their behavior in any detail. This has been done purposely in order to show the fundamental aspects and how they evolved quite naturally from elementary ac circuit theory. These concepts will be applied to numerous situations in succeeding chapters. It is, however, important for the student to have a basic understanding of the fundamentals before we proceed to other aspects of transmission lines. We shall derive these fundamental parameters once again in a more convenient form through the use of differential equations from which other basic characteristics of transmission lines can easily be obtained as well.

It is possible to analyze the entire behavior of transmission lines in terms of ac theory applied to small subsections of the line. However, this representation is that of a lumped-parameter transmission line, that is, discrete and separated $R, L, G,$ and C, whereas in reality such a line is a distributed-parameter line, that is, $R, L, G,$ and C are not discrete and distinguishably separated from each other. Thus the differential-equation approach is somewhat more realistic. Furthermore, the differential equations simplify the mathematical analysis and allow one to derive many important phenomena in a more convenient form. However, both the ac-theory and differential-equation approach are really approximate representations of wave propagation on an electrical line. Fortunately, these approximations are quite adequate for most applications. The exact solution for one general case of a strip line is detailed in Sec. 4.5 and the limitations of the approximate representations are presented in Sec. 4.6.

PROBLEMS

1-1. Calculate the resistance of an infinite ladder of 1-ohm resistors connected as in Fig. 1-3 with $L = C = 0$.

    Answer: 1.62 ohms.

1-2. Determine the input impedance in Fig. 1-3(a) when the shunt resistance losses G are not negligible.

1-3. Show that the circuit of Fig. 1-3(b) gives the same value for input impedance as that of (a) when G is not negligible.

1-4. Show that $e^{j\theta}$ is a vector of amplitude unity, inclined at an angle θ.

1-5. Show that $V_0 e^{j\omega t}$ is equivalent to a sinusoidal vector of amplitude V_0 and argument ωt.

1-6. Given a line for which the series resistance R is zero and shunt conductance G is small but not negligible, determine the expression for α and β.

   Answer: See Sec. 2.6.

1-7. Derive the input impedance of a simple series R, L, C circuit. Show that the resonant frequency is $1/\sqrt{LC}$ and that at this frequency, the input impedance equals R.

1-8. Derive the input impedance of a simple parallel R, L, C circuit. Show that the resonant frequency is $1/\sqrt{LC}$ and that at resonance, the input impedance equals R.

1-9. Show that at resonance, a series LC circuit has zero impedance, while a parallel LC circuit has infinite impedance.

REFERENCES

1.   Dwight, H. B.: "Tables of Integrals and Other Mathematical Data," 3rd ed., The Macmillan Company, New York, 1957.

2.   Lepage, W. R.: "Analysis of AC Circuits," McGraw-Hill Book Company, New York, 1952.

3.   Pierce, B. O., and R. M. Foster: "A Short Table of Integrals," 4th ed., Ginn and Company, Boston, 1956.

2 DIFFERENTIAL EQUATIONS AND GENERAL THEORY OF TRANSMISSION LINES

2.1 INTRODUCTION

In the previous chapter, we saw that a transmission line can be treated as a repeated array of small resistors, inductors, and capacitors. The analysis in terms of simple ac theory has provided us with some physical insights and an understanding of the fundamental behavior of such lines which are helpful as an introduction to this field. All transmission-line theory could be treated in terms of ac circuit analysis, but unfortunately the analyses become extremely involved for all but the simple cases. It is more convenient and useful to treat such lines in terms of differential equations which, as we shall see, develop quite naturally into the wave equation, which is of fundamental importance to all of electromagnetic theory.

From these differential equations the concepts of waves traveling on a line evolve, which is, in fact, the actual situation occurring in nature. The fundamental concepts derived in Chap. 1 are not changed but will be found to correlate exactly with the appearance of waves, with the phenomena of phase shift being generalized to the velocity of propagation of a traveling wave.

2.2 DIFFERENTIAL EQUATIONS FOR A UNIFORM LINE WITH AC EXCITATION

The differential equations are obtained from a simple ac circuit analysis of the equivalent circuit of Fig. 1-3 and then by letting the incremental section of line approach zero in length. Figure 2-1 shows the general equivalent circuit with notation of voltage and current at some general points located at x and $x + \Delta x$ along the line. As in Chap. 1, the parameters R, L, G, and C are taken as the per-unit-length values, i.e., as Δx is changed these parameters remain the same. We shall assume that the voltages and

25

currents are sinusoidal and that any point x, the time variation of voltage is $v_x = V_0 e^{j\omega t}$.

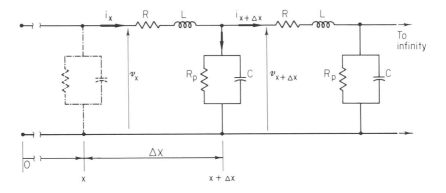

Fig. 2-1. Incremental equivalent circuit of an infinite transmission line.

Applying Kirchhoff's law around the first incremental loop in Fig. 2-1, we obtain

$$v_x = i_x R\Delta x + i_x j\omega L\Delta x + v_{x+\Delta x}$$

or

$$v_{x+\Delta x} - v_x = -i_x(R + j\omega L)\Delta x \qquad (2\text{-}1)$$

R and L were multiplied by Δx in order to get the actual resistance and inductance within the physical length Δx.

The total current i_x into the first increment at x minus the total current $i_{x+\Delta x}$ into the second incremental section at $x + \Delta x$ must be equal to the total current through the shunt capacitance C plus the parallel resistance R_p. In other words

$$i_x - i_{x+\Delta x} = \frac{v_{x+\Delta x}}{R_p \Delta x} + \frac{v_{x+\Delta x}}{1/j\omega C\Delta x} \qquad (2\text{-}2)$$

If we let $1/R_p = G$, the conductance per unit length, then Eq. (2-2) becomes

$$i_{x+\Delta x} - i_x = -v_{x+\Delta x}(G + j\omega C)\Delta x \qquad (2\text{-}3)$$

The per-unit-length parameters G and C are multiplied by Δx for the same reason as above, i.e., in order to get the total actual values within the increment Δx.

In Eq. (2-1), note that $v_{x+\Delta x} - v_x$ is just the incremental voltage drop along the line, referred to as Δv_x. Thus, dividing both sides by Δx (Eq. (2-1)) yields

$$\frac{\Delta v_x}{\Delta x} = -i_x (R + j\omega L) \tag{2-4}$$

In like manner, Eq. (2-3) becomes

$$\frac{\Delta i_x}{\Delta x} = -v_{x+\Delta x} (G + j\omega C) \tag{2-5}$$

Now if we let the incremental distance Δx become very small, then the incremental voltage or current change per incremental distance becomes the derivative. Thus, we arrive at the two fundamental differential equations for a uniform transmission line

$$\frac{dv_x}{dx} = -(R + j\omega L) i_x \tag{2-6}$$

$$\frac{di_x}{dx} = -(G + j\omega C) v_x \tag{2-7}$$

where all parameters are per unit distance.

In order to solve these equations, it is necessary to get each equation in terms of one unknown. An equation in terms of voltage can be obtained by taking the derivative of Eq. (2-6) with respect to x

$$\frac{d^2 v_x}{dx^2} = -(R + j\omega L) \frac{di_x}{dx} \tag{2-8}$$

Substituting Eq. (2-7) for di_x/dx in Eq. (2-8)

$$\frac{d^2 v_x}{dx^2} = (R + j\omega L)(G + j\omega C) v_x = \gamma^2 v_x \tag{2-9}$$

where

$$\gamma^2 = (R + j\omega L)(G + j\omega C) \qquad (2\text{-}10)$$

Similarly, an equation for current only can be obtained by differentiating Eq. (2-7) and substituting Eq. (2-6)

$$\frac{d^2 i_x}{dx^2} = (R + j\omega L)(G + j\omega C) i_x = \gamma^2 i_x \qquad (2\text{-}11)$$

It should be recalled that in deriving the equations, we assumed that the voltage and current were sinusoidal; thus they are vector quantities in Eqs. (2-9) and (2-11). The similarity between these two equations should be noted since it is a fundamental relationship governing wave propagation (as will become evident later), rather than pure coincidence.

The solution to these equations is obtained by the standard trial-and-error technique of guessing a solution and testing to see if it satisfies the equation. The solution to Eq. (2-9) is easily verified as

$$v_x = V_A e^{-\gamma x} + V_B e^{\gamma x} \qquad (2\text{-}12)$$

where γ is as previously defined and V_A and V_B are constants to be determined by boundary conditions at the input and output ends of the line. In a similar manner, Eq. (2-11) can be solved to give

$$i_x = I_A e^{-\gamma x} + I_B e^{\gamma x} \qquad (2\text{-}13a)$$

where I_A and I_B are also determined by the boundary conditions.

I_A and I_B are related to V_A and V_B; thus, Eq. (2-13) can be simplified as follows. From Eq. (2-6)

$$\frac{dv_x}{dx} = -(R + j\omega L) i_x = -\gamma V_A e^{-\gamma x} + \gamma V_B e^{\gamma x}$$

Thus

$$i_x = \frac{\gamma}{R + j\omega L} \left(V_A e^{-\gamma x} - V_B e^{\gamma x} \right) = \frac{1}{\sqrt{Z/Y}} \left(V_A e^{-\gamma x} - V_B e^{\gamma x} \right)$$

where

$$\sqrt{\frac{Z}{Y}} = \frac{R + j\omega L}{\gamma} = \sqrt{\frac{R + j\omega L}{G + j\omega C}}$$

so that

$$I_A = \frac{V_A}{\sqrt{Z/Y}} \qquad I_B = \frac{-V_B}{\sqrt{Z/Y}} \qquad (2\text{-}13b)$$

It should be recalled once again that sinusoidal variations of current and voltage were assumed in the derivation of the above equations and this fact is implicitly assumed. If the sinusoidal variation is represented explicitly, these equations become

$$v_x = e^{j\omega t}\left(V_A e^{-\gamma x} + V_B e^{\gamma x}\right) \qquad (2\text{-}14)$$

$$i_x = \frac{e^{j\omega t}}{\sqrt{Z/Y}}\left(V_A e^{-\gamma x} - V_B e^{\gamma x}\right) \qquad (2\text{-}15)$$

These two equations are the most fundamental and general equations of a uniform transmission line with sinusoidal excitation. Included within these equations are the concepts of traveling waves, standing waves, phase shift, and/or velocity of propagation, attenuation, and reflection coefficients as well as characteristic impedance.

The two constants V_A and V_B can be evaluated by noting that at $x = 0$, v_x must be the source voltage V_s minus the iR drop in the source resistance (see Fig. 2-2)

$$V_A + V_B = V_s - iR_s \qquad (2\text{-}16)$$

A second boundary condition is obtained by noting that at the output end of the line $x = \ell$ the voltage divided by current must equal the load impedance Z_ℓ

$$\left.\frac{v_x}{i_x}\right|_{x=\ell} = Z_\ell = Z_0 \frac{\left(V_A e^{-\gamma\ell} + V_B e^{\gamma\ell}\right)}{\left(V_A e^{-\gamma\ell} - V_B e^{\gamma\ell}\right)} \qquad (2\text{-}17)$$

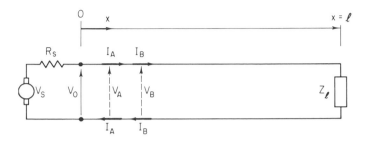

Fig. 2-2. Polarities and notation of parameters.

where

$$Z_0 = \sqrt{Z/Y}$$

The two equations are sufficient to determine the two unknowns; if $R_s = Z_0$, substitution of Eq. (2-13) into Eq. (2-16) and the result into Eq. (2-17) and solving gives

$$V_A = \frac{V_s}{2} \tag{2-18}$$

$$V_B = \frac{V_s}{2} e^{-2\gamma\ell} \left(\frac{Z_\ell - Z_0}{Z_\ell + Z_0} \right) \tag{2-19}$$

It should be understood that V_A and V_B are evaluated with their phase referenced to the source at $x = 0$. Some authors prefer to determine the solution in terms of distance measured from the load but this is of little consequence since all the analyses in the following sections are identical for either case, provided that one is consistent and defines all quantities properly.

The constants were evaluated in terms of the source voltage and load impedance. It should be apparent that they could have been evaluated in terms of any other two specified boundary conditions such as input current, input impedance, load voltage, load current, etc. Such forms are sometimes used where expedient but should cause no confusion.

We shall see later that terms with positive exponents in x, that is, $V_B e^{\gamma x}$, represent waves traveling in the negative x direction, while those with negative exponents in x represent waves traveling in the positive x direction. We shall also see that the sum of these positively and negatively traveling waves gives rise to standing waves on the line (the reflection coefficient simply being the ratio of the positive and negative traveling waves), and that the constant γ has an imaginary part which is a measure of the velocity of propagation of a traveling wave and a real part which is a measure of the attenuation (reduction in amplitude) of the wave as it travels along the line. Each of these concepts will be developed separately in the following sections and correlated where possible with the simple ac analysis of Chap. 1.

The expression given by Eq. (2-12) is not the only form of solution to Eq. (2-9); in fact, another solution in terms of hyperbolic functions is

$$v_x = E \sinh \gamma x + F \cosh \gamma x$$

as can easily be verified by substitution. The hyperbolic form is equivalent to the exponential form since the one can be derived from the other by simple substitution of the exponential form for the hyperbolic functions. If this is done, it will be found that $V_A = (F - E)/2$ and $V_B = (E + F)/2$. The hyperbolic equivalents of Eqs. (2-14) and (2-15) including time variation are thus

$$v_x = e^{j\omega t}(E \sinh \gamma x + F \cosh \gamma x)$$

$$i_x = \frac{e^{j\omega t}}{Z_0}(E \cosh \gamma x + F \sinh \gamma x)$$

The constants are determined as always by the boundary conditions, which can vary depending on what is assumed to be known.

These hyperbolic forms are frequently used in cases where a single frequency is used but the explicit designation of the positively and negatively traveling wave is lost. For cases where pulses (many frequencies) are present, the exponential form is more useful. We shall use the exponential form in all subsequent chapters.

2.3 TRANSMISSION-LINE EQUATIONS FOR A UNIFORM LINE WITH GENERAL EXCITATION

When the form of the excitation is not given explicitly, it is possible to obtain the transmission-line equations in terms of the time derivatives of voltage and current. The procedure is identical to that of Sec. 2.2 except that here we no longer have sinusoidal excitation. For the general case, using a uniform line, Eqs. (2-6) and (2-7) are obviously

$$\frac{\partial v_x}{\partial x} = -Ri_x - L\frac{\partial i_x}{\partial t} \tag{2-20}$$

$$\frac{\partial i_x}{\partial x} = -Gv_x - C\frac{\partial v_x}{\partial t} \tag{2-21}$$

Partial derivatives are now used since the unknowns are functions of both x and t. In order to obtain Eq. (2-20) in terms of v_x alone, it is first necessary to obtain the partial derivative of Eq. (2-20) with respect to x and of Eq. (2-21) with respect to t

$$\frac{\partial^2 v_x}{\partial x^2} = -R\frac{\partial i_x}{\partial x} - L\frac{\partial i_x}{\partial t \partial x} \tag{2-22}$$

$$\frac{\partial^2 i_x}{\partial x \partial t} = -G\frac{\partial v_x}{\partial t} - C\frac{\partial^2 v_x}{\partial t^2} \tag{2-23}$$

In Eq. (2-22), the first and second terms on the right are given by Eqs. (2-21) and (2-23), respectively. Substituting these and collecting terms

$$\frac{\partial^2 v_x}{\partial x^2} = RGv_x + (RC + LG)\frac{\partial v_x}{\partial t} + LC\frac{\partial^2 v_x}{\partial t^2} \tag{2-24}$$

In a similar manner, the equation of current can be obtained

$$\frac{\partial^2 i_x}{\partial x^2} = RGi_x + (RC + LG)\frac{\partial i_x}{\partial t} + LC\frac{\partial^2 i_x}{\partial t^2} \tag{2-25}$$

The above two equations are the more general forms of the transmission-line equations. If sinusoidal excitation is assumed for voltage and current, then these equations reduce to Eqs. (2-9) and (2-11).

It can be seen that for this general case, including losses, the equation is much more involved. Fortunately, in most cases of practical interest, the losses in the line are made relatively small so that to a first approximation they can be neglected. Inclusion of the losses will tend to obscure the essential, fundamental features of transmission lines. If the losses are assumed to be negligible, i.e., if $R = G = 0$, then Eq. (2-24) becomes

$$\frac{\partial^2 v_x}{\partial x^2} = LC \frac{\partial^2 v_x}{\partial t^2} \qquad (2\text{-}26)$$

with a similar equation for current

$$\frac{\partial^2 i_x}{\partial x^2} = LC \frac{\partial^2 i_x}{\partial t^2} \qquad (2\text{-}27)$$

This equation is known as the wave equation and is of fundamental importance in all wave phenomena, including electromagnetic theory and other
fields of science. As will become clear later, this equation describes a wave
or waves traveling in the x direction with a velocity determined by L and C.
Insight into this phenomenon can be obtained by a simple dimensional
analysis exercise, which provides a means for obtaining the velocity of a
traveling wave in the following manner. In Eq. (2-26), let us divide the
right-hand side by the left-hand side and also divide both sides by LC

$$\frac{1}{LC} = \frac{\partial^2 v_x}{\partial t^2} \frac{\partial x^2}{\partial v_x^2}$$

It is obvious that the right-hand side has dimensions of velocity squared;
thus, it follows that

$$\nu = \frac{1}{\sqrt{LC}} \qquad (2\text{-}28)$$

This represents the phase velocity or velocity of propagation of a sinusoid
traveling on an ideal line. This expression will occur many times and will
be considered more fully in Sec. 2.7.

2.4 CHARACTERISTIC IMPEDANCE OF A UNIFORM LINE

In Sec. 1.3 the input impedance of a uniform, lossless line was derived
from simple ac theory, and the same solution must be obtainable from the
previous differential equations developed in Sec. 2.2.

The impedance at a point x on the line is equal to the ratio of the voltage divided by the current at that point. This ratio can be obtained from

Eq. (2-6) in conjunction with the general solution given by Eq. (2-12) as follows. From Eq. (2-6)

$$i_x = -\frac{1}{R + j\omega L} \frac{dv_x}{dx} \tag{2-29}$$

The derivative on the right side can be obtained from Eq. (2-12)

$$\frac{dv_x}{dx} = \left(-\gamma V_A e^{-\gamma x} + \gamma V_B e^{\gamma x}\right) e^{j\omega t} \tag{2-30}$$

Substituting this into Eq. (2-29)

$$i_x = \frac{-\gamma}{R + j\omega L} \left(-V_A e^{-\gamma x} + V_B e^{\gamma x}\right) \tag{2-31}$$

The impedance is defined as the ratio of voltage to current at a given point

$$Z_0 = \frac{v_x}{i_x} = \frac{V_A e^{-\gamma x} + V_B e^{\gamma x}}{-V_A e^{-\gamma x} + V_B e^{\gamma x}} \frac{R + j\omega L}{(-\gamma)} \tag{2-32}$$

Now if we assume that the line is infinitely long, then it is necessary that the constant V_B be equal to zero; otherwise, as x becomes large, the terms with positive exponent become very large, an undesirable and unrealistic condition. As will become evident later, this is equivalent to stating that there are no waves traveling in the negative x direction on the line (which is a necessary condition). Thus, setting $V_B = 0$ and canceling appropriate terms

$$Z_0 = \frac{R + j\omega L}{\gamma} = \sqrt{\frac{R + j\omega L}{G + j\omega C}} \tag{2-33}$$

This is the so-called characteristic impedance of the line, since it depends on the characteristic parameters R, L, G, and C. It can be seen that for this general case the impedance is complex, containing resistive and reactive components, and that it is a function of the applied frequency, approaching a constant value of $\sqrt{L/C}$ for high frequency, assuming that all parameters are frequency-independent (which is not usually true).

If the line is lossless such that $R = G = 0$, then the characteristic impedance is

$$Z_0 = \sqrt{\frac{L}{C}} \tag{2-34}$$

which is a pure resistance, independent of frequency so long as L and C are independent of frequency. This result is identical to that derived in Sec. 1.3 from simple ac theory and is, of course, just as expected. The effect of small series losses on Z_0 is given by Eq. (1-11c).

2.5 PROPAGATION CONSTANT FOR A UNIFORM LINE

(a) Phase constant. In Sec. 1.5 we derived the amount of phase shift introduced by each small subsection of the line. As these subsections are reduced in length, it is apparent that the amount of phase shift introduced by each one approaches zero. However, the amount of phase shift introduced by a given physical length of line remains constant by the fact that if the length of any subsection is reduced, then there will be proportionally more such subsections in a given physical length of line. In other words, in choosing an equivalent circuit for the line, we are at liberty to make the subsections as small as we like so that a given length of line will have more such sections as their individual length is reduced. It is apparent from the results of Sec. 1.6 that the phase shift per unit length of a uniform line is a fixed value determined by the physical characteristics of the line, i.e., it is independent of location on the line. This phase constant can be determined simply by finding the total phase shift for a given length of line and dividing this by the given length. This is most easily accomplished with Eq. (2-12) and by using a length of line between $x = 0$ and $x = \ell$. Once again, as in Sec. 2.3, we will set $V_B = 0$, since we are assuming the line to be infinite and since we require all quantities to remain finite as x becomes large. At $x = 0$, the voltage is $v_0 = V_A$, while at $x = \ell$

$$v_\ell = V_A e^{-\gamma \ell}$$

Both of these are vector quantities and the phase difference is determined by their ratio

$$\frac{v_0}{v_\ell} = \frac{1}{e^{-\gamma \ell}} = e^{\gamma \ell}$$

Since y is a complex quantity, the phase shift is given by the imaginary part. The phase constant β is then this quantity divided by the length ℓ

$$\beta = \text{Im}\,\frac{(y\ell)}{\ell} = \text{Im}(y) \tag{2-35}$$

Using the definition of y from Eq. (2-10), we have

$$\beta = \text{Im}\,\sqrt{(R + j\omega L)(G + j\omega C)} \tag{2-36}$$

where the line parameters are per unit length. For the case of a lossless line where $R = G = 0$, the phase constant becomes

$$\beta = \omega\sqrt{LC} \tag{2-37}$$

with dimensions of radians per meter in RMKS units. This equation is identical to that developed from ac theory in Sec. 1.6. Note that in Sec. 2.3, it was found that the reciprocal of $\sqrt{LC}$ has dimensions of velocity. It follows from this and from Eq. (2-37) that the phase shift per unit length along the line is a measure of the velocity of propagation of a wave along the line

$$\nu = \frac{\omega}{\beta} = \frac{1}{\sqrt{LC}} \quad \frac{\text{meters}}{\text{sec}} \tag{2-38}$$

(b) Attenuation constant. In the preceding analysis, the phase shift along the line was obtained from the imaginary part of y. The real part of y gives the reduction in voltage or current along the line, as in Sec. 1.5. This quantity, when evaluated per unit length of line, is referred to as the attenuation constant

$$\alpha = \text{Re}(y) = \text{Re}\,\sqrt{(R + j\omega L)(G + j\omega C)} \tag{2-39}$$

For the case of a line with no losses, it is apparent that Eq. (2-39) gives $\alpha = 0$, which merely states that a line with no losses has no attenuation, as one might expect. This is analogous to the situation examined in terms of ac theory in Secs. 1.4 and 1.5, where it was shown that for a line with no losses, that is, $R = G = 0$, the voltage at any point along the line had the same amplitude as that at any other point but was shifted in phase by an amount equal to the total phase constant between points.

Thus, we can express the propagation constant as the sum of the attenuation and phase constant

$$\gamma = \alpha + j\beta \tag{2-40}$$

2.6 PROPAGATION CONSTANT FOR A LINE
WITH SMALL SERIES AND SHUNT LOSSES

In Sec. 1.6 the propagation constant for a line with small series losses was derived from ac theory. We now wish to extend this case to include small shunt losses, i.e., small G, as well. The propagation constant for a general uniform line is given by Eq. (2-10). A difficulty arises in that for the general case, the terms under the square root cannot be separated into real and imaginary parts with simple, closed-form expressions. However, if the losses are small, simple expressions can be obtained. In order to do this, it is expedient to make use of the identity

$$(1 \pm \nu)^{\frac{1}{2}} \approx 1 \pm \frac{\nu}{2} - \frac{\nu^2}{8} \quad \text{for small } \nu \tag{2-41}$$

Equation (2-10) can be put into this form by rewriting

$$\gamma = [RG - \omega^2 LC + j\omega(GL + RC)]^{\frac{1}{2}}$$

$$= \left\{ \omega^2 LC \left[\frac{RG}{\omega^2 LC} - 1 + \frac{j\omega}{\omega^2 LC}(GL + RC) \right] \right\}^{\frac{1}{2}}$$

$$= (-\omega^2 LC)^{\frac{1}{2}} \left\{ 1 - \left[\frac{RG}{\omega^2 LC} + j\omega \left(\frac{G}{\omega^2 C} + \frac{R}{\omega^2 L} \right) \right] \right\}^{\frac{1}{2}} \tag{2-42}$$

It is assumed that the losses are small but not necessarily negligible, which implies that $R \ll \omega L$ and $G \ll \omega C$. This being the case, Eq. (2-42) is of the form

$$\gamma = (-\omega^2 LC)^{\frac{1}{2}} [1 - \nu]^{\frac{1}{2}} \tag{2-43}$$

where $\nu \ll 1$ and the square root can be approximated by the identity of Eq. (2-41). If this is carried out, then after simplification it is found that

$$\gamma \approx \frac{R}{2\sqrt{L/C}} + \frac{G}{2}\sqrt{\frac{L}{C}} + j\omega\sqrt{LC}\left(1 + \frac{R}{8\omega^2 L^2} + \frac{G^2}{8\omega^2 C^2} - \frac{RG}{4\omega^2 LC}\right) \quad (2\text{-}44)$$

ERROR IN DIMENSION

It is apparent that the real part represents the attenuation and the imaginary part represents the phase constant

$$\alpha = \frac{R}{2\sqrt{L/C}} + \frac{G}{2}\sqrt{\frac{L}{C}} \quad (2\text{-}45)$$

$$\beta = \omega\sqrt{LC}\left(1 + \frac{R^2}{8\omega^2 L^2} + \frac{G^2}{8\omega^2 C^2} - \frac{RG}{4\omega^2 LC}\right)$$

$$= \omega\sqrt{LC} + \left(\frac{R}{2\sqrt{L/C}}\right)\frac{R}{4\omega L} + \left(\frac{G}{2}\sqrt{\frac{L}{C}}\right)\frac{G}{4\omega C} - \frac{RG}{4\omega\sqrt{LC}} \quad (2\text{-}46)$$

If $G = 0$, Eq. (2-45) is identical to Eq. (1-34), as is to be expected. It should be noted that the terms in parentheses in Eq. (2-46) are the attenuation terms appearing in Eq. (2-45). They are each multiplied by terms which are very small and since these two terms in parentheses are also small by assumptions then the losses have only a second-order effect on the phase constant. This is because the effect of losses on β comes from the $\nu^2/8$ term in Eq. (2-41). Thus, to a first approximation, a line with small losses has the same phase constant or same delay time as an ideal line. In other words

$$\beta \approx \omega\sqrt{LC} \quad (2\text{-}47)$$

to a first approximation. This is identical to Eq. (1-30) and is an important result, especially for lines operated with fast rise time pulses (high frequency) where the losses are generally small but not negligible.

2.7 TRAVELING WAVES

It is possible to demonstrate the existence of traveling waves and also to explore some of their characteristics in terms of the general wave equation given by Eq. (2-24) or Eq. (2-26). However, such general mathematical treatments often obscure the basic concepts. Since it is well known that any waveform, including single pulses, can be resolved into a Fourier series of sinusoidal components, it is expedient to derive wave phenomena assuming sinusoidal excitation, which we shall now do.

Equations (2-14) and (2-15) describe the general behavior of voltage and current on a uniform line with losses and sinusoidal excitation. We wish to show that these expressions describe two traveling waves, traveling in opposite directions with the same velocity of propagation and experiencing the same amount of attenuation. This is shown by substituting the expression for γ from Eq. (2-40) into Eq. (2-14)

$$v_x = e^{j\omega t}\left(V_A e^{-\alpha x}e^{-j\beta x} + V_B e^{\alpha x}e^{j\beta x}\right)$$

Combining terms with imaginary exponents

$$v_x = V_A e^{-\alpha x}e^{j(\omega t - \beta x)} + V_B e^{\alpha x}e^{j(\omega t + \beta x)} \qquad (2\text{-}48)$$

The essential concept required in order to prove that these two terms represent traveling waves is that the imaginary exponents represent the phase relation of a sinusoidal voltage at any point x on the line. Let us look at the behavior of any given phase point of v_x, which is equivalent to stating that the imaginary exponents in Eq. (2-48) equal a constant. Thus

$$\omega t - \beta x = K_1 = \text{constant} \qquad (2\text{-}49)$$

$$\omega t + \beta x = K_2 \qquad (2\text{-}50)$$

Taking the time derivatives of these, we get

$$\omega \pm \beta \frac{dx}{dt} = 0 \qquad (2\text{-}51)$$

where the $-$ and $+$ signs represent Eqs. (2-49) and (2-50), respectively. Solving

$$\frac{dx}{dt} = \mp \frac{\omega}{\beta} \qquad (2\text{-}52)$$

The dx/dt term is the velocity of a constant phase point and since it has already been shown (Eq. (2-38)) that the right-hand side equals $1/\sqrt{LC}$

$$\frac{dx}{dt} = \mp \frac{1}{\sqrt{LC}} = \nu \qquad (2\text{-}53)$$

where ν is the velocity of propagation as deduced form the wave equation.

The minus velocity is associated with Eq. (2-50) or with the second term of Eq. (2-48) and represents a wave traveling in the negative x direction. The positive attenuation exponent associated with this term indicates that the wave decreases in amplitude as it travels in the negative x direction (or increases in the positive x direction). Similarly, the first terms of Eq. (2-48) represent a wave traveling in the positive x direction and the negative attenuation exponent term indicates that this wave decreases in amplitude as it travels in the positive x direction. It is obvious from Eq. (2-53) that both these waves have the same velocity.

The concept of traveling waves is fundamental to an understanding of transmission lines as well as all wave phenomena. Further insight into this behavior can be gained if one considers some simple cases of sinusoidal waves on a lossless, uniform line. Such cases are treated in detail in Chap. 3 and the reader is referred to Sec. 3.2 for further treatment of a single frequency traveling on an ideal line.

2.8 STANDING WAVES

In the previous section, we have seen that waves can travel along a transmission line much as waves traveling in space do. Now we wish to examine the situation occuring when two traveling waves are present, but traveling in opposite directions. It will be shown that oppositely traveling waves give rise to so-called standing waves, which appear to be stationary in space, although the instantaneous amplitude varies with time.

Let us assume that there are two sinusoidal waves present on a given transmission line, traveling in opposite directions according to the relationship expressed by Eq. (2-48), except that we shall assume the line to be lossless, that is, $\alpha = 0$. For the present we will not concern ourselves with how these two waves were obtained; we will simply assume that they exist and will analyze their behavior. In vector space, i.e., complex plane, we know from ac theory that the exponential term $e^{j\theta}$ represents a circle of unit radius, the vector being inclined at the angle θ with respect to the positive real axis. Thus the time and distance variation of v_x is given by either the real or imaginary part of Eq. (2-48). If we take the imaginary part, the total instantaneous voltage at any x and t is (neglecting losses)

$$v_x = V_A \sin(\omega t - \beta x) + V_B \sin(\omega t + \beta x) \qquad (2\text{-}54)$$

Using a simple trigonometric identity for sin $(a \pm b)$, this may be expanded to

$$v_x = V_A \sin \omega t \cos \beta x - V_A \cos \omega t \sin \beta x + V_B \sin \omega t \cos \beta x$$

$$+ \, V_B \cos \omega t \sin \beta x$$

Collecting terms

$$v_x = (V_A + V_B) \sin \omega t \cos \beta x - (V_A - V_B) \cos \omega t \sin \beta x \qquad (2\text{-}55)$$

If v_x is plotted versus x for a fixed value of t, the curve will be sinusoidal; likewise if v_x is plotted versus t for a fixed value of x, the curve is again sinusoidal. We wish to show that this equation represents a standing wave with a peak amplitude of $V_A + V_B$ and a minimum amplitude of $V_A - V_B$. In order to show this, it is necessary to consider the variations with respect to x and t together. In order to illustrate the essential point, assume an actual situation in which there are two instruments available for observing the voltage along a given lossless transmission line. One instrument is an oscilloscope, which will display the entire waveform as a function of time for any value of x. The other instrument is a peak reading voltmeter, which reads only positive (or negative) peak values for any given x. The experimental observations to be expected from these two instruments are shown schematically in Fig. 2-3 for the case where $V_B = 0.5 \, V_A$ and both have the

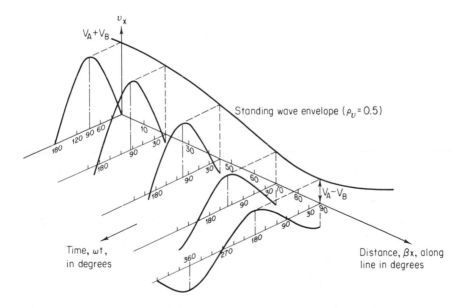

Fig. 2-3. Illustration of standing wave for $V_B = 0.5 \, V_A$; peaks of voltage vs. time at various points along line form standing wave envelope.

same phase relation at $x = 0$, that is, in phase at $x = 0$. The distance and time scales are actually in βx and ωt, respectively (both are measured in degrees in order to be more general). It is apparent that for any given x, the voltage is sinusoidal and would be observed on the oscilloscope just as shown.* The peak reading voltmeter would give a point-by-point plot of the "envelope" as a function of x, which is just a projection of the peak values of the time-varying sinusoids into the x-v plane, as shown.

A mathematical expression for this envelope can be obtained by evaluating the peak voltage at any given value of x. Since the value of t at which the peak voltage occurs (hereafter called t_{max}) varies with x, it is necessary to first determine t_{max} as a function of x. This can be done simply by taking the time derivative of Eq. (2-55) and equating it to 0

$$(V_A + V_B) \cos \beta x \cos \omega t + (V_A - V_B) \sin \beta x \sin \omega t = 0$$

or

$$\tan \omega t_{max} = -\frac{V_A + V_B}{(V_A - V_B) \tan \beta x} \tag{2-56}$$

Unfortunately, this is a transcendental equation and t_{max} cannot be obtained explicitly in terms of known parameters. The envelope is obtained by evaluating t_{max} for each value of x with the aid of Eq. (2-56); then, by substituting these values into Eq. (2-55), the peak voltage v_{max} can be evaluated

$$v_{max} = (V_A + V_B) \sin \omega t_{max} \cos \beta x - (V_A - V_B) \cos \omega t_{max} \sin \beta x \tag{2-57}$$

The envelope does not change with time but rather remains stationary; hence the name "standing wave." It is easily seen that the envelope has an identical curve in the $-v$ direction, obtained from the negative peak values of the time-varying sinusoids. If V_B were to be reduced to 0, the standing-wave envelope would degenerate into a straight line of amplitude V_A, that is, no standing wave present. If V_B were to be made larger than $0.5\,V_A$, it is apparent that the ratio of maximum to minimum value of the envelope would be greater. Standing waves can only occur when oppositely traveling waves are present and the ratio of maximum to minimum value of the envelope is a measure of the relative amounts of oppositely traveling waves. A useful figure of merit

---

*This assumes that the oscilloscope is triggered by the voltage at $x = 0$, so that the origin of the time scale for any x position is identical to that of the input voltage.

often employed in transmission-line work is the "standing wave ratio," abbreviated SWR. It is defined as the ratio of the absolute value of the maximum to the minimum value of the envelope

$$SWR = \frac{|V_A + V_B|}{|V_A - V_B|} \tag{2-58}$$

2.9 REFLECTION COEFFICIENT

In Sec. 2.8, we assumed that there were oppositely traveling waves present on a given transmission line without worrying about how such waves were generated. It is clear that all, or at least part, of the positively traveling wave could simply be the applied voltage; the negatively traveling wave must come from some other source. We now wish to consider a general source of negatively traveling waves, namely, reflections incurred from improper termination of a line. Before proceding with the mathematical analysis, we might consider the physical situation in order to better understand the nature of reflections, since they are of fundamental importance and will be used extensively in Chap. 5.

It has been shown in Chap. 1 that a lossless transmission line of infinite length looks like a pure resistance of value $Z_0 = \sqrt{L/C}$ to any excitation. If such a line is broken at any point and terminated in a resistor of this value, that is, $R_\ell = Z_0$ in Fig. 2-4(a), it will behave (externally) just as the infinitely long line. The power consumed by this load is simply

$$\frac{V_{rms}^2}{Z_0} \tag{2-59}$$

Suppose now that the terminating resistor R_ℓ is changed such as not to equal the characteristic impedance of the line. The power or energy which is traveling in the line before it reaches the termination is given by Eq. (2-59). The power dissipated in the new load resistor R_ℓ is

$$\frac{V_{rms}^2}{R_\ell} \tag{2-60}$$

Clearly, this is different from the power or energy which is incident on the load. If R_ℓ is greater than Z_0, there will be extra energy available at the load

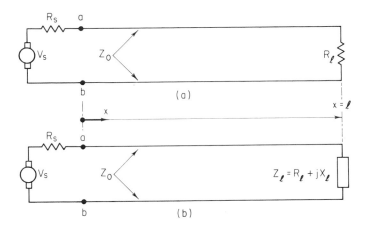

Fig. 2-4. Terminated transmission lines.

and since this energy cannot simply disappear, it is reflected back to the generator where it will be consumed in the internal impedance of the generator (assuming that this source impedance equals Z_0). If the load resistor R_ℓ is smaller than Z_0, the load will attempt to dissipate more energy than is available. Since it can only dissipate that which is available, a reflection will occur which is essentially a signal to the load to send more power. Both of these cases of improper termination will excite negatively traveling waves which will give rise to standing waves as was discussed in Sec. 2.8.

Since the amount of the negatively traveling wave clearly depends on the termination, it is desirable to find the ratio of the negative to the positive voltage in terms of the characteristic impedance and load impedance. We shall now derive this expression, known as the reflection coefficient, for the general case when the load impedance Z_ℓ is a complex quantity, i.e., resistive and reactive components, as in Fig. 2-4(b).

The reflection coefficient can easily be obtained from the previously derived equations for current and voltage

$$v_x = V_A e^{-\gamma x} + V_B e^{\gamma x} \tag{2-61}$$

$$i_x = I_A e^{-\gamma x} + I_B e^{\gamma x} \tag{2-62}$$

where I_A and V_A are amplitudes of the positively traveling components, I_B and V_B are amplitudes of the negatively traveling components, and x is the distance from the source generator. It is desirable to determine the voltage reflection coefficient at the load or the ratio of the negatively to the positively traveling wave at $x = \ell$. From Eq. (2-61) it is seen that at $x = \ell$, this ratio in both amplitude and phase is

$$\rho_v = \frac{V_B e^{\gamma \ell}}{V_A e^{-\gamma \ell}} = \frac{V_B}{V_A} e^{2\gamma \ell} \tag{2-63}$$

The constants V_A and V_B are given by Eqs. (2-18) and (2-19); substituting and simplifying yields

$$\rho_v = \left[e^{-2\gamma \ell} \left(\frac{Z_\ell - Z_0}{Z_\ell + Z_0} \right) \right] e^{2\gamma \ell} = \frac{Z_\ell - Z_0}{Z_\ell + Z_0} \tag{2-64}$$

An expression for the current reflection coefficient can be obtained in a very similar manner. From Eq. (2-62), it is apparent that at the load $x = \ell$

$$\rho_i = \frac{I_B e^{\gamma \ell}}{I_A e^{-\gamma \ell}} \tag{2-65}$$

I_A and I_B are given by Eq. (2-13b); substitution of these into Eq. (2-65) gives

$$\rho_i = -\frac{V_B}{V_A} e^{2\gamma \ell} = \frac{Z_0 - Z_\ell}{Z_\ell + Z_0} = -\rho_v \tag{2-66}$$

Thus, the current reflection coefficient is just the negative of the voltage reflection coefficient. It is apparent from Eqs. (2-64) and (2-66) that when $Z_\ell = Z_0$, both reflection coefficients are zero, as expected, since there can be no negatively traveling wave on a matched line. Also, it is easily seen that

$$0 \le |\rho_v| = |\rho_i| \le 1 \tag{2-67}$$

The reflection coefficient and Voltage Standing Wave Ratio (VSWR) are interdependent, as might be anticipated. This can easily be shown by rewriting Eq. (2-58)

$$\text{VSWR} = \frac{|1 + (V_B/V_A)|}{|1 - (V_B/V_A)|}$$

It is apparent in view of Eq. (2-63) that for a lossless line, $V_B/V_A = |\rho|$, so that

$$\text{VSWR} = \frac{1 + |\rho_v|}{1 - |\rho_v|} \tag{2-68}$$

Since $0 \leq |\rho_v| \leq 1$

$$1 \leq \text{VSWR} \leq \infty$$

The VSWR is a very useful concept when dealing with single frequencies and is often employed as a figure of merit in specifying commerical high-frequency equipment. It is a meaningless concept when pulses are propagating on a line, although the reflection coefficient can still be quite useful. The reason for the latter is that Eqs. (2-64) and (2-66) are quite general expressions. If Z_ℓ and Z_0 are both independent of frequency, then these equations give the amount of reflection experienced by a pulse as well as a single frequency. This concept will be used extensively in Chap. 5 as well as in other places throughout this book.

The reflection coefficient gives the amount of voltage (or current) which is reflected back, away from the load. From Eq. (2-12), the amount of voltage transmitted to the load is the sum of the incident plus reflected wave

$$v_x \Big|_{x=\ell} = V_\ell = V_A e^{-\gamma \ell} + V_B e^{\gamma \ell}$$

The ratio of the voltage at the load to the incident voltage, that is, $+x$ traveling wave, is defined as the transmission coefficient

$$\tau_v = \frac{V_\ell}{V_A e^{-\gamma \ell}} = 1 + \frac{V_B}{V_A} e^{2\gamma \ell} = 1 + \rho_v \tag{2-69}$$

In many cases, the load on a line is another transmission line; thus, Eq. (2-69) gives the voltage incident on the second line.

In a similar manner, the current transmission coefficient is

$$\tau_i = 1 + \rho_i \tag{2-70}$$

2.10 INPUT IMPEDANCE AS A FUNCTION OF TERMINATION AND LINE LENGTH FOR AC EXCITATION

It was previously shown in Secs. 1.3 and 2.4 that a lossless line terminated in its characteristic impedance will always have an input impedance which is a pure resistance of value $\sqrt{L/C}$. For such a case, a single wave traveling in the positive-x direction is present on the line, as in Sec. 2.8. If such a line is not terminated in its characteristic impedance, a reflection will occur, resulting in a standing wave, as shown in Sec. 2.9. When such standing waves are present on a line, the input impedance is no longer a pure resistance but rather can take on various values, including reactive components, depending on the termination and electrical length of the line. We will now derive the general expression for the input impedance of any line which is not properly terminated. More specifically, we shall determine the impedance of a line at any point x, looking toward the load with coordinates as in Fig. 2-4. This impedance is given simply by the ratio of the voltage to the current at that point. These can be obtained from Eqs. (2-14) and (2-15)

$$Z_x = \frac{v_x}{i_x} = Z_0 \frac{\left(V_A e^{-\gamma x} + V_B e^{\gamma x}\right)}{\left(V_A e^{-\gamma x} - V_B e^{\gamma x}\right)} = Z_0 \frac{\left(1 + (V_B/V_A)e^{2\gamma x}\right)}{\left(1 - (V_B/V_A)e^{2\gamma x}\right)} \qquad (2\text{-}71)$$

This can be expressed in terms of the reflection coefficient as follows. From Eq. (2-63)

$$\frac{V_B}{V_A} = \rho_v e^{-2\gamma \ell} \qquad (2\text{-}72)$$

Substitution of this into Eq. (2-71) yields

$$Z_x = Z_0 \frac{\left(1 + \rho_v e^{-2\gamma \ell} e^{2\gamma x}\right)}{\left(1 - \rho_v e^{-2\gamma \ell} e^{2\gamma x}\right)} \qquad (2\text{-}73)$$

If the input impedance of a line of length ℓ is desired, then Z_x is the impedance at the input terminals; thus, $x = 0$ and Eq. (2-73) becomes

$$Z_{\text{in}} = Z_0 \frac{\left(1 + \rho_v e^{-2\gamma \ell}\right)}{\left(1 - \rho_v e^{-2\gamma \ell}\right)} \qquad (2\text{-}74)$$

The above is an important equation which will be used in Sec. 2-11.

Equation (2-71) can also be expressed in terms of the load and characteristic impedances either by substituting Eqs. (2-18) and (2-19) for V_A and V_B or by substituting Eq. (2-64) into Eq. (2-73). In either case, the result is

$$Z_x = Z_0 \left[\frac{Z_\ell + Z_0 \left(e^{-\gamma x} - e^{-2\gamma\ell}e^{\gamma x}\right)\left(e^{-\gamma x} + e^{-2\gamma\ell}e^{\gamma x}\right)^{-1}}{Z_0 + Z_\ell \left(e^{-\gamma x} - e^{-2\gamma\ell}e^{\gamma x}\right)\left(e^{-\gamma x} + e^{-2\gamma\ell}e^{\gamma x}\right)^{-1}} \right] \quad (2\text{-}75)$$

where ℓ is the length of the line. Equation (2-75) represents the most general case of the impedance of a line of length ℓ looking toward the load from some point x away from the generator. If we let x go to zero, i.e., determine input impedance of total line, then Eq. (2-75) simplifies to

$$Z_{in} = Z_0 \left[\frac{Z_\ell + Z_0 \tanh \gamma\ell}{Z_0 + Z_\ell \tanh \gamma\ell} \right] \quad (2\text{-}76)$$

For a lossless line, $\gamma\ell$ is just a measure of the total phase shift of the given line, since $\gamma = j\beta$. For a given transmission line, β is a function of frequency, so that the input impedance of the line depends on the applied frequency and physical length of line in addition to the characteristic and termination impedance. This is seen by letting $\gamma = j\beta = j2\pi/\lambda_w$; Eq. (2-76) becomes

$$Z_{in} = Z_0 \frac{[Z_\ell + jZ_0 \tan(2\pi/\lambda_w)\ell]}{[Z_0 + jZ_\ell \tan(2\pi/\lambda_w)\ell]} \quad (2\text{-}77)$$

If the line is shorted so that $Z_\ell = 0$, then the input impedance is obviously

$$Z_{in} = Z_0 \tanh \gamma\ell \quad (2\text{-}78)$$

or for a lossless line

$$Z_{in} = Z_0 \tanh j\beta\ell = jZ_0 \tan \frac{2\pi}{\lambda_w} \ell \quad (2\text{-}79)$$

From the above, it is seen that if the line length equals exactly one-fourth of the wavelength of the applied frequency, that is, if $\ell = \lambda_w/4$, then

$$Z_{in} = jZ_0 \tan \frac{\pi}{2} = \infty$$

The input impedance of a quarter-wavelength line, short-circuited at its load, looks like an infinite impedance to the source generator. This is a very important result which is used in numerous ways.

In a similar manner, if the line is open-circuited so that Z_ℓ is infinite, then the input impedance for an ideal line becomes

$$Z_{in} = Z_0 \coth j\beta\ell = -jZ_0 \cot \frac{2\pi}{\lambda_w} \ell \qquad (2\text{-}80)$$

Once again, if $\ell = \lambda_w/4$, then the above gives zero input impedance. Thus, a quarter-wavelength, open line looks like a short circuit, which is also an important result.

For the general case when Z_ℓ is not zero (when it is real or imaginary or both), the input impedance will contain real and imaginary parts. Evaluation of these parts becomes a tedious task which can be greatly simplified by the use of normalized (universal) charts. These are discussed in the next section.

2.11 TRANSMISSION-LINE CHARTS

In the following, only ideal lines will be considered. For an ideal line terminated in a resistance equal to its characteristic impedance, the input impedance will always be resistive and equal to Z_0 for any length of line. However, in many cases, a line is not terminated in its characteristic impedance either through necessity or, as is often the case, on purpose. For the general case of resistive plus reactive termination, it is apparent that solution of Eq. (2-74) or Eq. (2-77) results in both real and imaginary components to the input impedance which depend on Z_ℓ as well as the length and phase constant of the line. Although the solution of this equation is straightforward in principle, the required computations become extensive and tedious for involved problems. Thus, it is desirable to devise some simple means of evaluation. There are several such means available, two of which are the rectangular* and Smith chart. The latter is the most useful and is the better known, and so we shall concentrate on it.

The input impedance of a line of any length ℓ is given by Eq. (2-74)

$$\frac{Z_{in}}{Z_0} = R_n + jX_n = \frac{1 + \rho e^{-2j\beta\ell}}{1 - \rho e^{-2j\beta\ell}} \tag{2-81}$$

where R_n and X_n are the normalized real and imaginary parts of the impedance looking toward the load.

It should be kept in mind that in general ρ can be complex, so let us simply represent it by a vector in the complex plane, i.e., let

$$w = u + jv = \rho e^{-2j\beta\ell} \tag{2-82}$$

Substitution into Eq. (2-81) gives

$$R_n + jX_n = \frac{1 + u + jv}{1 - (u + jv)} \tag{2-83}$$

The above is easily separated into real and imaginary parts by multiplying numerator and denominator by $1 - u + jv$, which, after simplification, yields

$$R_n = \frac{1 - u^2 - v^2}{(1 - u)^2 + v^2} \tag{2-84}$$

$$X_n = \frac{2v}{(1 - u)^2 + v^2} \tag{2-85}$$

Let us specify that R_n and X_n are constants and determine the contours in the u-v plane. If Eq. (2-84) is rearranged and if the term $R_n^2/(1 + R_n)^2$ is added to both sides to complete the square

$$\left(u - \frac{R_n}{1 + R_n}\right)^2 + v^2 = \frac{1}{(1 + R_n)^2} \tag{2-86}$$

The above is obviously the equation of a circle of radius $(1 + R_n)^{-1}$ with

---

*See [2, p. 435] or [1, p. 251].

center at $u = R_n/(1 + R_n)$, $v = 0$. In a similar manner, it can be shown that Eq. (2-85) simplifies to

$$(u - 1)^2 + \left(v - \frac{1}{X_n}\right)^2 = \frac{1}{X_n^2} \tag{2-87}$$

which is also a circle of radius $1/X_n$ and center $u = 1$, $v = 1/X_n$. Thus, the contours for any preassigned values of R_n and X_n can easily be constructed; several such contours are shown in Fig. 2-5. In the usual form of

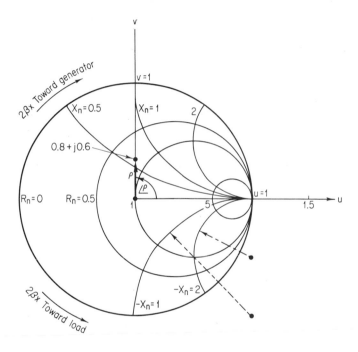

Fig. 2-5. Construction of Smith chart showing reflection coefficient of $0.333\underline{/90°}$ and normalized impedance values.

a Smith chart, many such contours are given with the scales of the u and v axes removed. In order to avoid confusion, it is helpful to keep in mind that such a chart is essentially a superposition of two quantities as follows: (1) the $w = u + jv$ plane with $|u| = |v| = 1$ at the outer edge (largest circle) is a vector or polar plot of the magnitude and phase of the reflection coefficient at the load; (2) the values of R_n and X_n corresponding to each uv

point in the plane are shown as circles of constant R_n or X_n. Thus, for any value of load impedance R_n, X_n, and $x = 0$, the reflection coefficient from Eq. (2-82) is determined as the uv vector from the origin to the intersection of the R_n and X_n circle. For example, for a load impedance (normalized) of $0.8 + j\ 0.6$, the reflection coefficient from Fig. 2-5 is found to be $0.333\underline{/90°}$. The angle is read directly as the angle in the u-v plane. However, if it is desired to determine the reflection coefficient as measured from some point a distance $x = \ell$ away from the load, then it is necessary to move on a circle of constant radius $|\rho|$ through an angle $2\beta\ell$ in a clockwise direction, as specified by Eq. (2-82). For instance, if ℓ is one quarter-wavelength long, then $2\beta\ell = \pi$, so that the reflection coefficient given above, seen looking into the line at this point, would be $0.333\underline{/-90°}$, i.e., proceed from ρ in Fig. 2-5 a distance π in the clockwise direction (this angle $2\beta\ell$ is usually specified around the outer periphery of the chart and the direction toward generator or toward load is also specified).

Because of the way the chart is constructed, if any two of the quantities, input impedance, line length, reflection coefficient, VSWR, or load impedance are specified, the others can be obtained from the chart. For instance, suppose that the input impedance and length of line are given as $0.5 + j\ 0.5$ and ℓ is one-eight of a wavelenght ($2\beta\ell = \pi/2°$). The input impedance is immediately determined as the intersection of the two circles $R_n = 0.5$ and $X_n = 0.5$, as shown in Fig. 2-6. The reflection coefficient seen looking in at that point ($x = \ell$) is $0.45\underline{/116.5°}$. The reflection coefficient at the load and the load impedance are obtained by tracing a circle of constant radius through

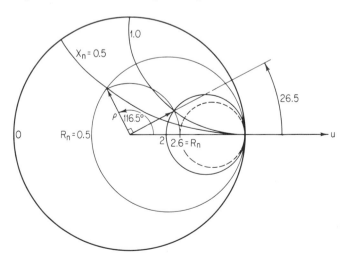

Fig. 2-6. Example of use of Smith chart for $2\beta\ell = \pi/2$ and $\rho = 0.45\underline{/26.5°}$ at load.

an angle of 90° in the clockwise direction to read $\rho = 0.45\underline{/26.5°}$ and load impedance of $2 + j\ 1$. The VSWR is obtained as 2.6 by reading the value of R_n given by the intersection of the reflection coefficient circle with the u-axis between $R_n = 1$ and infinity, while the position of voltage maximum or current minimum is then 26.5°/2 or about 0.075 wavelengths from the load (proof is left as an exercise).

It should be apparent that complex problems can be solved for lines with various stub interconnections and many others ([2, p. 439], or [3, p. 349]). The most serious drawback is that such charts are only useful for a single frequency. An applied pulse will contain many frequencies and the line behavior is best understood in terms of the concepts developed in Chap. 5. Nevertheless, the Smith chart* remains an important aid in understanding transmission lines and can be used, for instance, to understand the devices described in footnote 8 of [4] in Chap. 7, which use quarter-wavelength lines to achieve the necessary impedances.

2.12 IMPEDANCE MATCHING WITH TRANSMISSION-LINE SECTIONS

It is often necessary to match a line of one impedance Z_1 to another line of a different impedance Z_2. For a single applied frequency, this can often be done by using a quarter-wavelength line as a matching transformer, as in Fig. 2-7(a). In order to understand how this is accomplished, consider the equivalent circuit of Fig. 2-7(b). The impedance looking into terminals a-b is given by Eq. (2-77), or for this case, assuming ideal lines

$$Z_{in} = \left[\frac{Z_2 + jZ_0 \tan(2\pi/\lambda_w)\ell}{Z_0 + jZ_2 \tan(2\pi/\lambda_w)\ell}\right] Z_0 \qquad (2\text{-}88)$$

If $\ell = \lambda_w/4$, then $\tan(\pi/2)$ approaches infinity so

$$Z_{in} = \frac{Z_0^2}{Z_2} \qquad (2\text{-}89)$$

But Z_{in} must equal Z_1 for a proper match and if Z_1 and Z_2 are known, the

---

*Accurately scaled, normalized charts are available from most graphic suppliers.

impedance of the quarter-wave section must be

$$Z_0 = \sqrt{Z_1 Z_2} \qquad\qquad (2\text{-}90)$$

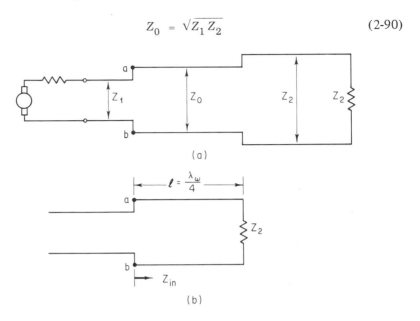

Fig. 2-7. Impedance matching using quarter-wavelength line.

While the above gives a perfect impedance match, it is often difficult in practice to obtain lines of the proper impedance. For instance, suppose that it is desired to match a 50-ohm line to a 93-ohm line, both of which are common commercial coaxial lines. From Eq. (2-90), this would require a matching section of impedance $(50 \times 93)^{1/2}$ or 68 ohms. The nearest commercial line is 73 ohms, so that unless construction of a line is undertaken. some mismatch must be tolerated.

It should be apparent that complex impedance matching can also be done by using sections of different length; the Smith chart is particularly suited for solving such problems.

2.13 NONUNIFORM TRANSMISSION LINES

In the previous analysis, it has been assumed that the transmission lines were uniform such that the line parameters R, L, G, and C were independent of the position x along the line. However, it is possible for these parameters to be a function of x either through necessity or by design. Nonuniform lines can be designed to have practical value, but for the present, we shall be concerned with the fundamental general equations for such a line.

In the derivation of the general equation of a nonuniform line, the incremental circuit Eqs. (2-20) and (2-21) serve as the starting point

$$\frac{\partial v_x}{\partial x} = -i_x R - L \frac{\partial i_x}{\partial i} \tag{2-91}$$

$$\frac{\partial i_x}{\partial x} = -v_x G - C \frac{\partial v_x}{\partial t} \tag{2-92}$$

where R, L, G, and C are now functions of x. In order to obtain an equation in terms of one variable, we must take the derivative of Eq. (2-91) with respect to distance x and the derivative of Eq. (2-92) with respect to time t. Thus

$$-\frac{\partial^2 v_x}{\partial x^2} = R \frac{\partial i_x}{\partial x} + i_x \frac{\partial R}{\partial x} + L \frac{\partial^2 i_x}{\partial x \partial t} + \frac{\partial i_x}{\partial t} \frac{\partial L}{\partial x} \tag{2-93}$$

$$\frac{\partial^2 i_x}{\partial t^2} = -G \frac{\partial v_x}{\partial t} - C \frac{\partial^2 v_x}{\partial t^2} \tag{2-94}$$

Substituting Eqs. (2-92) and (2-94) into Eq. (2-93)

$$-\frac{\partial^2 v_x}{\partial x^2} = R\left(-Gv_x - C \frac{\partial v_x}{\partial t}\right) + i_x \frac{\partial R}{\partial x} + L\left(-G \frac{\partial v_x}{\partial t} - C \frac{\partial^2 v_x}{\partial t^2}\right) + \frac{\partial i_x}{\partial t} \frac{\partial L}{\partial x} \tag{2-95}$$

The above equation still contains two variables but can be simplified by assuming R and G to be negligibly small, which is usually the case. Thus, by allowing R and G to approach zero and making use of Eq. (2-91) we get

$$\frac{\partial^2 v_x}{\partial x^2} = LC \frac{\partial^2 v_x}{\partial t^2} + \frac{1}{L} \frac{\partial L}{\partial x} \frac{\partial v_x}{\partial x} \tag{2-96}$$

In a similar manner, it can be shown that

$$\frac{\partial^2 i_x}{\partial x^2} = LC \frac{\partial^2 i_x}{\partial t^2} + \frac{1}{C} \frac{\partial C}{\partial x} \frac{\partial i_x}{\partial x} \tag{2-97}$$

These represent the general equations of a lossless line with position-dependent parameters. If the parameters do not vary with x, then Eqs. (2-96) and (2-97) reduce to the usual wave Eqs. (2-26) and (2-27) as expected.

PROBLEMS

2-1. Show that the characteristic impedance of a line is given by $Z_0 = \sqrt{Z_{sc} Z_{oc}}$ where Z_{sc} is the input impedance with a short-circuited load and Z_{oc} is the input impedance with open-circuited load.

2-2. Show that the propagation constant of a line is given by $\gamma = \sqrt{Z_{sc}/Z_{oc}}$ with parameters as defined in the previous problem.

2-3. Show that, in general, a standing wave is composed of two waves of the same frequency traveling in opposite directions. (This is an important fundamental concept, applicable to many physical problems.)

2-4. Explain why a large number of quarter-wavelength, shortcircuited stubs projecting from a long, terminated transmission line, as shown, have no effect on the impedance of the line. Note that in the limit of a very large number of such stubs such as to give a continuous rectangular box, the transmission line still has an impedance not equal to zero, although its propagation characteristics change; the line has then become a wave guide and TEM modes can no longer propagate.

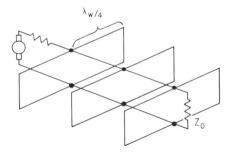

Fig. 2P-1.

REFERENCES

1. Jordan, E. G.: "Electromagnetic Waves and Radiating Systems," Prentice Hall, Inc., New York, 1950.
2. Kraus, J. C.: "Electromagnetics," McGraw-Hill Book Company, New York, 1953.
3. Skilling, H. H.: "Electric Transmission Lines," McGraw-Hill Book Company, New York, 1951.

3 VELOCITY OF PROPAGATION

3.1 INTRODUCTION

An understanding of the velocity of propagation of traveling waves is important in the study of transmission lines, especially at high frequencies, in radio-wave propagation using the ionosphere (Heaviside layer) to reflect sky waves back to earth, and in wave mechanics where the group velocity corresponds to the velocity of a particle which is guided by a wave. Even though it is usually possible to find some meaningful measure of the velocity of a wave in practical cases, this is nevertheless one of the concepts most often confused by a lack of understanding of the basic principles as well as of what certain concepts do and do not mean. We will approach this subject from a fundamental viewpoint, starting with very simple situations and proceeding to the more complex. The most simple case is that of a single frequency at steady state on a line, which leads to the concept of phase velocity, an obvious classification in light of Secs. 2.4 and 2.6. Two different frequencies present simultaneously on a line complicate the situation somewhat and give rise to the so-called group velocity, which is characteristic of "beat" phenomena associated with modulated carrier waves. For the transient analysis of a single frequency or a pulse (many frequencies) on a dispersive line, the situation becomes more involved and the velocity of propagation loses its simple meaning. We shall investigate each of these cases in order to clarify the various possibilities and the meaning of the term "velocity of a wave."

Before proceeding, some general statements should be made to help orient the reader and give a better understanding of the analysis to come. It is a fundamental law of nature that no energy or, what amounts to the same thing, no intelligible signal can ever be propagated in any medium faster than the velocity of light ($c = 3 \times 10^8$ m/sec). This concept is a fundamental assumption in Einstein's special theory of relativity, which sets this upper limit on all of nature and which has been experimentally verified. It will be

57

shown that in dispersive media, i.e., where characteristics are frequency-dependent, the group velocity can exceed the velocity of light but the actual signal or energy velocity must be less than c. The group velocity can have almost any value, positive, negative, or infinite, in a dispersive medium. When the media are not dispersive, the various terms take on their simple meaning for all cases, i.e., it is the dispersive quality of a medium which gives rise to distortion of the wave. A simple way of understanding this for a wave composed of several or many frequencies is to consider each frequency component of the wave separately. We shall use this approach whenever possible.

3.2 SINGLE FREQUENCY PRESENT ON A TRANSMISSION LINE—PHASE VELOCITY

In Chaps. 1 and 2 we developed the idea of phase shift and velocity of a traveling wave on a line for a single excitation frequency. The phase velocity is given by Eq. (2-38), which for an ideal line is

$$\nu = \frac{1}{\sqrt{LC}} = \frac{\omega}{\beta} \tag{3-1}$$

where β is the phase constant. The physical interpretation of phase velocity is illustrated in Fig. 3-1, in which the voltage as a function of distance along a line is plotted for several values of time. As pointed out in Sec. 2.7, the phase velocity is simply the velocity of a given phase point of the sinusoid, i.e., how fast a given point on the wave travels along the line.

3.3 TWO FREQUENCIES SIMULTANEOUSLY PRESENT ON A DISPERSIONLESS LINE—GROUP VELOCITY

Two or more frequencies must be present on a line in order to give rise to wave groups and group velocity. The distinction between phase and group velocity has been known for many years and was discussed by Lord Rayleigh in his study of the theory and propagation of sound. In order to understand how group velocity arises, let us consider two sinusoidal voltage waves of different frequency and therefore different phase constants to be linearly added on a transmission line. The two waves do not necessarily

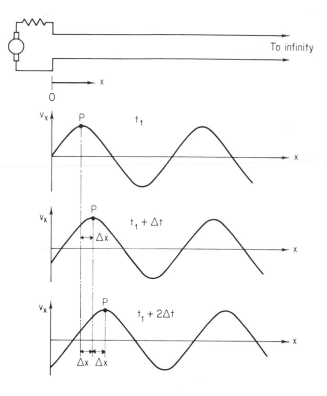

Fig. 3-1. Illustration of phase velocity for a single frequency traveling on an ideal line.

have the same amplitude; they are given by

$$v_1 = V_1 \cos(\omega_1 t - \beta_1 x)$$
$$v_2 = V_2 \cos(\omega_2 t - \beta_2 x)$$

$$(3\text{-}2)$$

Assume for simplicity's sake that $V_2 > V_1$ and $\omega_2 > \omega_1$. If these two voltages are added together linearly

$$v = v_1 + v_2 = V_1 \cos(\omega_1 t - \beta_1 x) + V_2 \cos(\omega_2 t - \beta_2 x) \qquad (3\text{-}3)$$

A more convenient form is

$$\frac{v}{V_1} = \cos(\omega_1 t - \beta_1 x) + (1 + m) \cos(\omega_2 t - \beta_2 x)$$

$$(3\text{-}4)$$

$$= \cos(\omega_1 t - \beta_1 x) + \cos(\omega_2 t - \beta_2 x) + m \cos(\omega_2 t - \beta_2 x)$$

where

$$m = \frac{V_2}{V_1} - 1 \qquad (3\text{-}5)$$

Making use of the identity

$$\cos\theta_1 + \cos\theta_2 = 2\cos\left(\frac{\theta_1 + \theta_2}{2}\right)\cos\left(\frac{\theta_2 - \theta_1}{2}\right) \qquad (3\text{-}6)$$

Eq. (3-4) becomes

$$\frac{v}{V_1} = m\cos\theta_2 + 2\cos\left(\frac{\theta_2 + \theta_1}{2}\right)\cos\left(\frac{\theta_2 - \theta_1}{2}\right)$$

or

$$v = (V_2 - V_1)\cos(\omega_2 t - \beta_2 x)$$

$$+ 2V_1 \cos\left[\frac{(\omega_2 + \omega_1)t - (\beta_2 + \beta_1)x}{2}\right]\cos\left[\frac{(\omega_2 - \omega_1)t - (\beta_2 - \beta_1)x}{2}\right]$$

$$(3\text{-}7)$$

This is the general form obtained with two frequencies of unequal amplitude present on a line. In order to illustrate the essential idea, we simplify the above expression by assuming that $V_2 = V_1$

$$v = 2V_1 \cos\left[\frac{(\omega_2 + \omega_1)}{2}t - \frac{(\beta_2 + \beta_1)}{2}x\right]\cos\left[\frac{(\omega_2 - \omega_1)}{2}t - \frac{(\beta_2 - \beta_1)}{2}x\right]$$

$$(3\text{-}8)$$

This represents a voltage containing the average of the sum and difference of the two applied frequencies.

If ω_2 is very nearly equal but decidedly different from ω_1, then Eq. (3-8) becomes a so-called modulated carrier. This is more readily seen by letting

$$\varphi_0 = \frac{\omega_2 + \omega_1}{2} \qquad \Delta\omega = \frac{\omega_2 - \omega_1}{2}$$

$$\beta_0 = \frac{\beta_2 + \beta_1}{2} \qquad \Delta\beta = \frac{\beta_2 - \beta_1}{2}$$

(3-9)

Substituting these, Eq. (3-8) becomes

$$v = 2V_1 \cos(\omega_0 t - \beta_0 x) \cos(\Delta\omega t - \Delta\beta x) \qquad (3\text{-}10)$$

In a sufficiently small time interval, the second cosine term in Eq. (3-10), that is, $\cos[\Delta\omega t - \Delta\beta x]$, does not change appreciably while the first term may undergo many cycles. This results from the fact that we assumed ω_2 to be nearly equal to ω_1, or $\Delta\omega \ll \omega_0$. Thus, ω_0 represents the carrier while $\Delta\omega$ represents the slower modulation frequency, as illustrated in Fig. 3-2,* where

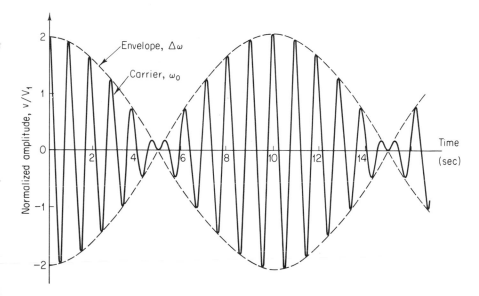

Fig. 3-2. Modulated carrier obtained from the linear addition of two nearly equal frequencies $f_2 = 1.1 f$, $f_1 = 1$ Hz.

*We obtained the modulated carrier by linearly adding two distinct frequencies which had very nearly the same frequency, resulting in a carrier and two sidebands. This is different in detail but similar in principle to the method of carrier modulation usually used in AM radio, where a high-frequency carrier is mixed with a low-frequency modulator (the two widely separated frequencies are essentially multiplied together rather than linearly added). If the two widely different frequencies were linearly added, then a modulated carrier would not result: see Prob. 3-1.

$\omega_2 = 1.1\omega_1$. The carrier ω_0 is essentially sinusoidal, i.e., single-frequency, with an amplitude which varies essentially sinusoidally, but at a very slow rate. This envelope will travel along with the carrier and both will preserve their nearly sinusoidal shape. Thus, it is possible to associate a velocity of propagation with the constant phase points of the envelope which is commonly known as the group velocity, since this measures the velocity of the hills and valleys (groups) of the waveform. In other words, group velocity is the phase velocity of the groups or envelope. The constant phase points of the envelope can be obtained from Eq. (3-10) by setting the argument of the modulating term equal to a constant or zero. Thus

$$\Delta\omega t - \Delta\beta x = 0$$

or group velocity is

$$u = \frac{x}{t} = \frac{\Delta\omega}{\Delta\beta} \tag{3-11}$$

The phase velocity of the carrier can be obtained from Eq. (3-1)

$$\mathcal{v} = \frac{\omega_0}{\beta_0} \tag{3-12}$$

For the line we have been considering, the velocity of propagation is independent of frequency (dispersionless), so that ω_1 and ω_2 propagate at the same velocity. Figure 3-3, in which the phase constant is plotted as a

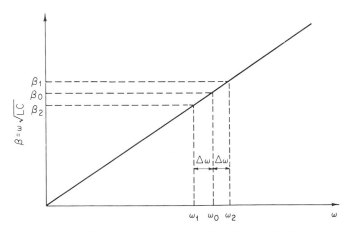

Fig. 3-3. Phase constant vs. frequency for a dispersionless line.

function of frequency showing the relative values of ω_1, ω_2, ω_0, and $\Delta\omega$, illustrates this. Since this curve is a straight line, both the group and phase velocity are given by the slope, which is a constant. Thus, the group and phase velocity are identical and no ambiguity arises with two very nearly equal frequencies on a dispersionless line.

In the above case, when ω_1 and ω_2 were nearly equal so that $\omega_0 \gg \Delta\omega$, a modulated carrier was obtained. Now if ω_1 and ω_2 are not equal, so that $\Delta\omega$ is on the order of ω_0, the result is quite different. If ω_1 and ω_2 are not integer multiples of one another, then the linear addition of two very different frequencies gives a waveform which is not sinusoidal and which in general is very erratic in amplitude. The waveform of the total wave will nevertheless still be periodic, as will be the envelope. The envelope consists of the peaks of the waveform and will not be sinusoidal. The situation where a linear summation of two frequencies of equal amplitude is plotted for $\omega_2 = 1.8\omega_1$ is shown in Fig. 3-4. It is apparent from the lack of sinusoidal representation of the "carrier" or "modulation" (envelope) that a constant phase point of either is more difficult to describe mathematically than for the previous case when ω_1 was nearly equal to ω_2. Nevertheless, phase points can be identified and if the voltage is measured at two points separated by a known distance ℓ by means of a dual trace oscilloscope, for instance, the difference in time between corresponding phase points such as P_1 and P_2 or P_1' and P_2' will give the delay time one way. Dividing the distance ℓ by the delay time will give the velocity of propagation of point P_1 (or an integer multiple of it if the line is more than one wavelength long). Since the groups no longer contain many cycles of the carrier wave, the group velocity as associated with Fig. 3-2 loses its simple meaning and becomes synonymous with phase velocity.* The phase velocity of the carrier of Fig. 3-4 is identical to the velocity of propagation of the two individual frequency components.

3.4 MANY FREQUENCIES PRESENT ON A DISPERSIONLESS LINE—GROUP VELOCITY

It is easily seen from the complexity of Fig. 3-4 for two frequencies that simple visualization of the composite waveform behavior is difficult to achieve when many frequencies are linearly superimposed. If the frequency spectrum of interest becomes continuous, it is necessary to use Fourier integrals to analyze the behavior.

---

*This follows from the fact that the envelope can no longer be identified. This can be seen by connecting the peaks in Fig. 3-4 with straight lines: The envelope so formed is meaningless.

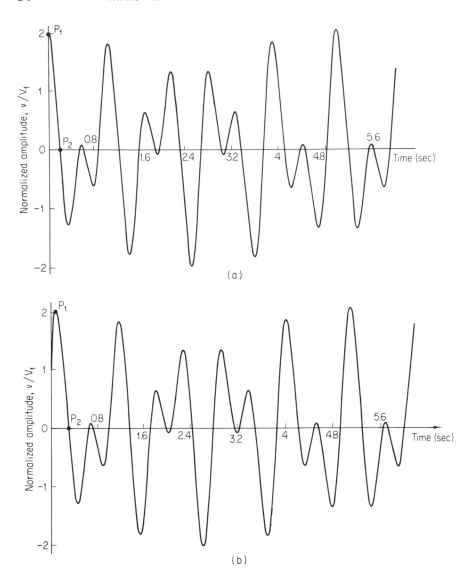

Fig. 3-4. Linear addition of two widely different frequencies $f_1 = 1$, $f_2 = 1.8 f_1$; $V_1 = V_2$ and equal phase velocity of 1 m/sec; $\beta_1 \ell = \pi/4$, $\beta_2 \ell = 1.8 \pi/4$. (a) input at $x = 0$; (b) output at $x = \ell$.

Let us consider the case of a continuous spectrum of frequencies $g(\omega)$ present on a line and determine the group velocity. The Fourier integrals for a traveling wave yield

$$g(\omega) = \int_{-\infty}^{\infty} v(x, t) \exp[-j(\omega t - \beta x)] \, dt \tag{3-13}$$

$$v(x, t) = \frac{1}{2\pi} \int_{-\infty}^{\infty} g(\omega) \exp[j(\omega t - \beta x)] \, d\omega \tag{3-14}$$

If we are given the frequency spectrum, we can find $v(x, t)$ or vice versa. We will not consider a specific case, as this is unnecessary, but will specify that the frequency spectrum $g(\omega)$ has a significant amplitude only within a region of $2\delta\omega$, as shown in Fig. 3-5. This is equivalent to specifying that $\Delta\omega$ is small or that ω_1 is nearly equal to ω_2, as was done in Sec. 3.3. Thus, we can expect to have a modulated carrier which is given by

$$v(x, t) = \frac{1}{2\pi} \int_{\omega_0 - \delta\omega}^{\omega_0 + \delta\omega} g(\omega) \exp[j(\omega t - \beta x)] \, d\omega \tag{3-15}$$

where we have simply replaced the limits in Eq. (3-14) to correspond with our assumption that $g(\omega)$ is negligible outside the interval of $2\delta\omega$ about ω_0.

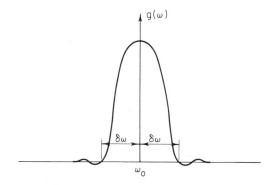

Fig. 3-5. Continuous frequency spectrum with significant portion extending only $2\delta\omega$ about ω_0.

We have assumed that the transmission line is dispersionless; thus the phase constant as a function of frequency is a straight line, as in Fig. 3-3

$$\beta(\omega) = \beta_0 + \frac{\partial\beta}{\partial\omega}(\omega - \omega_0) \tag{3-16}$$

Using this expression, the exponent in Eq. (3-15) can be rewritten

$$\omega t - \beta x = \omega_0 t + (\omega - \omega_0)t - \left[\beta_0 + \frac{\partial\beta}{\partial\omega}(\omega - \omega_0)\right]x$$

$$= \omega_0 t - \beta_0 x + (\omega - \omega_0)\left[t - \frac{\partial\beta}{\partial\omega}x\right] \tag{3-17}$$

Substituting this into Eq. (3-15)

$$v(x, t) = \frac{1}{2\pi} \int_{\omega_0 - \delta\omega}^{\omega_0 + \delta\omega} g(\omega) \exp[j(\omega_0 t - \beta x)] \exp\left\{ j\left[(\omega - \omega_0)\left(t - \frac{\partial\beta}{\partial\omega} x \right) \right] \right\} d\omega$$

(3-18)

It is apparent that this represents a modulated carrier with carrier frequency ω_0 and modulation frequency $\delta\omega$, very similar to the situation in Fig. 3-2. The slower time variation represents the groups or envelope and is given by the second exponential term in Eq. (3-18). It is apparent that the amplitude of the envelope will be constant at points defined by

$$(\omega - \omega_0)\left(t - \frac{\partial\beta}{\partial\omega} x \right) = \text{constant}$$

These represent constant phase points of the envelope; thus, the group velocity is obviously given by $t - x\partial\beta/\partial\omega = 0$ which yields

$$u = \frac{x}{t} = \frac{1}{\partial\beta/\partial\omega}\bigg|_{\omega=\omega_0}$$

(3-19)

This is, of course, essentially the same as Eq. (3-11) for the case when just two frequencies are present, as would be expected provided that $\delta\omega$ is small.

In the above derivation of group velocity, it was necessary to assume that $\delta\omega$ was small in order to obtain a slow time varying wave modulating a higher-frequency carrier. This allows a simple distinction between the groups or envelope and the carrier. However, in a dispersionless medium, group and phase velocity are always identical, so that it is not necessary to restrict $\delta\omega$ to be small. Nevertheless, as $\delta\omega$ becomes larger, the groups become less well defined (as in Fig. 3-4), so that only phase velocity has any significance. In Sec. 3.7 we shall consider the same problem, though in a dispersive medium, where we will find that $\delta\omega$ must be small in order for us to have any definition of group velocity at all. Otherwise, the velocity of the groups is ambiguous.

The reader is cautioned against placing too much emphasis on the concept of group and phase velocity, since they merely represent the relative arrangement of phases of the various frequency components of a steady-state wave. As we shall see later, meaningful velocities of propagation must be determined for a transient rather than a steady-state wave, for which the group and phase velocity have no direct bearing on the signal and wavefront velocity.

3.5 DISPERSION IN MATERIALS

The remainder of this chapter will deal with dispersive lines for which the propagation constant of a given frequency on the line depends on the frequency.* We will first digress to discuss the phenomenon of dispersion and later will consider the effect on propagation properties of dispersive lines. Those readers not interested in this subject can skip to Sec. 3.6 with no loss of continuity.

Since the ambiguity concerning the velocity of propagation of a signal stems primarily from the phenomenon of dispersion, it is important to have a clear understanding of the general character of dispersion. If no dispersion is present, i.e., no losses and no parameters are frequency-dependent, then group velocity and phase velocity become identical and no ambiguity can arise. However, all materials are dispersive to some degree, being more so at certain frequencies than at others.

Historically, the phenomenon of frequency-dependent dispersion has been known for many years, having been associated primarily with optical phenomena and the index of refraction. We are all familiar with the separation of white light which is passed through a glass prism into its component colors as a result of the dispersive nature of glass. The higher frequencies, e.g., violet, are bent more than the lower frequencies, e.g., red, since the index of refraction is larger for the higher frequencies. This is just another way of stating that the higher frequencies travel more slowly or have a smaller phase velocity than the lower frequencies. This is generally true for optical materials when they are excited sufficiently far from an absorption band. Since dispersion curves were first compiled for such cases, this became known as "normal" dispersion. The general situation is illustrated in Fig. 3-6(a). The index of refraction is proportional to the reciprocal of the phase velocity

$$n = \frac{c}{v} \tag{3-20}$$

*In addition to frequency-dependent dispersion, there is another type, namely, amplitude-dependence, which arises from a variation in the transmission-line parameters with amplitude of the applied voltage or current. This type is not nearly so common but is important in some applications. We will not consider this type. There is also dispersion caused by skin effect, treated in Sec. 5.11.

(n increases with frequency while the phase velocity decreases with frequency).

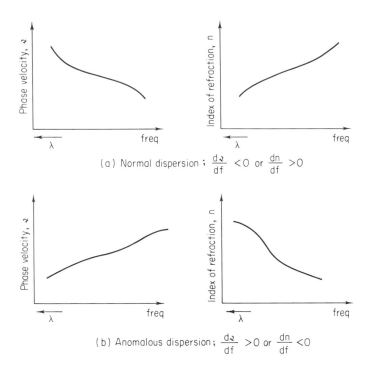

(a) Normal dispersion ; $\dfrac{d\vartheta}{df} <0$ or $\dfrac{dn}{df} >0$

(b) Anomalous dispersion; $\dfrac{d\vartheta}{df} >0$ or $\dfrac{dn}{df} <0$

Fig. 3-6. General behavior of phase velocity and index of refraction for normal and anomalous dispersion.

It is possible for the slope of the curves of Fig. 3-6(a) to have signs opposite those shown, in which case the curves might resemble those of Fig. 3-6(b). Since such materials differed from the normal case, they were arbitrarily classified as anomalous. This arbitrary classification is unfortunate in many respects and is often the source of much confusion since, in general, anomalous dispersion is much more common than normal dispersion. This can be understood from very simple concepts. For most nonmagnetic materials, the relative permeability is unity (permeability equal to that of free space), so that the index of refraction is

$$n = \sqrt{\epsilon_r} \qquad (3\text{-}21)$$

where ϵ_r is the relative dielectric constant. It is well known that, in general, the dielectric constant of materials decreases with increasing frequency, eventually approaching that of free space. However, the dielectric constant is not well behaved throughout the entire frequency spectrum, but rather the same value can exist for several different frequencies as a result of absorption bands. The general case of the relative dielectric constant as a function of frequency for a hypothetical material is shown in Fig. 3-7.*

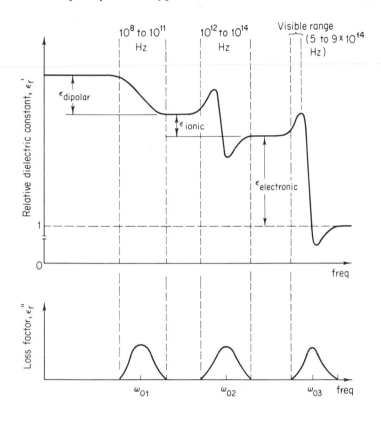

Fig. 3-7. Complex relative dielectric constant $\epsilon_r = \epsilon'_r - j\epsilon''_r$ of a hypothetical material.

It can be seen that in this case there are three major contributions to ϵ'_r, namely, dipolar, ionic, and electronic polarization. Since the slope of this curve essentially determines the group velocity, as will be shown in Sec. 3.6, the dispersion can be normal or anomalous, depending on the frequency being used. Points where $d\epsilon_r/df > 0$, that is, positive slope, correspond to

*Derivation and further discussions concerning this type of relationship can be found in [3, Chap. 6] and [7, Chap. 7].

normal dispersion, while the relationship $d\epsilon_r/df < 0$, that is, negative slope, indicates anomalous dispersion. It can be seen that the general trend of the curve is downward (anomalous) with only small regions of normal dispersion. Thus, such a material could not be generally classified as either anomalous or normal.

A typical example of materials of this type is glass. In the visible frequency range, glass exhibits normal dispersion since it refracts (slows down) the higher frequencies more than the lower frequencies. However, in the engineering range of interest, from 0 to about 10^{11} Hz, the dielectric constant, although nearly constant, protrays a distinct decrease with increasing frequency. The latter is also true for other engineering materials which are commonly used for transmission lines. The dielectric constant as a function of frequency is shown in Fig. 3-8* for several materials commonly encountered in transmission lines: ϵ_r' for teflon and quartz is a constant at 2.1 and 3.78, respectively, through 10^{10} Hz, while the others have a small but measurable decrease with frequency. All the materials show an increase in loss tangent (dielectric losses) as the frequency increases, but generally these are still quite small for most engineering purposes.

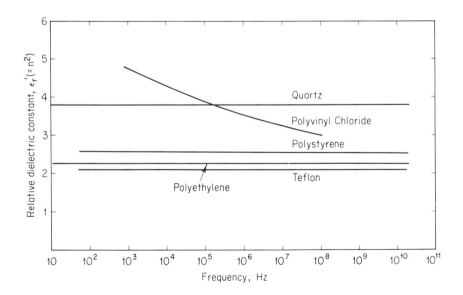

Fig. 3-8. Dielectric constant vs. frequency for several commonly used dielectrics.

*Data taken from [12]. For additional information, see [16].

In order to be more general, it would be necessary to extend the curves of Fig. 3-8 to much higher frequencies. If this were done, absorption bands would be found to exist much as illustrated in Fig. 3-7. The nature of these absorption bands is very important in determining the propagation characteristics of any given material, as will be seen in Sec. 3.8. A brief discussion and derivation of the dielectric constant near an absorption band is given in Appendix 3A, where it is shown that ω_{0n} are the characteristic frequencies of oscillation of the charged particles.

A metal, or any material with relatively large conductivity through which an electromagnetic wave is propagating, is generally considered to exhibit anomalous dispersion. In order to understand this, consider a voltage wave propagating on a transmission line which has an "insulator" exhibiting a large conductivity, that is, G is not negligible. If we assume that the series resistance of the line is negligible, then the propagation constant as given by Eq. (1-54) or Eq. (2-10) is

$$\gamma = \sqrt{j\omega L (G + j\omega C)} \qquad (3\text{-}22)$$

If the conductivity is sufficiently large (as in a metal), we can neglect the capacitance, so that Eq. (3-22) becomes

$$\gamma = \sqrt{j\omega LG} \qquad (3\text{-}23)$$

The parameters of any transmission line are given by

$$L = \mu K \qquad G = \frac{\sigma}{K}$$

where

K = geometry or form factor*
μ = permeability of the medium
σ = conductivity of the medium

Using these and the identity

$$\sqrt{j} = \frac{1 + j}{\sqrt{2}} \qquad (3\text{-}24)$$

---

*For a strip line with large width-to-separation ratio, for instance, K would be S/W; see Chap. 8 for further details.

Eq. (3-23) becomes

$$\gamma = (1 + j) \sqrt{\frac{\omega\mu\sigma}{2}} = \alpha + j\beta \qquad (3\text{-}25)*$$

The attenuation α and the phase constant β have equal amplitude. The phase velocity is easily found

$$\nu = \frac{\omega}{\beta} = \frac{\omega}{\sqrt{\omega\mu\sigma/2}} = \sqrt{\frac{2\omega}{\mu\sigma}} \qquad (3\text{-}26)$$

It is apparent from the above that $\partial\nu/\partial\omega$ is positive; thus, the conductor must exhibit anomalous dispersion. This can be more clearly shown by a comparison of group and phase velocity. The group velocity which would exist for a narrow band of frequencies as given by Eq. (3-19) is

$$u = \frac{1}{\partial\beta/\partial\omega} = 2\sqrt{\frac{2\omega}{\mu\sigma}} = 2\nu \qquad (3\text{-}27)$$

Thus, the group velocity is greater than the phase velocity as it should be in an anomalously dispersive medium.

It should be noted that even though conductors satisfy the requirement of $\partial\nu/\partial\omega > 0$ for anomalous dispersion, a wave propagating in such a medium will undergo severe distortion due to the large attenuation factor α. Each frequency component will experience a different amount of attenuation so that the voltage wave at a large distance from the input terminals may bear no resemblance to the applied voltage. For such cases, group velocity cannot be determined and must therefore be limited to distances for which the total attenuation is not significant.

3.6 TWO FREQUENCIES PRESENT ON A DISPERSIVE LINE

When two voltage waves which are sufficiently close in frequency are linearly added on a dispersive line, a modulated carrier which exhibits either normal or anomalous dispersion results. We pointed out in Sec. 3.5 that if the higher frequency travels with a slower phase velocity than the lower

---

*This is identical to the result that one would obtain by means of the application of Maxwell's equations to a plane wave in a conducting medium; see [8, p. 396] or [11, p. 307].

frequency, the dispersion is said to be normal and group velocity of the envelope is less than the phase velocity of the carrier wave. If the higher frequency travels faster than the lower frequency, the group velocity exceeds the phase velocity and the dispersion is termed anomalous. In order to clearly show the distinction between the significance of these two types of dispersion, we will consider an example consisting of two voltages of nearly the same frequency, linearly added first on a normally, and then on an anomalously dispersive line. The losses are assumed to be negligible for simplicity's sake, although in reality they have a significant effect, as discussed in Sec. 3.8. Before we proceed, it is helpful to digress for a moment and consider the matter of identifying the dispersion and determining the group velocity in terms of known parameters.

There are numerous parameters used to describe traveling waves and, therefore, also numerous methods and criteria for the evaluation and distinction of normal and anomalous dispersion. Since the phenomena arise as a result of the frequency-dependence of the parameters, it is reasonable to consider frequency as the independent variable. If the phase velocity v, index of refraction n, or phase constant β is known as a function of frequency (or wavelength), then the type of dispersion and value of group and phase velocity can easily be determined. For instance, suppose β versus frequency is known. If the line is well behaved, there is no dispersion at the lower frequencies. At the higher frequencies where dispersion occurs, there are numerous possibilities, as shown in Fig. 3-8. If the line is excited by a modulated carrier such as that shown in Fig. 3-2, then the phase velocity of the carrier is equal to the absolute value of $2\pi f/\beta$ at any given f and will always be a positive number. The group velocity is equal to the reciprocal of the slope of the curve at any given point. At all points where the slope (tangent) of the curve equals the slope of the line passing from that point through the origin, there is no dispersion, as illustrated by points Q_1 and Q_2 in Figs. 3-9(a) and (b). The dispersion on either side of these points will be normal or anomalous as shown. Since the group velocity equals the reciprocal of the slope of these curves, it is apparent that it can take on positive and negative values, as well as infinity at points where the slope is zero, as shown. It is possible for materials, especially when they are excited near absorption bands, to exhibit shapes similar to those shown in Fig. 3-9, for example, as in Fig. 3-7.

Suppose that instead of the phase constant, the phase velocity as a function of frequency is known. If the line is again well behaved, the velocity is independent of frequency in the lower frequency range and group velocity equals phase velocity. At the higher frequencies where dispersion appears, the curve might look like that in Fig. 3-9(c). The distinction

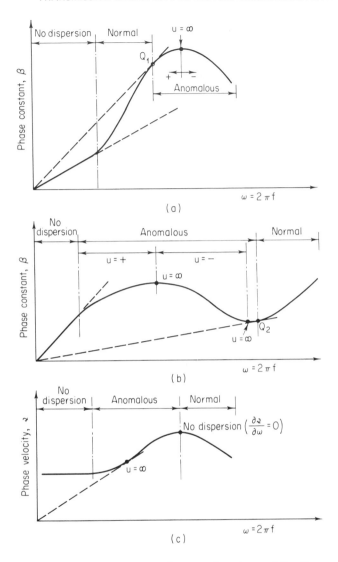

Fig. 3-9. Frequency variation of propagation characteristics for a general case indicating (a) regions of normal and anomalous dispersion, (b) β vs. ω, (c) phase velocity vs. ω.

between anomalous and normal dispersion is more easily portrayed than that in Fig. 3-9(a) and (b) but the value of group velocity is not as simple to determine. The group velocity can be obtained in terms of υ and ω by using Eqs. (3-19) and (3-1)

$$u = \frac{1}{\partial\beta/\partial\omega} \qquad \beta = \frac{\omega}{\upsilon}$$

Taking the derivative of β gives

$$\frac{\partial \beta}{\partial \omega} = \frac{\nu - \omega(\partial \nu/\partial \omega)}{\nu^2} \tag{3-28}$$

The group velocity is then

$$u = \frac{\nu^2}{\nu - \omega(\partial \nu/\partial \omega)}$$

It can be seen from Eq. (3-28) that for a given modulated carrier, u depends on ν, ω, and the slope of the curve at the carrier frequency. Thus, group velocity can be positive, negative, or infinite; for instance, u is infinite when the denominator of Eq. (3-28) is zero or when

$$\frac{\partial \nu}{\partial \omega} = \frac{\nu}{\omega} \tag{3-29}$$

This criterion for infinite u is identical to that associated with the curve of β versus frequency, i.e., the slope of the curve at that point must equal the slope of a line passing from that point through the origin.

From Eq. (3-28), it is apparent that when the slope of Fig. 3-9 is zero, that is, when $\partial \nu/\partial \omega = 0$, then $u = \nu$ and no dispersion occurs for narrow frequency bands, i.e., small $\Delta\omega$, about such points. Now we shall consider an example which illustrates the waveforms to be expected for various cases of dispersion. For this example, the parameters were arbitrarily chosen as follows

$$f_1 = 1, \quad f_2 = 1.1 \text{ Hz} \quad \text{so} \quad f_2 = 1.1 f_1$$

$$\nu_1 = 1 \text{ m/sec}, \quad \text{line length } \ell = \frac{\pi}{4\beta_1} = \frac{\pi}{4}\frac{\nu_1}{2\pi f_1} = \frac{1}{8} \text{ meter}$$

$$\text{Amplitudes } V_1 = V_2 = 1$$

The various possibilities were obtained by keeping ν_1 constant and varying the phase velocity of f_2, that is, by varying ν_2 and therefore $\beta_2\ell$. The value of ν_2 was chosen such that it represents the cases of no dispersion, normal, anomalous with u a positive value, and anomalous with u a negative number

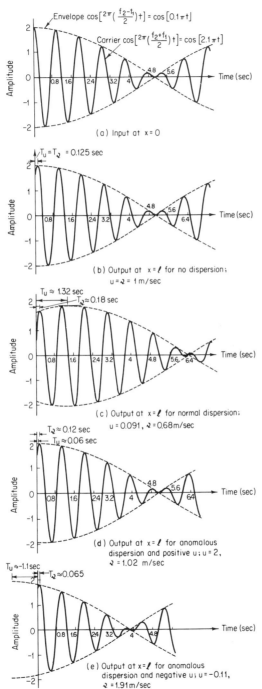

Fig. 3-10. Waveforms illustrating group and phase velocity for two nearly equal frequencies $f_2 = 1.1 f_1$ added linearly and traveling in a dispersive medium for a distance $\beta_1 \ell = \pi/4$.

1. No dispersion

$$\nu_2 = \nu_1 = 1 \text{ m/sec}$$

$$\beta_1 \ell = \frac{\pi}{4} \qquad \beta_2 \ell = 1.1 \frac{\pi}{4}$$

2. Normal

$$\nu_2 = 0.52 \text{ m/sec}$$

$$\beta_1 \ell = \frac{\pi}{4} \qquad \beta_2 \ell = 2.1 \frac{\pi}{4}$$

3. Anomalous $(+u)$

$$\nu_2 = 1.05 \text{ m/sec}$$

$$\beta_1 \ell = \frac{\pi}{4} \qquad \beta_2 \ell = 1.05 \frac{\pi}{4}$$

4. Anomalous $(-u)$

$$\nu_2 = 11.0 \text{ m/sec}$$

$$\beta_1 \ell = \frac{\pi}{4} \qquad \beta_2 \ell = 0.1 \frac{\pi}{4}$$

These two frequencies are linearly added

$$v(x, t) = \cos(\omega_1 t - \beta_1 x) + \cos(\omega_2 t - \beta_2 x) \qquad (3\text{-}30)$$

and are plotted point by point, as in Fig. 3-10, to get the actual waveform. At the input to our hypothetical line, that is, at $x = 0$, the waveform is as shown in Fig. 3-10(a). The carrier has a frequency $(f_2 + f_1)/2 = 1.05$, while the envelope is $(f_2 - f_1)/2 = 0.05$ Hz. The output of the line at $x = \ell$ is shown in Fig. 3-10(b), (c), (d), and (e) for the four cases specified above. The phase and group velocity can be determined either directly from the output waveforms or analytically. If Fig. 3-10 is used directly, then the phase and group velocity are respectively

$$\nu = \frac{\omega_2}{\beta_2} = \frac{2\pi f_2}{2\pi f_2 T_2 / \ell} = \frac{\ell}{T_2} = \frac{1}{8 T_2} \text{ m/sec} \qquad (3\text{-}31)$$

and

$$u = \frac{\Delta \omega_u}{\Delta \beta_u} = \frac{2\pi f_u}{2\pi f_u T_u / \ell} = \frac{1}{8 T_u} \quad \text{m/sec} \tag{3-32}$$

In other words, u and v are inversely proportional to the delay times T_u and T_v, respectively; T_u and T_v are expressed in seconds and can be read directly from the waveforms. A more exact analytical method yields

$$v = \frac{\omega_v}{\beta_v} = \frac{\omega_2 + \omega_1}{\beta_2 + \beta_1} = \frac{2.1(2\pi)}{\beta_2 + \beta_1} \tag{3-33}$$

$$u = \frac{\partial \omega_u}{\partial \beta_u} = \frac{\Delta \omega_u}{\Delta \beta_u} = \frac{\omega_2 - \omega_1}{\beta_2 - \beta_1} = \frac{0.1(2\pi)}{\beta_2 - \beta_1} \tag{3-34}$$

The values of β_2 and β_1 were specified for the various cases; substituting these values gives u and v, as indicated in Fig. 3-10. If the values of T_u and T_v and Eqs. (3-31) and (3-32) are used instead, the group and phase velocity will be found to agree with the analytical values; verification of this is left as an exercise for the reader.

These waveforms are essentially those which would actually be observed, for instance, on an oscilloscope which is triggered by the input waveform, i.e., where all time scales are referenced to the input waveform. Since the applied waveform is a steady-state sinusoid, T_u and T_v are just the relative delay time or phase relation between these waveforms; as a result, T_v must always be positive, i.e., the carrier output phase must always lag the input, but this need not be true for the envelope. Let us consider each case separately.

For no dispersion, as in Fig. 3-10(b), the peak of the envelope is delayed by the same amount as the peak of the carrier, hence $u = v$.

For normal dispersion, as in Fig. 3-10(c), the lower velocity of the higher frequency component has caused the peaks of the envelope to be delayed more than the peak of the carrier, so that from Eqs. (3-31) and (3-32) it is apparent that $u < v$. For anomalous dispersion, just the opposite has occurred; thus, $u > v$, as in Fig. 3-10(d). If the anomalous dispersion, that is, v_2, is sufficiently large, the peak of the envelope on the output end can occur at the same point as the peak of input, that is, $T_u = 0$; for such a case, u is obviously infinite. If v_2 is made still larger, T_u can become negative, resulting in a negative value of u, as shown in Fig. 3-10(e).

This example is significant in that it illustrates some very fundamental and important concepts, namely, that phase and group velocity merely represent the arrangement of the frequency components with respect to one another. An infinite or negative group velocity has no significance except in specifying the value of T_u, as in Fig. 3-10. Since the velocity of propagation of energy or intelligible signal cannot be negative or infinity, it is apparent that group velocity has little significance in dispersive media.

3.7 MANY FREQUENCIES PRESENT ON A DISPERSIVE LINE

In Sec. 3.4, the group velocity was derived for a dispersionless transmission line excited with a continuous frequency spectrum over a narrow frequency band of $2\delta\omega$. We now wish to consider an identical situation, except that the line is dispersive, so that β as a function of frequency might have the general shape shown in Fig. 3-11. The traveling wave voltage on any

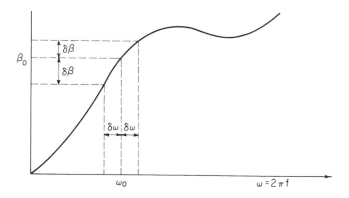

Fig. 3-11. Frequency dependence of phase constant for a dispersive line with an increment of $2\delta\omega$ of applied frequencies.

line can be represented by Eq. (3-15). The propagation constant as a function of frequency can be expressed in terms of a Taylor series expansion about ω_0

$$\beta(\omega) = \beta_0 + \left(\frac{\partial\beta}{\partial\omega}\right)_{\omega_0} (\omega - \omega_0) + \frac{1}{2}\left(\frac{\partial^2\beta}{\partial\omega^2}\right)_{\omega_0} (\omega - \omega_0)^2 + \text{etc.} \quad (3\text{-}35)$$

It is apparent that if $\delta\omega$ is sufficiently small, then only the first two terms are necessary in the above, i.e., we can approximate the curve by a straight line for the small interval $\delta\omega$. Equation (3-35) then reduces to Eq. (3-16) and since the remaining analysis is the same, the group velocity is given by Eq. (3-19). For a dispersive line such as that of Fig. 3-11

$$\left.\frac{\partial\beta}{\partial\omega}\right|_{\omega_0} \neq \frac{\beta_0}{\omega_0}$$

Thus, the group velocity (reciprocal of the former) does not equal the carrier phase velocity (reciprocal of the latter).

Since the group velocity is proportional to the reciprocal of the slope of the curve of Fig. 3-11, it can be seen that u can assume almost any value, depending upon the location of ω_0, as was discussed in Sec. 3.6.

In the above analysis, it was necessary to assume that $\delta\omega$ was small in order to approximate the β-vs.-ω curve by a straight line. If $\delta\omega$ is still much less than ω_0, so as to give an identifiable carrier with a lower frequency modulation, though large enough to require the three terms in Eq. (3-35), then the expression for group velocity will contain the additional term. This can be seen by including the additional terms as in Eq. (3-17)

$$\omega t - \beta x = \omega_0 t + (\omega - \omega_0)t - \left[\beta_0 + \left(\frac{\partial\beta}{\partial\omega}\right)_{\omega_0}(\omega - \omega_0) + \frac{1}{2}\left(\frac{\partial^2\beta}{\partial\omega^2}\right)_{\omega_0}(\omega - \omega_0)^2\right]x$$

$$= \omega_0 t - \beta_0 x + (\omega - \omega_0)\left\{t - \left[\left(\frac{\partial\beta}{\partial\omega}\right)_{\omega_0} + \frac{1}{2}\left(\frac{\partial^2\beta}{\partial\omega^2}\right)_{\omega_0}(\omega - \omega_0)\right]x\right\}$$

The equivalent expression of Eq. (3-18) for this case will have a constant amplitude and therefore constant phase points for the envelope when

$$t - \left[\left(\frac{\partial\beta}{\partial\omega}\right)_{\omega_0} + \frac{1}{2}\left(\frac{\partial^2\beta}{\partial\omega^2}\right)_{\omega_0}(\omega - \omega_0)\right]x = 0$$

or

$$u = \frac{x}{t} = \left[\left(\frac{\partial\beta}{\partial\omega}\right)_{\omega_0} + \frac{1}{2}\left(\frac{\partial^2\beta}{\partial\omega^2}\right)_{\omega_0}\delta\omega\right]^{-1} \tag{3-36}$$

It is obvious that the expression has become more complex and $\delta\omega$ is still restricted to reasonably small values. If $\delta\omega$ becomes large, then we can no longer identify a modulated carrier (Fig. 3-4). The waveform is greatly distorted as it travels down the line because of the dispersion. Thus, group velocity loses all meaning, i.e., we cannot identify a constant phase point on a group, since the groups do not exist in any simple form and, furthermore, the shape changes drastically as one proceeds down the line.

Thus, the concept of group velocity in a dispersive medium can be identified only for a narrow band of frequencies; this restriction applies also to nondispersive media. Another way of visualizing this is to recognize that the group velocity is really nothing more than the phase velocity of the envelope. If the frequency band becomes large, the envelope loses its significance and the group velocity therefore becomes meaningless.

3.8 SIGNAL VELOCITY AND FORERUNNERS—TREATMENT OF SOMMERFELD AND BRILLOUIN

In this section, it will be shown that in addition to phase and group velocity, three other velocities can be defined, namely, wavefront, signal, and (average) energy velocity. The concept of forerunners, those infinitesimally small vibrations which always travel at the speed of light regardless of the medium and precede all transient waves, evolves quite naturally from the mathematics of the treatment and will be seen to be associated with the wavefront velocity. It will further be seen that the signal velocity depends entirely upon the sensitivity of the detector which one uses to measure the velocity in an actual experiment. As the sensitivity of the detector increases, the signal velocity becomes identical to the forerunner or wavefront velocity, which always equals the velocity of light c in any medium. The energy velocity gives the average rate of energy propagation and is less than or equal to c, the former always being the case in dispersive media. We will not consider energy velocity but will limit our discussion to the other types of velocity, and will give some physical interpretation of their meaning.

The historical development of the basic ideas of propagation velocity is very interesting [1]. A few of the more important historical aspects deserve mention here. The first concept of group velocity dates back at least to 1839 [5], and the early concepts were well known to Lord Rayleigh in his study of the propagation of sound waves in 1877. He held that the propagation velocity of energy or signals was identical to group velocity. In 1905 Einstein made known his special theory of relativity, which assumes that no energy (signal) can be propagated faster than the

velocity of light.   We have seen in Secs. 3.6 and 3.7 that it is possible for
the group velocity to exceed that of light under certain conditions.   If one
assumes that energy velocity is identical to group velocity, the contradic-
tions with the theory of relativity are obvious.

Many objections of this sort were presented in the years following the
announcement of the theory of relativity to "prove" that it was incorrect.
Fortunately, the entire situation was clarified by Sommerfeld in 1912 [14]
and again in 1914 [15] when he showed that no signal could be propagated
faster than c in any medium with either normal or anomalous dispersion.   Although he formulated the problem correctly and could draw some definite
conclusions, Sommerfeld did not derive the complete shape of the signal
waveform in a dispersive media.   Brillouin [2], extended the methods
of complex integration and obtained a complete solution to the problem.
By choosing the paths of integration properly, one can obtain integrals
which can be evaluated, and these give rise to the forerunners as well as to
the main body of the signal, which will be discussed later.

It is easy, in retrospect, to understand why so much confusion over
the velocity of propagation existed in the early 1900s and in fact still exists
to some extent today.   The results of the preceding sections, particularly
Secs. 3.6 and 3.7, show the reason for this rather vividly.   In general, the
propagation of useful signals necessitates the presence of many frequencies
over a frequency spectrum.   In dispersive media, the parameters which determine the propagation velocity of individual frequency components will depend on the frequency itself.   In principle, it is possible simply to linearly
add these different components at various points, as was done in Sec. 3.3;
from the shape of the waveforms one can determine the velocity of propagation.   There are two difficulties with this procedure: First, we have
already seen the difficulty encountered in identifying group velocity with
just two frequencies on a line; the presence of several components would
greatly compound the situation.   Furthermore, such a combination of discrete frequency components would apply only to the steady-state solution.
In order to be meaningful, the analysis must consider the propagation of a
wave which is discrete rather than continuous with respect to space (distance)
and time.   The wave must propagate into a medium where there was initially
no wave at all and which therefore requires a transient analysis.

The problem originally treated by Sommerfeld and Brillouin consisted
of a single-frequency wave traveling in free space.   This wave strikes the
surface of a dispersive medium at time $t = 0$, with a reflection and transmission of energy at the boundary resulting.   The reflection can be neglected;
the transient analysis of the propagation into the material is desired.   In
principle, the same situation can be realized by an infinite transmission line

constructed from a dispersive dielectric and fed from a sinusoidal source, as in Fig. 3-12. The switch is closed at $t = 0$; it is desirable to find the voltage at some distance $x = \ell$ along the line as a function of time from $t = 0$ to infinity, i.e., to perform the transient analysis. The situation is identical both in principle and in detail to the original case considered by Sommerfeld except for some practical difficulties with transmission lines at frequencies beyond the microwave region, e.g., with visible light.

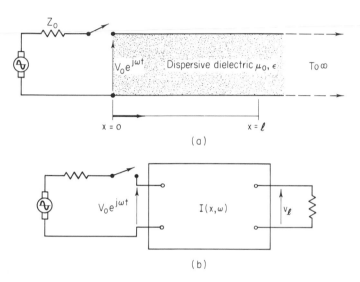

Fig. 3-12. Physical arrangement for determining velocity of propagation in a dispersive medium. (a) Dispersive transmission line; (b) equivalent circuit in terms of transfer function of a length $x = \ell$.

This problem can be solved in several ways: one is to simply apply the general form of the wave equation

$$\frac{\partial^2 v(x, t)}{\partial x^2} = \mu\epsilon \frac{\partial^2 v(x, t)}{\partial t^2}$$

to the line and obtain the complementary function and particular integral, i.e., to use transient analysis, as one would for any circuit. However, two problems occur; first, there are two variables x and t which ordinarily would present no difficulty since a separation of variables is possible in most cases (see [6, p. 289]). However, ϵ is not a constant but varies with time; unfortunately, the time variation of ϵ is not known, but in general, ϵ as a function of frequency is known. Thus, in order to obtain a solution, it is expedient to work in the frequency domain, i.e., with the aid of Laplace transforms. This approach is identical to that of Stratton ([13, p. 334]), except that the problem will be put in terms understandable to electrical engineers and others familiar with system analysis. This approach differs from that of Sommerfeld and Brillouin only with respect to the use of Laplace instead of Fourier transforms.*

Referring to Fig. 3-12(a), since the line is infinitely long and only the portion from $x = 0$ to ℓ is of interest we can assume that it is properly terminated and can represent the line as in Fig. 3-12(b). The transfer function of the line is just the steady-state solution for v_ℓ / V_0 in terms of frequency. Since the line is (assumed) properly terminated, there can be no negatively traveling wave, so that from (2-14)

$$\frac{v_\ell}{V_0} = \exp[j\omega t - \gamma x] = I(x, \omega) \tag{3-37}$$

The above equation represents the transfer function or impulse response of the line. In order to obtain the response of the line (in the frequency domain) to any other input, it is necessary to multiply the Laplace transform of the input by the impulse response. The Laplace transform of the applied sinusoid is

$$\mathcal{L}(e^{j\omega t}) = V_0 \int_0^\infty e^{j\omega t} e^{-st} dt = \frac{V_0}{s - j\omega} \tag{3-38}$$

Thus the transient solution in the frequency domain is

$$\mathcal{V}(x, \omega) = \frac{V_0}{s - j\omega} I(x, \omega) = V_0 \frac{\exp[j\omega t - \gamma x]}{s - j\omega} \tag{3-39}$$

In order to get the solution in the time domain, it is necessary to perform the inverse Laplace transform

$$v(x, t) = \frac{1}{2\pi j} \oint \frac{V_0 \exp[j\omega t - \gamma x]}{j\omega - s} ds \tag{3-40}$$

*It is desirable to work with Laplace transforms because the Fourier integral leads to certain convergence problems which are circumvented by restriction of the integration in the complex plane, which then essentially reduces the Fourier integral to the Laplace transform. For an introduction to the Fourier integral and Laplace transform, see [6, p. 266] or [13, p. 310].

where $x = \ell$, and $\gamma = \sqrt{\mu\epsilon}$ and is a function of frequency. An important decision is required now in regard to just what analytical expression should be used for γ versus frequency. For all good dielectrics, the permeability is that of free space, so that $\mu = \mu_0$. It was shown in Sec. 3.5 that, in general, ϵ varies greatly with frequency, having several absorption bands, as shown in Fig. 3-7. It is these absorption bands which give rise to the dispersion and it is thus essential that they be included. Appendix 3A provides an expression for ϵ_r which is somewhat oversimplified but is nevertheless reasonably accurate. It will be assumed that the dispersive dielectric has only one characteristic frequency ω_0, that is, one absorption band, over the entire frequency range and that it can be represented by Eq. (3-44)

$$\epsilon_r(s) = 1 + \frac{a^2}{s^2 + \rho c + \omega_0{}^2} \tag{3-41}$$

where $a^2 = 4\pi Ne^2/m$ and $j\omega$ has been replaced by s, since the transient response of the dielectric must be included. The use of this simplified expression and only one absorption band has little bearing on the character of the final result, i.e., a more refined analysis would change the details but not the overall concepts and conclusions. Thus the problem is reduced to

$$v(x, t) = \frac{1}{2\pi j} \oint \frac{\exp\left(st - \dfrac{\sqrt{\epsilon_r(s)}}{c}x\right)}{s - j\omega} \, ds \tag{3-42}$$

where ω is the applied frequency. It is easily seen that contour integration in the complex plane is required which contains many subtle details. The integration must be broken up into several different paths and, as might be expected, certain of these paths near the poles and zeros of ϵ_r give rise to different components of the total solution. We shall not attempt this evaluation but will instead present the conclusion and physical interpretations of the results.

Let us analyze the physical situation in a qualitative manner. When the switch in Fig. 3-12 is closed, the voltage impressed on the line begins to increase in valué and at the same time to travel down the line. In order for the presence of the dielectric to be felt, it is necessary that the charged particles respond to the applied field, which requires a transient buildup of the displacement of the particles which were initially at rest (except for thermal motion, which is random and can be neglected). Needless to say, this transient response of the dielectric constitutes the entire problem, since an ideal, lossless transmission line with constant μ and ϵ has no transient response. It is obvious that a finite response time is required for the particles to begin oscillating, so that some of the energy of the applied wave gets through unimpeded, as if the medium were free space. This represents the so-called first forerunner, which always travels with the velocity c in all media. The first forerunner starts with zero amplitude, and period approximately given by

$$\tau \approx \frac{\pi^2 c}{a^2 x} \tag{3-43}$$

This period is independent of the period of the applied frequency as well as the characteristic frequency ω_0 of the oscillating particles, and is determined only by the distance of travel x and the dispersive power of the material, that is, $a \propto Ne^2/m$. The amplitude and period gradually increase until the period approaches the characteristic frequency ω_0, at which time the amplitude quickly decreases as a result of absorption of energy. If ρ is small, the amplitude decreases as $(t - x/c)e^{-2\rho x}$. Subsequently, a second disturbance begins, known as the second forerunner, traveling with a velocity $\omega_0 c \times \left(\omega_0^2 + a^2\right)^{-\frac{1}{2}}$, provided that $\omega < \omega_0$. Its period is initially large, gradually decreasing, as shown in Fig. 3-13. The main body of the wave then suddenly

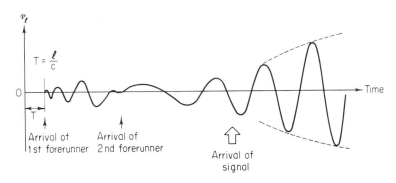

Fig. 3-13. Approximate theoretical waveform at point $x = \ell$ along a dispersive line when switch is closed at $t = 0$ in Fig. 3-12.

begins to arrive, traveling with signal velocity w, and builds up to the steady state value with frequency ω. The exact point of arrival of this main portion, and hence signal velocity w, is difficult to determine except by analytical methods. It is seen that a measurement of "wavefront velocity" is meaningless since (assuming that no other noise is present), as the sensitivity of the detector, e.g., oscilloscope, is increased, the velocity increases and eventually approaches the velocity of light, which represents the arrival of the first forerunner. In practice, such measurements are always impeded by significant noise, and since the amplitudes of these forerunners are very small, their detection is extremely difficult. As an example, Brillouin has estimated the amplitudes of the first and second forerunner (see [1, pp. 55, 70, 73]), to be roughly 2×10^{-3} and $1/30$ times that of the applied signal, respectively, or intensities of 4×10^{-6} and 10^{-3} times that of the incident intensity, assuming that $\omega = 4 \times 10^{15}$ (visible light), $\omega_0 = 4 \times 10^{16}$ rad/sec, $x = 1$ cm, and $a \approx \omega_0$.

This problem has been studied with the assumption that the dielectric has only one characteristic frequency or one absorption band. If several absorption bands were present, it would be necessary to represent ϵ_r by Eq. (3-47), and it seems likely that there would be additional forerunners with periods lying between the bands.

When two or more very nearly equal frequencies are present to give rise to group velocity, the signal velocity is almost equal to group velocity, provided that the applied frequencies are very different from the characteristic frequency ω_0 of the dielectric. An analytical definition of signal velocity is given on p. 74 of [1] as follows: "The arrival of the signal can be arbitrarily defined as the moment the path of integration reaches the pole ω" (where $\omega = 2\pi f$, the applied frequency). This definition is applicable even when the applied signal falls in the region of anomalous dispersion. Using this definition for signal velocity and assuming that two or more frequencies in the neighborhood of ω are present, the phase, group, and signal velocity as a function of ω are as shown in Fig. 3-14 in the region near the absorption band of the dielectric; the phase and group velocity are obtained as the steady-state values, since they are meaningless during the transient, as already seen. If we use Eq. (3-44) for ϵ_r, then if the damping term ρ is very small, the phase velocity is approximately

$$ \vartheta = \frac{c}{\sqrt{\epsilon_r}} $$

If the applied frequency is larger than ω_0 of the dielectric such that $\omega_0^2 - \omega^2 = -\Delta\omega^2$, that is, in the anomalous region, then

$$ \epsilon_r = 1 - \frac{a^2 \Delta\omega^2}{\Delta\omega^4 + \rho^2\omega^2} $$

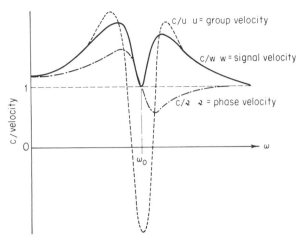

Fig. 3-14. Various velocities as a function of midband frequency in the neighborhood of the absorption band ω_0 for a narrow band of applied frequencies.

Thus ϵ'_r can be less than unity (see Fig. 3-7), and ν can therefore be greater than c, but only in the anomalous range as shown.

The significant points concerning velocity of propagation can be summarized as follows. Group velocity u and phase velocity ν can only be defined at steady state; furthermore, in order for u to exist, at least two frequencies that are very nearly equal must be linearly added together. In nondispersive media, u and ν are always equal and less than or equal to c. In dispersive media, u can have any value, positive, negative, or zero, and can be greater than or less than c. In dispersive media with no losses, we must have $\epsilon_r > 1$, so that phase velocity must always be less than c. In more typical dielectrics, however, losses are significant near the absorption band, which causes ν to be greater than c in the anomalous region, but ν must always be positive. For transient propagation, the wavefront velocity increases as the detector sensitivity increases, approaching c in the limit. A signal velocity can be defined analytically, but its measurement is ambiguous.

APPENDIX 3A

CLASSICAL THEORY OF ABSORPTION BANDS IN DIELECTRICS

In general, the permittivity is complex. It is represented by $\epsilon = \epsilon' - j\epsilon''$ or relative values of $\epsilon_r = \epsilon'_r - j\epsilon''_r$; ϵ'_r is the part usually thought of as being the dielectric constant and gives rise to capacitance, while ϵ''_r represents the loss term. The dielectric constant will vary with frequency, as shown in Fig. 3-7 for a general material with several absorption bands. According to the elementary classical theory of dielectric constant, a material is composed of charged particles (ions, atoms, molecules, electrons, etc.), held in place by elastic restoring forces so that each particle can undergo harmonic oscillation about its equilibrium position. If a sinusoidal field $E_0 e^{j\omega t}$ is applied, the equation of motion for the displacement y of a mass m of charge e is (see [3, p. 154])

$$m \frac{d^2 y}{dt^2} + m\rho \frac{dy}{dt} + m\omega_{0n}^2 y = eE_0 e^{j\omega t}$$

where ω_{0n} represents the characteristic frequency at each absorption band. The permittivity is given by $\epsilon = \epsilon_0 + P/E$ where P is the dipole moment per unit volume or polarization, given by $P = ey/N$. If it is assumed that there is only one characteristic frequency ω_0, then substitution of the steady-state solution for y in terms of ω and ω_0 in the above equations gives

$$\epsilon_r = \frac{\epsilon}{\epsilon_0} = 1 + \frac{Ne^2}{m\epsilon_0} \frac{1}{\omega_0^2 - \omega + j\rho\omega} \tag{3-44}$$

in RMKS units where ρ is the damping factor, N is the number of particles/ volume, and ω_0 is the characteristic frequency. Separation of the above into its real and imaginary parts yields

$$\epsilon_r' = 1 + a^2 \frac{\omega_0^2 - \omega^2}{\left(\omega_0^2 - \omega^2\right)^2 + \rho^2\omega^2} \tag{3-45}$$

$$\epsilon_r'' = a^2 \frac{\rho\omega}{\left(\omega_0^2 - \omega^2\right)^2 + \rho^2\omega^2} \tag{3-46}$$

where

$$a^2 = \frac{Ne^2}{m\epsilon_0}$$

These are plotted as a function of frequency near an absorption band in Fig. 3-15. Note that ϵ_r'' is directly proportional to ρ, so that if there is no damping, there is no loss term. This is similar to a series RLC circuit with ρ analogous to R: If $R = 0$, there can be no losses.

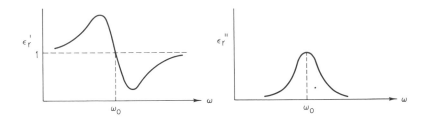

Fig. 3-15. Dielectric constant of a material with one characteristic frequency ω_0.

Equation (3-44) is good only for materials with low density N, i.e., gases, so that the internal field approximately equals the applied field. For materials with high densities, i.e., solids, the Lorentz expression for internal field must be used. This leads to a result identical to Eq. (3-44) except that $\omega_0{}^2$ must be replaced by $\omega_0{}^2 - a/3$. Thus, the form of ϵ_r is identical in either case. Another important assumption in the above is that the wavelength of the applied field must be much larger than the distance between particles; otherwise the polarization is not constant over a neighborhood of a given particle and an average dielectric constant cannot be defined. This assumption begins to break down approximately in the ultraviolet region for gases and in the x-ray region for solids.

For materials with several characteristic frequencies, the complex dielectric constant becomes

$$\epsilon_r = 1 + \sum_n \frac{a_n{}^2}{\omega_{0n}^2 - \omega^2 + j\rho_n\omega} \tag{3-47}$$

PROBLEMS

3-1. In ordinary AM radio transmission, a modulated carrier wave is obtained by "mixing" a low frequency (audio v_a) with a high frequency (carrier v_c) signal. Show that the linear addition of two such frequencies, as shown, into a resistance load does not lead to a modulated carrier, but that by using a series diode which has a nonlinear square law IV characteristic given by $i = a_1 v + a_2 v^2$, the v^2 term multiplies the two frequencies to give a modulated carrier (plus other components).

3-2. Show that in the previous problem, if the two frequencies are close together, the diode is not needed in order to obtain a modulated carrier.

3-3. Show that a diode with square-law characteristics as in Prob. 3-1 can be used as a detector or demodulator.

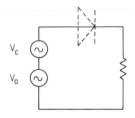

Fig. 3P-1.

REFERENCES

1. Brillouin, Leon: "Wave Propagation and Group Velocity," Academic Press, New York, 1960.

2. *Ibid.,* Über die Fortpflanzung des Lichtes in Dispergierenden Medien, *Ann. Physik,* vol. 44, p. 203, 1914.

3. Decker, A. J.: "Solid State Physics," Prentice-Hall, Inc., Englewood Cliffs, N. J., 1957.

4. Einstein, A.: Zur Elektrodynamik bewegter Körper (On Electrodynamics of Moving Bodies), *Ann. Physik,* ser. 4, vol. 17, p. 891, 1905.

5. Hamilton, W. R.: Researches Respecting Vibration Connected with the Theory of Light, *Proc. Roy. Irish Acad.,* vol. 1, pp. 341-349, 1836-1840, also published in A. W. Conway and A. J. McConnell (eds.), "The Mathematical Papers of Sir William Rowan Hamilton," vol. 2, p. 578, Cambridge University Press, New York, 1940.

6. Johnson, W.: "Mathematical and Physical Principle of Engineering Analysis," McGraw-Hill Book Company, New York, 1944.

7. Kittel, C.: "Introduction to Solid State Physics," 2nd ed., John Wiley & Sons, Inc., New York, 1956.

8. Kraus, J. C.: "Electromagnetics," McGraw-Hill Book Company, New York, 1953.

9. Lord Rayleigh: "The Theory of Sound," vol. 1, Dover Publications, Inc., New York, 1945 (first published by Macmillan, London, 1877).

10. *Ibid.*, "Scientific Papers," vol. 1, p. 537, Dover Publications, Inc., New York, 1964 (first published by Cambridge University Press, New York, 1899).

11. Ramo, S. and J. R. Whinnery: "Fields and Waves in Modern Radio," 2nd ed., John Wiley & Sons, Inc., New York, 1953.

12. International Telephone and Telegraph Corp., "Reference Data for Radio Engineers," American Book-Stratford Press, Inc., New York, 1956.

13. Stratton, J. A.: "Electromagnetic Theory," p. 334, McGraw-Hill Book Company, New York, 1941.

14. Sommerfeld, A.: "Festschrift zum 70. Geburtstage von Heinrich Weber," Teubner, Leipzig, 1912.

15. *Ibid.*, Über die Fortpflanzung des Lichtes in dispergierenden Medien, *Ann. Physik,* vol. 44, pp. 177-202, 1914.

16. von Hippel, A. R.: "Dielectrics and Waves," John Wiley & Sons, Inc., New York, 1954.

4 SKIN EFFECT—NORMAL AND ANOMALOUS

4.1 INTRODUCTION

The phenomena of skin effect and penetration of an electromagnetic wave into a conductor are classical subjects which are treated in numerous books. However, there are a number of characteristics of these phenomena which are often neglected or not brought together in a coherent form, such as the relationship between conductor resistance, internal inductance, surface impedance, and simultaneous propagation of a wave along the surface of a conductor as well as into a conductor. Also neglected in the usual college courses is the phenomenon of anomalous skin effect, which is becoming increasingly important. We shall cover these subjects in this chapter and demonstrate the close correlation between analyses of a transmission line as a circuit and as a field theory problem. Before we proceed, it is desirable to provide some orientation for the reader in order that he can understand the organization of the various topics to follow.

All skin-effect phenomena result from the fact that the current and/or magnetic field distribution in a conductor is frequency-dependent and also from the fact that current in any real conductor gives rise to an electric field as given by Ohm's law. The student exposed to skin effect for the first time learns that it causes current to flow on the surface of a conductor, thus increasing its effective resistance. In elementary circuits courses, the student learns, quite apart from the phenomenon of resistance change, that a conductor has an internal inductance due to magnetic flux penetration in the conductor. This internal inductance arises in addition to the external or circuit inductance. However, these two effects arise from the same phenomena, so that when the effective resistance changes, the internal inductance also changes. Both result from the more fundamental notions of penetration effects, and a much better understanding of skin effect is obtained if an electromagnetic wave penetrating or propagating into a conductor is pictured. As we shall see, it does not matter whether the wave is traveling along the surface, as in a transmission line, or is incident at a right angle to the surface; the same result is obtained for both. From such a picture, it is easily seen that current (ac) in a conductor will always *concentrate on the surface that is nearest the wave which creates the current.* Thus, for a general system of conductors, e.g., three wires of odd shapes, it is not possible to say *a priori* on which surfaces the currents will concentrate until the total physical configuration as well as means of excitation have been specified.

Since normal skin effect results from the frequency-dependence of the current distribution in a conductor, the most fundamental way of investigating the subject is by application of Maxwell's equations and by finding the proper differential equations for E and H. Application of boundary conditions gives the fields and current distribution as a function of frequency for any given geometry. From this the attenuation and phase constant or effective resistance and inductance can easily be determined. However, these same quantities can be determined from more elementary concepts of Faraday's law of induced voltages and Kirchhoff's law that the sum of voltage rises and drops around a closed loop equals zero. Thus, to see the correspondence between ordinary circuit and field analysis, we shall first calculate the resistance and inductance of a solid round wire from elementary principles.

It is appropriate to start with the simplest case, namely, that of low frequency, for which the current is uniform over the cross-sectional area. Such a case provides a low-frequency limit check on more general analyses and quickly shows the effect of frequency on the parameters of interest. Afterwards, we will proceed to determine the current distribution and surface impedance as a function of frequency for a round wire, using approximations of low losses.

A similar analysis can be done for a strip line, but it is of interest to obtain the exact solution for a wave propagating between two strip conductors. The penetration of the wave into the conductors as it propagates along the surface causes a reduction in phase velocity, an increase in series impedance, and an attenuation. The solution for low losses will be found to be identical to that of the transmission line with low losses in Secs. 1.6 and 2.6, with the added feature that the exact solution gives the series resistance and inductance of the line as well as penetration of the field into the surface of the conductor.

Some of the more interesting general results to be shown are as follows. For a solid round wire of radius r, the surface resistance R and reactance ωL are generally not equal. At dc, R is inversely proportional to r^2 while L is independent of r. As the frequency increases, R and ωL both change, but at different rates. At sufficiently high frequency where the skin depth is much smaller than r, we will find $R = \omega L$. This latter equality is also true for a thick-plane conductor, e.g., a strip line conductor, but at all frequencies, not just at high frequency, and the plane-conductor analysis can be used for round wires, provided that the frequency is sufficiently high as might be expected.

After consideration of the classical theory, a survey of the anomalous skin effect will be presented with emphasis on the basic concepts which have bearing on practical considerations. We shall examine why this effect occurs, what the practical consequences are, and where and under what conditions it is to be taken into consideration. A number of useful curves and rules of thumb will be presented.

4.2 LOW—FREQUENCY RESISTANCE AND INDUCTANCE OF A SOLID ROUND WIRE

Let us consider any ordinary wire of circular cross section and determine its internal impedance at low frequency. It is assumed that the applied frequency is sufficiently low so that the current is uniformly distributed over the cross-sectional area of the conductor. This is obviously true at dc and we will determine the frequency limits for this assumption later.

It is apparent that for the wire of Fig. 4-1, the resistance per unit length R is the resitivity $1/\sigma$ divided by the cross-sectional area

$$R = \frac{1}{\sigma \pi r^2} \tag{4-1}$$

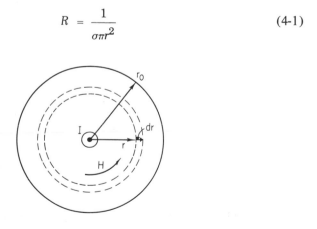

Fig. 4-1. Circular wire with uniform current $I = \pi r_0^2 J$.

The determination of the internal inductance is not quite so simple but is easily obtained with the aid of the principle of flux linkages, that is, $L = \lambda/I$, where λ is total flux linkage. Referring to Fig. 4-2, since the current density is known and is a constant equal to **J**, the magnetic field at any point within the conductor is obtained from Ampere's circuital law

$$\oint \mathbf{H} \cdot dl = H 2\pi r = I = J \pi r^2 \tag{4-2}$$

or

$$\mathbf{B} = \mu \mathbf{H} = \frac{\mu J r}{2} \tag{4-3}$$

In order to determine the inductance, we must find the flux linkage per unit current. Since there are partial flux linkages in this case, it is necessary to multiply the total flux contained within the wall of an incremental cylinder of radius r and wall of thickness dr by the fraction of the total current linked inside the cylinder and sum all these flux linkages from $r = 0$ to r_0.

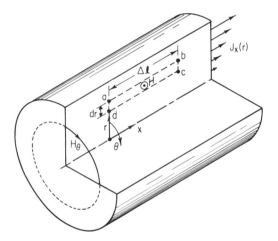

Fig. 4-2. Section of circular wire with non-uniform current.

The total flux per unit length in the incremental cyclinder wall dr is

$$\frac{\phi}{\ell} = \frac{\mathbf{B} \cdot dA}{\ell} = \mathbf{B} \cdot dr \tag{4-4}$$

Since the fraction of total current within any elemental cyclinder is $\pi r^2 / \pi r_0^2$, the total flux linkages are

$$\lambda = \int_0^{r_0} \frac{r^2}{r_0^2} \, \mathbf{B} \cdot dr = \frac{\mu J}{2r_0^2} \int_0^{r_0} r^3 \, dr = \frac{\mu J r_0^2}{8} \qquad (4\text{-}5)$$

But since $J = I/\pi r_0^2$, substitution yields

$$L = \frac{\lambda}{I} = \frac{\mu}{8\pi} = 0.5 \times 10^{-7} \mu_r \frac{\text{henries}}{\text{meter}} \qquad (4\text{-}6)$$

This is the internal inductance of the wire and is independent of all geometrical parameters, provided that the wire is round. We will use these results in Sec. 4.4 when the resistance and inductance are determined for all frequencies.

4.3 CURRENT DISTRIBUTION IN SOLID CIRCULAR WIRE WITH AC EXCITATION

In general, when ac is applied to the circular wire of Fig. 4-1, the current will not be uniform over the cross section. We shall determine the current distribution and also the frequency range where various simplified expressions can be used. The differential equation governing the current will be obtained from the elementary principles of Kirchhoff's circuit law and Faraday's induction law. It is assumed that the wire is long in the axial direction and that there are no variations of current or field quantities in the axial or circumferential direction. The situation where a section of the wire has been cut away to show the fields and currents inside is illustrated in Fig. 4-2.

The assumption that there are no variations of the fields or current in the axial direction implies that there is no wave propagation in this direction. Strictly speaking, this assumption is not valid for long wires, but we can assume that we are concerned with only a small section of line, for which this is a reasonable assumption. This question will be considered in detail in Sec. 4.5, where an exact solution for a plane conductor will be presented.

The current density will be in the x direction only and the magnetic field in the θ direction only, both vectors being functions only of r. In order to relate the current and magnetic field to the conductivity of the metal, it is necessary to recognize that for any imaginary incremental loop such as $abcd$ in Fig. 4-2, the difference between the voltage drop along the sides ab and cd must be the voltage induced by the time changing flux within the loop, which is just another statement of Kirchhoff's law. The voltage induced in the loop is given by Faraday's induction law

$$v_{ind} = \frac{d\phi}{dt} = \frac{\partial}{\partial t} \mathbf{B} \cdot \Delta\ell \, dr$$

The voltage induced along side $c\,d$ of the incremental loop is

$$v_{cd} = iR = \mathbf{J} \frac{\Delta\ell}{\sigma}$$

For the voltage drop along side $a\,b$, we know that $\mathbf{J}$ varies with r, but if dr is sufficiently small, we can use the first two terms of a Taylor expansion for $\mathbf{J}$ in terms of r to get

$$v_{ab} = (i + di)\,R = \left(i + \frac{\partial i}{\partial r} dr\right) R = \frac{\Delta\ell}{\sigma}\left(\mathbf{J} + \frac{\partial \mathbf{J}}{\partial r} dr\right)$$

Thus, using Kirchhoff's law

$$v_{ind} = v_{ab} - v_{cd}$$

or

$$\Delta\ell \, dr \frac{\partial \mathbf{B}}{\partial t} = \frac{\Delta\ell}{\sigma} \frac{\partial \mathbf{J}}{\partial r} dr$$

Thus we get

$$\frac{1}{\sigma} \frac{\partial \mathbf{J}}{\partial r} = \mu \frac{\partial \mathbf{H}}{\partial t} \tag{4-7}$$

where $\mathbf{B} = \mu \mathbf{H}$.

In order to get the above differential equation in terms of $\mathbf{J}$ only, we can use Ampere's circuital law to relate H to J

$$\int \mathbf{H} \cdot d\mathbf{l} = I = \int \mathbf{J} \cdot d\mathbf{A}$$

H and **J** are uniform around the circumference, so that

$$dl = 2\pi r \quad \text{and} \quad dA = 2\pi r dr$$

(l is not axial length here). Substituting these gives

$$r\mathbf{H} = \int r\mathbf{J}\, dr$$

To eliminate the integral, we can take the derivative of both sides, recalling that **H** is a function of r

$$\mathbf{H}\frac{\partial}{\partial r}(r) + r\frac{\partial \mathbf{H}}{\partial r} = r\mathbf{J}$$

To get the final expression, we can take the time derivative of the above equation and the derivative with respect to r of Eq. (4-7)

$$\frac{1}{r}\frac{\partial \mathbf{H}}{\partial t} + \frac{\partial^2 \mathbf{H}}{\partial r\, \partial t} = \frac{\partial \mathbf{J}}{\partial t} \quad \frac{1}{r}\frac{\partial^2 \mathbf{J}}{\partial r^2} = \mu\frac{\partial^2 \mathbf{H}}{\partial t\, \partial r} \tag{4-8}$$

Substitution of appropriate factors and rearranging results in the final differential equation for the circular wire

$$\frac{\partial^2 \mathbf{J}}{\partial r^2} + \frac{1}{r}\frac{\partial \mathbf{J}}{\partial r} = \mu\sigma\frac{\partial \mathbf{J}}{\partial t} \tag{4-9}*$$

In order to solve this equation, we will assume sinusoidal time variation of the form $e^{j\omega t}$. The equation to be solved is thus

$$\frac{d^2 \mathbf{J}}{dr^2} + \frac{1}{r}\frac{d\mathbf{J}}{dr} = j\omega\mu\sigma \mathbf{J} = \frac{-2}{j\delta^2}\mathbf{J} \tag{4-10}$$

where the substitution $\delta^2 = 2/\omega\mu\sigma$ has been made. This equation can be solved by the usual technique of assuming a solution and checking to see if it fits: For this case, a power series of the form

$$\mathbf{J} = a_0 + a_1 r + a_2 r^2 + \cdots a_n r^n$$

---

*The same equation could have been obtained much more simply from Maxwell's equations, which essentially take care of the details of incremental areas implicitly.

will be found to work, provided that the constants are properly chosen. It is easily verified that the solution is a convergent series. It is so important that it has been well tabulated; it is known as the Bessel function of zero order and first kind and is given by

$$J_0(x) = 1 - \frac{x^2}{2} + \frac{(x/2)^4}{(2!)^2} + \text{etc.} = \sum_{n=0}^{\infty} \frac{(-1)^n (x/2)^{2n}}{(n!)^2} \tag{4-11}$$

The more general solution to the differential equation is

$$J = CJ_0\left(\frac{r}{\delta}\sqrt{\frac{2}{j}}\right) + DH_0^{(1)}\left(\frac{r}{\delta}\sqrt{\frac{2}{j}}\right) \tag{4-12}$$

where J_0 is the Bessel function (not current density) of zero order and first kind and $H_0^{(1)}$ is a linear combination of Bessel functions of the first and second kind, called the Hankel function.* We need not worry about the latter, since H_0 becomes infinite as r goes to 0;[†] thus, the boundary conditions require that the constant D equal 0. The term in parentheses is the argument of the Bessel function. We need only evaluate the constant C, which can be done in terms of current density J_s on the surface of the conductor. We do not yet know the value of J_s, but in most cases, its evaluation is not necessary, since usually the ratio of current at some radius r to that on the surface is sufficient. J_s can be evaluated in various ways but we will do this later when the surface impedance is determined.

At the surface of the wires, it is necessary that

$$J(r = r_0) = J_s$$

From the previous equation, it is apparent that

$$C = \frac{J_s}{J_0\left(\frac{r_0}{\delta}\sqrt{\frac{2}{j}}\right)} \quad \text{so} \quad J = \frac{J_s}{J_0\left(\frac{r_0}{\delta}j\sqrt{2j}\right)} J_0\left(\frac{r}{\delta}j\sqrt{2j}\right) \tag{4-13}$$

*For a complete discussion of Bessel and related functions, see [21].
†See [8, p. 576-581].

Since the argument of the Bessel function is complex, it is desirable to separate the expression into its real and imaginary parts. This can be done easily with the aid of the following identities*

$$J_0(uj\sqrt{j}) = \text{ber}(u) + j\,\text{bei}(u) \tag{4-14}$$

where, for our case

$$u = \frac{r\sqrt{2}}{\delta} \tag{4-15}$$

and the real and imaginary parts of the Bessel function are

$$\text{ber}(u) = 1 - \frac{(u/2)^4}{(2!)^2} + \frac{(u/2)^8}{(4!)^2} - \text{etc.} \tag{4-16}$$

$$\text{bei}(u) = \frac{(u/2)^2}{(1!)^2} - \frac{(u/2)^6}{(3!)^2} + \frac{(u/2)^{10}}{(5!)^2} - \text{etc.} \tag{4-17}$$

Fortunately, these functions are well tabulated,[†] and hence they need not be evaluated by the infinite series. The final solution for current in a solid circular wire is

$$\mathbf{J} = \mathbf{J}_s \frac{\text{ber}\left(\frac{r}{\delta}\sqrt{2}\right) + j\,\text{bei}\left(\frac{r}{\delta}\sqrt{2}\right)}{\text{ber}\left(\frac{r_0}{\delta}\sqrt{2}\right) + j\,\text{bei}\left(\frac{r_0}{\delta}\sqrt{2}\right)} \tag{4-18}$$

For the present, we are interested only in the ratio of current at any point r to that at the surface, as a function of the ratio r/δ. It can be seen that since $\mathbf{J}$ is a complex quantity, it will vary in phase and amplitude as a function of r. Curves showing this ratio of amplitude and phase from the center of the wire out to the surface are plotted in Fig. 4-3 for several values of r_0/δ. It can be seen that as r_0/δ becomes very large, i.e., high frequency, most of the current concentrates on the surface. Under such conditions, it is reasonable to expect that a circular wire can be approximated by a plane

---

*See [6, p. 184] for complete definitions of Bessel functions.
†See [6, p. 276].

surface. We will calculate the penetration in a plane surface later and a comparison will show this approximation to be justified.

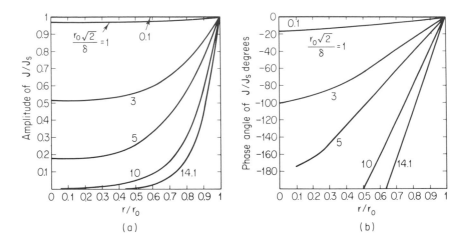

Fig. 4-3. Theoretical current distribution in a solid circular wire.

4.4 SURFACE IMPEDANCE OF A SOLID CIRCULAR WIRE WITH AC EXCITATION

The internal impedance of a conductor is commonly referred to as the surface impedance and is simply obtained as the ratio of the voltage drop along the surface to the total current enclosed. We will always be interested in the impedance per unit length $\Delta \ell$

$$Z_{surf} = \frac{1}{\Delta \ell} \frac{V_{surf}}{I} = \frac{E_s}{I} \frac{ohms}{meter} \qquad (4\text{-}19)$$

We know that the surface E is related to the surface J by Ohm's law

$$E_s = \frac{1}{\sigma} J_s \qquad (4\text{-}20)$$

Thus, to find the impedance, it is only necessary to evaluate the total current in terms of the current density at the surface. This can be done by integrating the current density over the cross-sectional area of the wire, but a simpler way would be to use Ampere's circuital law at the surface

$$\oint \mathbf{H}_s \cdot dl = 2\pi r_0 \mathbf{H}_s = I \tag{4-21}$$

$\mathbf{H}_s$ can be found by using Eq. (4-7) and assuming sinusoidal excitation

$$\frac{1}{\sigma} \frac{\partial \mathbf{J}}{\partial r} = \mu \frac{\partial \mathbf{H}}{\partial t} = j\omega\mu\mathbf{H}$$

or

$$\mathbf{H}_s = \frac{1}{j\omega\mu} \frac{\partial \mathbf{E}_s}{\partial r} \tag{4-22}$$

Substituting this for $\mathbf{H}_s$ in Eq. (4-21) yields

$$I = \frac{2\pi r_0}{j\omega\mu} \frac{\partial \mathbf{E}_s}{\partial r} = 2\pi r_0 \frac{\partial \mathbf{J}_s}{\partial r} \tag{4-23}$$

The current as a function of r is given by Eq. (4-13)

$$\frac{\partial \mathbf{J}_s}{\partial r} = \mathbf{J}_s \frac{\dfrac{\partial}{\partial r} J_0 \left(\dfrac{r_0}{\delta} j\sqrt{2j} \right)}{J_0 \left(\dfrac{r_0}{\delta} j\sqrt{2j} \right)} \tag{4-24}$$

Substituting Eqs. (4-23) and (4-24) into Eq. (4-19) and simplifying

$$Z_{\text{surf}} = \frac{j\omega\mu}{2\pi r_0} \frac{J_0 \left(\dfrac{r_0}{\delta} j\sqrt{2j} \right)}{\dfrac{\partial}{\partial r} J_0 \left(\dfrac{r_0}{\delta} j\sqrt{2j} \right)} \tag{4-25}$$

Since the Bessel functions are complex quantities, the surface impedance will have a resistive and reactive part, as expected. To separate these parts, we can use the identities of Eq. (4-14) to get

$$\frac{\partial}{\partial r} J_0 (uj\sqrt{j}) = \frac{d}{du} [\text{ber}\, u + j\,\text{bei}\, u] \frac{du}{dr} = [\text{ber}'u + j\,\text{bei}'u] \frac{du}{dr} \tag{4-26}$$

where

$$u = \frac{r_0 \sqrt{2}}{\delta} \tag{4-27}$$

Using these identities, the surface impedance becomes

$$Z_{surf} = \frac{j\omega\mu}{2\pi r_0} \frac{(\mathrm{ber}\, u + j\,\mathrm{bei}\, u)}{(\sqrt{2}/\delta)(\mathrm{ber}'u + j\,\mathrm{bei}'u)} = R + j\omega L_{in} \tag{4-28}$$

where

$$\delta = \left(\frac{1}{\pi f \mu \sigma}\right)^{1/2} \tag{4-29}$$

Elimination of the complex quantity in the denominator yields

$$R = \frac{1}{r_0}\left(\frac{\mu f}{2\pi\sigma}\right)^{1/2}\left[\frac{\mathrm{ber}\, u\, \mathrm{bei}'u - \mathrm{bei}\, u\, \mathrm{ber}'u}{(\mathrm{ber}'u)^2 + (\mathrm{bei}'u)^2}\right]\frac{\mathrm{ohms}}{\mathrm{meter}} \tag{4-30}$$

$$\omega L_{in} = \frac{1}{r_0}\left(\frac{\mu f}{2\pi\sigma}\right)^{1/2}\left[\frac{\mathrm{ber}\, u\, \mathrm{ber}'u + \mathrm{bei}\, u\, \mathrm{bei}'u}{(\mathrm{ber}'u)^2 + (\mathrm{bei}'u)^2}\right]\frac{\mathrm{ohms}}{\mathrm{meter}} \tag{4-31}$$

As the frequency approaches 0, the above expressions should reduce to those of Sec. 4.2. This can easily be checked: as f approaches 0, δ goes to infinity or u approaches 0. For small arguments, the Bessel functions can be represented by the first terms of their series expansions, which are

$$
\begin{array}{ll}
\mathrm{ber}\, u \approx 1 & \mathrm{bei}\, u \approx \dfrac{u^2}{4} \\[2mm]
\mathrm{ber}'u \approx \dfrac{u^3}{16} & \mathrm{bei}'u \approx \dfrac{u}{2}
\end{array}
\qquad \text{for small } u \tag{4-32}
$$

Substitution of these approximations for R and L_{in} gives

$$R_{dc} = \frac{1}{\sigma\pi r_0^2} \qquad L_{in} = \frac{\mu}{8\pi} \tag{4-33}$$

just as was previously obtained.

At sufficiently high frequencies, δ becomes very small or u is very large. For such a case, the Bessel functions can be approximated by simplified expressions* and it can easily be shown that for the resistance term

$$\frac{\text{ber}\,u\,\text{bei}'u - \text{bei}\,u\,\text{ber}'u}{(\text{ber}'u)^2 + (\text{bei}'u)^2} = \frac{1}{\sqrt{2}} \tag{4-34}$$

for large u. The same result is obtained for the Bessel function part of the reactive term. Thus, at high frequencies, the surface impedance becomes

$$R_{\text{hf}} = \frac{1}{r_0}\left(\frac{\mu f}{2\pi\sigma}\right)^{1/2}\frac{1}{\sqrt{2}} = \frac{1}{2\pi r_0\,\sigma\delta} = \omega L_{\text{in}} \quad \frac{\text{ohms}}{\text{meter}} \tag{4-35}$$

or

$$Z_{\text{surf}} = R_{\text{hf}}(1 + j) \tag{4-36}$$

The resistive and reactive terms thus become equal at high frequencies and are given by greatly simplified expressions.

The significance of δ can now be deduced from R_{hf}. At high frequency, δ will become very small and r_0/δ very large; Eq. (4-35) states that the effective resistance is the same as if all the current were uniformly distributed over a conductor of circumference $2\pi r_0$ and thickness δ, that is, the dc resistance of the outer layer of the conductor of thickness δ. It is apparent that the current will in fact be nonuniformly distributed over a thickness greater than δ, but the resistance appears "as if" it were as stated above. This parameter δ is referred to as the penetration or skin depth and will appear again in connection with plane conductors. In fact, we shall find that plane conductors have a surface impedance identical to Eq. (4-36) for all frequencies, not just high frequency.

Intermediate between dc and high frequency, the surface impedance is a complicated function of the parameters. In the penetration depth analysis of transmission lines, the ratio of actual (effective) resistance or inductance to the dc values is usually of interest since this is a measure of the increase in attenuation and phase constant. The dc resistance and inductance are given by Eq. (4-33), so that Eq. (4-30) yields

$$\frac{R}{R_{\text{dc}}} = \frac{r_0}{\delta\sqrt{2}}\left[\frac{\text{ber}\,u\,\text{bei}'u - \text{bei}\,u\,\text{ber}'u}{(\text{ber}'u)^2 + (\text{bei}'u)^2}\right] \tag{4-37}$$

*See [6, p. 184].

Multiplying Eq. (4-31) by ω gives

$$\frac{L_{in}}{L_{dc}} = \frac{4}{\sqrt{2}} \frac{\delta}{r_0} \left[\frac{\text{ber}\, u \,\text{ber}'u + \text{bei}\, u \,\text{bei}'u}{(\text{ber}'u)^2 + (\text{bei}'u)^2} \right] \tag{4-38}$$

These normalized curves are shown in Fig. 4-4 as a function of r_0/δ or normalized frequency. It should be remembered that the reactive curve must be multiplied by ω to get actual inductive impedance and that this impedance increases with increasing r_0/δ, as shown by the dashed curve. At sufficiently large r_0/δ the real and reactive curves will approach each other as indicated.

Curves of the actual skin-effect resistance of pure copper wires as a function of frequency from very low to very high frequencies are shown in Fig. 4-5 for several values of wire radius. The smaller wires are less affected at low frequencies since the skin depth is larger than the wire radius, i.e., appreciable frequencies are necessary before the current density becomes nonuniform.

It should be recalled that in all the calculations thus far, it has been assumed that there was no variation of any of the vector quantities in the axial direction. It has not been shown that this is, in fact, valid but we have been concerned with only incremental distances $\Delta \ell$ in the axial direction. It is reasonable to expect that a range of parameters must exist for which the fields change by a negligible amount over $\Delta \ell$ compared with changes in the radial direction into the conductor. This is indeed the case, and in the next section the exact solution for a plane conductor will be found, showing the relationship between a wave traveling along the surface of a conductor while simultaneously penetrating into the conductor.

4.5 EXACT GENERAL SOLUTION FOR PROPAGATION PROPERTIES OF A WAVE BETWEEN THE SURFACES OF THICK, PLANE STRIP-LINE CONDUCTORS

In the preceding sections, the internal conductor impedance parameters were calculated using the approximation of small losses in the axial direction. These parameters could then be used in the transmission-line analyses presented in Chaps. 1 and 2 to determine the characteristic impedance, phase velocity, attenuation, or other properties. This approach is the simplest and most widely used, but it is of interest to calculate the exact general solution of a wave propagating on a plane conducting surface without resorting to the previous transmission line analyses. This problem will point out the

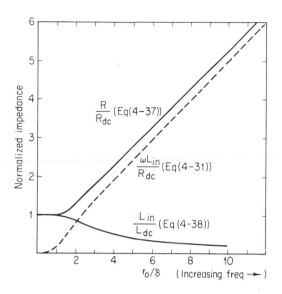

Fig. 4-4. Skin impedance components for a solid circular wire.

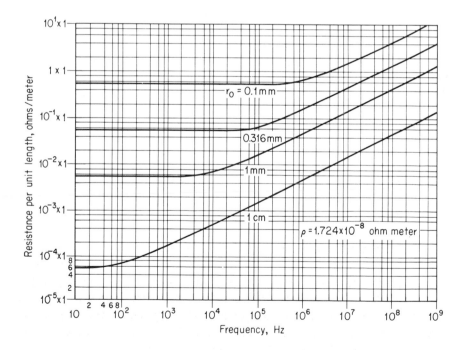

Fig. 4-5. Skin resistance of solid circular wire of pure copper at 20°C.

significance of the assumptions used in Sec. 4.4 and will lead to a clearer appreciation of the relationship between field and circuit analysis. We will find that as the fields (wave) propagate along the surface, they also propagate into the conductor, and the latter causes a slowing down as well as attenuation of the wave.

The geometry to be analyzed is shown in Fig. 4-6 and consists of two strip conductors, each assumed to be very thick, having a separation S and a width W, and assumed to be very long. A sinusoidal current of the form $J_0 e^{j\omega t}$ is applied with positive polarities taken as shown. It is assumed that only the vector components J_x, H_y, E_x, and E_z exist as shown. There is no variation of any of these vectors in the y direction (W/S is assumed to be very large), but only in x and z direction. The problem is to determine the x and z variation of the field vectors.

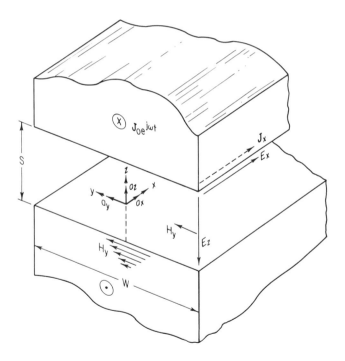

Fig. 4-6. Coordinate system of infinitely thick strip line.

In order to do this, the following relationships will be used (the reader is assumed to be familiar with them*).

---

*It is possible to proceed from more elementary considerations, as in Sec. 4.3, to arrive at the same basic differential equations, but the use of Maxwell's equations simplifies the steps.

Maxwell's equations

$$\nabla \times \mathbf{E} = -\frac{\partial \mathbf{B}}{\partial t} \qquad (4\text{-}39)$$

$$\nabla \times \mathbf{H} = \mathbf{J}_s + \frac{\partial \mathbf{D}}{\partial t} \qquad (4\text{-}40)$$

$$\nabla \cdot \mathbf{D} = \rho_v \qquad (4\text{-}41)$$

$$\nabla \cdot \mathbf{B} = 0 \qquad (4\text{-}42)$$

Ohms' law

$$\mathbf{J}_x = \sigma \mathbf{E}_x \qquad (4\text{-}43)$$

We further assume linear isotropic materials for both conductor and dielectric

$$\mathbf{D} = \epsilon \mathbf{E} \qquad (4\text{-}44)$$

$$\mathbf{B} = \mu \mathbf{H} \qquad (4\text{-}45)$$

The general relations for any material will first be derived. Then application of physical facts such as conduction current but no displacement current in conductors and displacement but no conduction current in insulators will simplify the differential equations and allow a general solution. Application of boundary conditions will give all constants and the desired solution.

The differential equation governing the electric field can be obtained by taking the curl of both sides of Eq. (4-39)

$$\nabla \times \nabla \times \mathbf{E} = -\frac{\partial}{\partial t} \nabla \times \mathbf{B} \qquad (4\text{-}46)$$

or

$$\nabla (\nabla \cdot \mathbf{E}) - \nabla^2 \mathbf{E} = -\mu \frac{\partial}{\partial t} \nabla \times \mathbf{H} \qquad (4\text{-}47)$$

There is no storage of volume charge density, so that

$$\nabla \cdot \mathbf{D} = \nabla \cdot \mathbf{E} = 0$$

Substituting Eqs. (4-40) and (4-43) into Eq. (4-47) gives

$$\nabla^2 \mathbf{E} = \mu \frac{\partial}{\partial t} \left(\sigma \mathbf{E} + \epsilon \frac{\partial \mathbf{E}}{\partial t} \right) \tag{4-48}$$

Since we have assumed sinusoidal time variation, Eq. (4-48) becomes

$$\nabla^2 \mathbf{E} = (j\omega\mu\sigma - \omega^2\mu\epsilon) \mathbf{E} \tag{4-49}$$

An identical equation for **H** can be found in similar fashion, though we will not need it.

The problem is now to use Eqs. (4-39) through (4-45) and Eq. (4-49) to find H_y, E_x, and E_z in terms of constants which can be evaluated from the boundary conditions. From Eq. (4-49), since $E_y = 0$

$$\nabla^2 \mathbf{E} = \frac{\partial^2 E_x}{\partial x^2} + \frac{\partial^2 E_x}{\partial z^2} + \frac{\partial^2 E_z}{\partial x^2} + \frac{\partial^2 E_z}{\partial z^2} = (j\omega\mu\sigma - \omega^2\mu\epsilon)(E_x + E_z) \tag{4-50}$$

It is apparent that the differential equation will require exponential-type solutions. Since it is known from experience that a wave will propagate on such lines, it is assumed that all field variations in the x direction are of the form $e^{-\gamma x}$, where γ can be complex (negative exponent only, since a positive exponent will give an infinite field at x equal to infinity, an impossible situation).* It is apparent that propagation of all vector quantities both in the dielectric and conductor will be determined by γ, since these vectors are not independent.

First consider the dielectric region. Here $\sigma = 0$;† thus Eq. (4-12) gives for the x component

$$\frac{\partial^2 E_x}{\partial x^2} + \frac{\partial^2 E_x}{\partial z^2} = -\omega^2\mu\epsilon E_x \tag{4-51}$$

The z variation of E_x must be exponential or harmonic; referring to Fig. 4-6, it is seen that at $z = S/2$, E_x must be positive, while at $z = -S/2$, E_x

---

*Sinusoidal variation is implicit, so that the fields actually vary as $\exp(j\omega t - \gamma x)$.
†Or at least $\omega\mu\sigma \ll \omega^2\mu\epsilon$ in the dielectric.

must be negative and of equal amplitude. It is apparent that E_x must vary sinusoidally with z; thus, assume that

$$E_x = C_1 e^{-\gamma x} \sin \Gamma_1 z \tag{4-52}$$

Substituting into Eq. (4-51) gives

$$\gamma^2 - \Gamma_1^2 = -\omega^2 \mu \epsilon \tag{4-53}$$

The constants γ, Γ_1, and C_1 are to be determined.

Since there is only a y component of H, Eq. (4-40) gives

$$\nabla \times \mathbf{H} = \mathbf{a}_z \frac{\partial H_y}{\partial x} - \mathbf{a}_x \frac{\partial H_y}{\partial z} = j\omega\epsilon(\mathbf{a}_x E_x + \mathbf{a}_z E_z) \tag{4-54}$$

with $\mathbf{a}_x$ and $\mathbf{a}_z$ being the unit vectors. Equating like components yields for the x component

$$-\frac{\partial H_y}{\partial z} = j\omega\epsilon E_x$$

or

$$H_y = -\int j\omega\epsilon E_x \, dz \tag{4-55}$$

Substituting Eq. (4-52) gives

$$H_y = j\omega\epsilon \, e^{-\gamma x} \frac{C_1}{\Gamma_1} \cos \Gamma_1 z \tag{4-56}$$

For the z component Eq. (4-54) yields

$$\frac{\partial H_y}{\partial x} = j\omega\epsilon E_z \tag{4-57}$$

Substituting Eq. (4-56) gives

$$E_z = -\gamma \, e^{-\gamma x} \frac{C_1}{\Gamma_1} \cos \Gamma_1 z \tag{4-58}$$

Now consider the conductor, and proceed in a manner similar to that used with the dielectric. Since the z component of E in the dielectric must terminate on charges at the surface of the conductor, E_z must be 0 inside the conductor. It is assumed that displacement current is negligible as compared to conduction current

$$\omega^2 \mu\epsilon \ll \omega\mu\sigma$$

Thus Eq. (4-50) gives for the x component

$$\frac{\partial^2 E_x}{\partial x^2} + \frac{\partial^2 E_x}{\partial z^2} = j\omega\mu\sigma E_x \tag{4-59}$$

An exponential variation is the z direction is necessary; thus, assume that

$$E_x = C_2 e^{-\Gamma_2 z} e^{-\gamma x} \tag{4-60}$$

(negative exponents only, since positive values give infinite fields as x increases). Substitution into Eq. (4-59) yields

$$\gamma^2 + \Gamma_2{}^2 = j\omega\mu\sigma \tag{4-61}$$

For the magnetic field, since $E_z = 0$

$$\nabla \times \mathbf{H} = J_x = \sigma E_x$$

or

$$\nabla \times \mathbf{H} = -\frac{\partial H_y}{\partial z} = \sigma E_x \tag{4-62}$$

Substitution of Eq. (4-60) and integration gives

$$H_y = \frac{\sigma C_2}{\Gamma_2} e^{-\Gamma_2 z} e^{-\gamma x} \tag{4-63}$$

The constants $C_1, C_2, \Gamma_1, \Gamma_2,$ and γ can be evaluated from boundary conditions. At the dielectric conductor boundary $z = S/2$, the tangential components of E, that is, E_x, must be equal, or from Eqs. (4-52) and (4-60)

$$C_1 \, e^{-\gamma x} \sin \Gamma_1 \frac{S}{2} = C_2 \, e^{-\Gamma_2 S/2} \, e^{-\gamma x}$$

or

$$C_2 = C_1 \frac{\sin \Gamma_1 \, S/2}{e^{-\Gamma_2 S/2}} \tag{4-64}$$

In like manner at $z = S/2$, the magnetic field must be continuous so from Eqs. (4-56) and (4-63)

$$j\omega\epsilon \, e^{-\gamma x} \frac{C_1}{\Gamma_1} \cos \Gamma_1 \frac{S}{2} = \sigma \frac{C_2}{\Gamma_2} e^{-\Gamma_2 S/2} \, e^{-\gamma x}$$

Substituting Eq. (4-64) and simplifying gives

$$\tan \Gamma_1 \frac{S}{2} = \frac{j\omega\epsilon}{\sigma} \frac{\Gamma_2}{\Gamma_1} \tag{4-65}$$

Γ_1 and Γ_2 are known in terms of γ and physical parameters. Substitution of Eqs. (4-53) and (4-61) gives

$$\tan \left[\frac{S}{2} (\omega^2 \mu\epsilon + \gamma^2)^{1/2} \right] = \frac{j\omega\epsilon}{\sigma} \left(\frac{j\omega\mu\sigma - \gamma^2}{\omega^2 \mu\epsilon + \gamma^2} \right)^{1/2} \tag{4-66}$$

This equation can be solved for γ and represents the general case of propagation on such a line. However, the equation is transcendental, making simple expressions impossible. In many cases of practical interest, the losses, that is, E_x, are sufficiently small so that simple expressions are possible. This is similar to the case in Secs. 1.6 and 2.6 where low-loss cases were easily solved. We shall now show the correspondence between the present, more exact field solution and previous circuit-type solutions.

   If the losses are reasonably low, then it will be found that $|\gamma|^2 \ll \omega\mu_c \sigma$ (μ_c is conductor permeability). This assumes, or implies, that the fields are reasonably uniform in the x direction for a much greater distance than in the z direction. Let us assume this and check it in Sec. 4.6. Thus Eq. (4-61) simplifies to

$$\Gamma_2 \approx (\omega\mu\sigma)^{1/2}\left(\frac{1+j}{\sqrt{2}}\right) = \frac{1+j}{\delta} \qquad (4\text{-}67)$$

where

$$\delta = (\pi f \mu\sigma)^{-1/2}$$

This δ is identical to that used in Secs. 4.3 and 4.4 for circular wires. Continuing the analysis, we have from Eq. (4-53)

$$\Gamma_1{}^2 = \omega^2\mu\epsilon + \gamma^2 \qquad (4\text{-}68)$$

Since it has already been assumed that the losses are small, then E_x must be small at the conductor surfaces $z = \pm S/2$. From Eq. (4-52) it can be seen that this can only be true if $\Gamma_1 S/2$ is small; therefore

$$\tan\Gamma_1\frac{S}{2} \approx \Gamma_1\frac{S}{2}$$

Thus, for low losses, the general form of Eq. (4-65) with the use of Eq. (4-68) becomes

$$\Gamma_1{}^2 = \frac{j\omega\epsilon\,\Gamma_2}{\sigma\,S/2} = \omega^2\mu\epsilon + \gamma^2 \qquad (4\text{-}69)$$

Substitution of Eq. (4-67) gives

$$\gamma^2 = -\omega^2\mu\epsilon - (1-j)\frac{\omega\epsilon}{S/2}\left(\frac{\omega\mu}{2\sigma}\right)^{1/2} \qquad (4\text{-}70)$$

But

$$\frac{1}{S/2}\left(\frac{1}{2\omega\sigma\mu}\right)^{1/2} = \frac{\delta}{S}$$

so

$$\gamma^2 = -\omega^2 \mu\epsilon \left(1 + \frac{\delta}{S} - j\frac{\delta}{S}\right) = -\omega^2 \mu\epsilon(1 + \nu) \tag{4-71}$$

or

$$\gamma = j[\omega^2 \mu\epsilon(1 + \nu)]^{\frac{1}{2}} \tag{4-72}$$

Since the losses are assumed small, it is necessary that

$$\nu = \frac{\delta}{S}(1 - j) \ll 1 \tag{4-73}$$

But*

$$(1 + \nu)^{\frac{1}{2}} \approx 1 + \frac{\nu}{2} \quad \text{for} \quad \nu \ll 1 \tag{4-74}$$

So

$$\gamma = j\omega\sqrt{\mu\epsilon}\left(1 + \frac{\delta}{2S} - j\frac{\delta}{2S}\right) = \alpha + j\beta \tag{4-75}$$

$$\beta = \omega\sqrt{\mu\epsilon}\left(1 + \frac{\delta}{2S}\right) \tag{4-76}$$

$$\alpha = \omega\sqrt{\mu\epsilon}\,\frac{\delta}{2S} \tag{4-77}$$

Thus, the wave propagates along the surface with a phase constant which is greater than that of the bulk dielectric by an amount proportional to $\delta/2S$, that is, the wave is slowed down by the conductor, and is attenuated as it travels in the $+x$ direction. As the wave propagates along the surface, it also propagates into the conductor, the effective penetration depth being δ.

It is interesting to make a comparison of the attenuation of this case with the low-loss case of Secs. 1.6 and 2.6. An approximate expression for low losses was previously found to be

$$\alpha = \frac{R}{2\sqrt{L/C}} \tag{4-78}$$

*See Eq. (1-25).

For the strip line under consideration here, it is apparent that the total series resistance for two conductors is, from Eq. (4-1)

$$R = 2\frac{1}{W\sigma\delta} \tag{4-79}$$

It has already been assumed that W/S is very large, so that if we neglect the internal inductance of the conductor, then for the strip line (see Sec. 8.5)

$$L = \mu\frac{S}{W} \qquad C = \epsilon\frac{W}{S} \tag{4-80}$$

Substituting Eqs. (4-79) and (4-80) into Eq. (4-78) yields

$$\alpha = \frac{R}{2\sqrt{L/C}} = \frac{\sqrt{\epsilon/\mu}}{S\sigma\delta} = \omega\sqrt{\mu\epsilon}\frac{\delta}{2S} \tag{4-81}$$

This is identical to Eq. (4-77); thus the low-loss field analysis of this strip line yields the same result as the circuit-type analysis of Secs. 1.6 and 2.6. The advantage of the field analysis is that it not only yields the form of the solution, but also evaluates the line parameters directly, as a function of frequency, as is desired.

An approximate expression for β for the low-loss case is given by Eq. (2-46) for $G = 0$

$$\beta = \omega\sqrt{LC}\left(1 + \frac{R^2}{8\omega^2 L^2}\right) \tag{4-82}$$

In order to show that this is equivalent to Eq. (4-76), it is necessary to use the total inductance of the line, i.e., the external circuit inductance plus internal inductance, the former being given by Eq. (8-20) and the latter by Eq. (4-35) or Eq. (4-103)

$$L_T = L_{ex} + L_{in} = \mu\frac{S}{W} + 2\left(\frac{1}{\omega\delta\sigma W}\right) \tag{4-83}$$

In the expression for the attenuation which was previously derived, i.e., Eq. (4-81), the internal inductance was neglected. The reason is simply that the conductor inductance is a second-order effect and therefore usually has negligible effect on the attenuation. However, the phase constant is influenced only by second-order effects and as shall now be shown, if the internal inductance is negligible, the phase constant is not influenced by the losses.

In Sec. 2.6 it was necessary to assume that $R/\omega L \ll 1$ in order to obtain the simplified expression of Eq. (4-82). However, if this assumption is true, then $R^2/8\omega^2 L^2$ certainly can be neglected; thus

$$\beta = \omega\sqrt{CL_T} \tag{4-84}$$

where L_T is given by Eq. (4-83). Therefore

$$\beta = \omega\sqrt{CL_{ex}}\left(1 + \frac{L_{in}}{L_{ex}}\right)^{1/2} \tag{4-85}$$

If it is assumed that $L_{ex} \gg L_{in}$, that is, the inductance between the conductors is far greater than that of the conductors themselves, but L_{in} is not negligible, then, with the aid of Eq. (1-25)

$$\left(1 + \frac{L_{in}}{L_{ex}}\right)^{1/2} \approx 1 + \frac{1}{2}\frac{L_{in}}{L_{ex}} = 1 + \frac{1}{2}\frac{2}{\omega\delta\sigma\mu S} \tag{4-86}$$

Thus, using the fact that $\delta^2 = 2/\omega\mu\sigma$, and using Eqs. (4-80) for L_{ex} and C, the phase constant becomes

$$\beta = \omega\sqrt{\mu\epsilon}\left(1 + \frac{\delta}{2S}\right) \tag{4-87}$$

This is identical to Eq. (4-76), which thus indicates the equivalence between the lumped-circuit and field-theory analysis for the case of low losses.

4.6 ANALYSIS OF ASSUMPTIONS USED IN TRANSMISSION-LINE ANALYSES

We will consider both the approximate range of validity of the various assumptions used in the exact analysis of Sec. 4.5 and the assumptions implicit in the incremental circuit approach of Chaps. 1 and 2 and their relationship to the exact analysis for the general case.

In the exact analysis of the strip line of Sec. 4.5, the assumptions used to obtain the general expression of Eq. (4-66) were that we may neglect conduction current in dielectrics and displacement currents in conductors. These are commonly used assumptions which are generally true. Let us examine the range of validity. First, in the dielectric, it was assumed

$$\omega \mu \sigma \ll \omega^2 \mu \epsilon \quad \text{or} \quad \frac{\sigma}{\epsilon} \ll \omega$$

In a good dielectric such as polyethylene or teflon, $1/\sigma$ is about 10^{15} ohm-meters (dc volume resistivity at room temperature)* and ϵ is about $2\epsilon_0$. Substituting these values, we get

$$\omega \gg 10^{-4} \text{ Hz}$$

The assumption is thus essentially valid over all frequencies of practical interest. However, as the frequency gets very high, the dielectric losses become important, i.e., the imaginary (lossy) part of ϵ, can no longer be neglected, so that the general expression is $\sigma \ll \omega(\epsilon' - j\epsilon'')$ or

$$\sigma + j\omega\epsilon'' \ll \omega\epsilon'$$

We have already shown that for good dielectrics, σ can be neglected; thus it is necessary that $\epsilon' \ll \epsilon''$ or

$$\frac{\epsilon''}{\epsilon'} \ll 1$$

Losses are generally given in terms of $\tan \delta = \epsilon''/\epsilon'$. It is thus necessary that $\tan \delta \ll 1$[†] for the analysis to be accurate. For polyethylene and teflon at $f = 25 \times 10^9$ Hz, $\tan \delta = 0.0006$, which satisfies the requirements, showing the wide range of frequency covered by the assumption.

In the conductor, it is necessary that $\omega \mu \sigma \gg \omega^2 \mu \epsilon$ or $\sigma \gg \omega\epsilon$. For copper at room temperature, $\sigma = 1/1.7 \times 10^{-8}$ (ohm-meters)$^{-1}$, and $\epsilon = \epsilon_0$; thus it is necessary that

$$\omega \ll \frac{\sigma}{\epsilon} = 6.7 \times 10^{18} \text{ Hz}$$

---

*Surface conduction current is usually more significant than bulk current, reducing the effective resistivity sometimes by a few orders of magnitude, depending on actual surface conditions.
† δ is the angle between ϵ'' and ϵ', not the skin depth.

The above inequality will certainly hold true for any line of practical interest.

Only the above two assumptions were required in order to obtain the general solution with γ, the propagation contant in the direction of propagation, expressed in the form of Eq. (4-66). This transcendental equation can only be solved approximately and simple, closed-form solutions are possible only for the assumption of low losses, as was the case in Sec. 4.5. This raises a very important point: Since the exact solution to general transmission-line propagation can only be obtained approximately, the solution for γ and therefore also for Z_0 obtained from the incremental-circuit approach in Chaps. 1 and 2 must be only approximate and have certain implicit assumptions.   This is, in fact, true and we will now consider what these approximations are.

In Chaps. 1 and 2, the general propagation constant for a good dielectric with $G = 0$ was found to be

$$\gamma^2 = (R + j\omega L_T) j\omega C \tag{4-88}$$

This equation is always valid, as long as R, L_T, and C are known. In the derivation of this equation, it was just implicitly assumed that the parameters R, L_T, and C could be calculated (or measured) at the frequency of interest and a simple lumped-equivalent circuit would represent the line exactly. This is the difficulty because the parameters are not simply evaluated for the general case and in fact only approximate expressions for R, L_T, and C are possible for the general case. In order to see this, let us ask ourselves what R and L_T are in terms of physical parameters for the general case. When the losses are large as we now assume, the total inductance per unit length must include the internal conductor inductance and external circuit inductance. The latter is easily obtained but the former can only be obtained approximately. Also, R can only be obtained approximately. If certain assumptions are made concerning the losses, then simple closed form solutions of R and L_T are given by Eqs. (4-79) and (4-83). C is given by Eq. (4-80). If these are substituted into the lumped-equivalent-circuit expression of Eq. (4-88), the result, after some manipulation, is

$$\gamma^2 = \frac{j2\omega\epsilon}{S\sigma\delta} - \omega^2\mu\epsilon - \frac{2\omega\epsilon}{S\sigma\delta} = -\omega^2\mu\epsilon\left(1 + \frac{\delta}{S} - j\frac{\delta}{S}\right) \tag{4-89}$$

This is identical to Eq. (4-71), which was obtained from the transcendental Eq. (4-66) using only the two assumptions

$$\tan \Gamma_1 \frac{S}{2} = \Gamma_1 \frac{S}{2} \tag{4-90}$$

$$\gamma^2 \ll \omega \mu_c \sigma \tag{4-91}$$

μ_c is permeability of the conductor and is usually equal to μ_0. In order words, the above assumptions are required in order that R and L_T can be given simple expressions. Let us now see what limitations the above expressions place on the frequency for which R and L_T can be expressed in simple terms.

The assumption of Eq. (4-90) reduces to

$$\Gamma_1 \frac{S}{2} \ll 1 \text{ radian}$$

or, from Eq. (4-69)

$$\frac{S}{2}(\omega^2 \mu \epsilon + \gamma^2)^{1/2} \ll 1 \tag{4-92}$$

Squaring both sides gives

$$\omega^2 \mu \epsilon + \gamma^2 \ll 4/S^2$$

But γ^2 is given by Eq. (4-71) and we are interested only in an approximate amplitude comparison. Thus, $(j - 1) \approx 1$ in γ^2, so that, after collecting terms, the above becomes

$$\omega^{3/2} \ll \frac{4}{S \epsilon} \left(\frac{\sigma}{2\mu}\right)^{1/2} \tag{4-93}$$

where μ and ϵ are the permeability and permittivity of the dielectric.

As an example, copper at room temperature has $1/\sigma = 1.7 \times 10^{-8}$ ohm-meters. If we assume that the dielectric is air, then the above becomes approximately

$$\omega^{3/2} \ll \frac{24 \times 10^{17}}{S} \tag{4-94}$$

If $S = 10^{-3}$ meters, then the approximate expression is

$$f \ll 10^{13} \text{ Hz}$$

This frequency limitation is quite high, well beyond that normally encountered.

The assumption of Eq. (4-91) can be analyzed in a similar manner; substitution of Eqs. (4-69) and (4-61) and simplification and collection of terms yields

$$\frac{\epsilon}{S}\left(\frac{2\omega}{\sigma\mu}\right)^{1/2}(j-1) \ll \sigma + \omega\epsilon_d \qquad (4\text{-}95)$$

where it has been assumed that $\mu_c = \mu_d = \mu_0$, which is usually true (subscripts c and d refer to conductor and dielectric). We are once again interested only in an approximate amplitude comparison, so that $(j-1) \approx 1$; also, if we assume that ϵ_d is not greatly different from ϵ_0, that is, $\epsilon_d/\epsilon_0 \approx 2$ to 4, then $\sigma \gg \omega\epsilon_d$, since displacement current is negligible. The above equation therefore reduces to

$$\omega^{1/2} \ll \frac{S\sigma}{\epsilon_d}\left(\frac{\sigma\mu}{2}\right) \qquad (4\text{-}96)$$

Using copper at room temperature and $\epsilon_d = \epsilon_0$, this becomes approximately

$$\omega^{1/2} \ll 0.5S \times 10^{20}$$

If S is 10^{-3} meters, then $\omega \ll 25 \times 10^{32}$ Hz. This frequency is so high that it will always hold true.

Equation (4-93) indicates that in order to keep the frequency limit high, σ should be as large, and S as small as possible. Equation (4-96) indicates that both σ and S should be large. Thus, we see that σ should always be as large as possible in order to extend the frequency range of applicability of the assumptions.

The general conclusion is that the high-frequency limit imposed by these assumptions is high enough that Eqs. (4-79) and (4-83) are good for nearly all cases of practical interest, with γ given by Eq. (4-71). For the more general case, Eq. (4-66) is necessary.

The expression for γ given by Eq. (4-71) or Eq. (4-88) is still not quite as simple as is desired. In Secs. 1.6 and 2.6, we derived simple expressions for α and β and an identical expression in Sec. 4.5 from Eq. (4-71). The additional assumption required was that $R \ll \omega L$ in Secs. 1.6 and 2.6, and $\delta/S \ll 1$ in Sec. 4.5. For plane conductors, these two assumptions are identical, as can be seen if we substitute the expression for R and L of a plane conductor in the former to get

$$\frac{2}{W\sigma\delta} \ll \omega\mu\frac{S}{W}$$

But since $\delta^2 = 2/\omega\mu\sigma$, the above gives $\delta/S \ll 1$ or $f \gg 1/(\pi\mu\sigma S^2)$. It is interesting to note that this assumption gives a low-frequency limit and no high-frequency limit. This point is often overlooked or not clearly understood. Obviously, σ should be as large as possible in order to extend the low end of the bandwidth. This low-frequency limit really comes about because we have been concerned with an infinitely thick conductor: as the frequency becomes very low the depth of penetration and hence the internal conductor inductance increases. In the limit of f approaching zero, R must equal ωL for a plane conductor, and obviously the assumption $S \gg \delta$ is no longer valid. For a thin conductor, the internal inductance is usually (but not always) negligible. Thus the necessary requirement is $S \gg T$ if the thickness T is smaller than the penetration depth, and again σ should be as large as possible so as to keep the losses small.

The high-frequency limits imposed on these simple expressions for α and β are given by Eqs. (4-90) and (4-91), which are the same ones imposed on the simplified expressions for R and L.

4.7 SURFACE IMPEDANCE OF A THICK, PLANE CONDUCTOR

It is of interest to determine the surface impedance of either of the plane conductors of Fig. 4-6, since this will give the internal inductance and effective resistance. If W/S is very large, then the geometrical (external) inductance is negligible and the surface impedance per unit length is the voltage drop per unit length on the surface divided by the total current enclosed*

$$Z_{\text{surf}} = \frac{E_x\,(z = \pm S/2)}{I_T} \qquad (4\text{-}97)$$

At the conductor surface $z = S/2$, the electric field is given by Eq. (4-60)

$$E_x = C_2\,e^{-\Gamma_2 S/2}\,e^{-\gamma x} \qquad (4\text{-}98)$$

---

*A simplified and sometimes more expedient expression for obtaining the surface impedance is derived in Appendix 4A.

The total current is obtained from

$$\int H_y \cdot d_y = I_T$$

Since H_y is uniform in the y direction, then as long as W/S is very large, no field exists outside the conductors

$$I_T = WH_y\big|_{z=S/2} \tag{4-99}$$

Substituting Eq. (4-63) for H_y gives

$$I_T = \frac{\sigma C_2}{\Gamma_2} e^{-\Gamma_2 S/2} e^{-\gamma x} \tag{4-100}$$

Substituting Eqs. (4-98) and (4-100) in Eq. (4-97) yields

$$Z_{surf} = \frac{\Gamma_2}{\sigma W} \tag{4-101}$$

Let us assume that the losses are small, as was done in Sec. 4.5; then Γ_2 is given by Eq. (4-67) and

$$Z_{surf} = \frac{1 + j}{\delta \sigma W} = R + j\omega L \quad \frac{\text{ohms}}{\text{meter}} \tag{4-102}$$

where

$$R = \omega L = \frac{1}{\delta \sigma W} = \frac{1}{W}\left(\frac{\pi f \mu}{\sigma}\right)^{1/2} \tag{4-103}$$

These values represent the impedance per unit length for the given width W. As W changes, these values change. The surface impedance can be made independent of the dimensions of the strip by multiplying both sides by W

$$Z_{surf} = \frac{1 + j}{\delta \sigma} = (1 + j)\left(\frac{\pi f \mu}{\sigma}\right)^{1/2} \quad \frac{\text{ohms}}{\text{square}} \tag{4-104}$$

Note that this is not ohms per unit length squared but ohms per square of any size, as long as the length equals the width, i.e., the impedance of a surface square of thickness δ; f = Hz, and σ = dc conductivity in (ohm-meters)$^{-1}$. It should be noted that the above represents the impedance of one conductor only. The total surface impedance per unit length of the entire strip line is twice this value.

It is apparent that for the low-loss case, the effective resistance and internal inductance of a plane conductor are equal at all frequencies. Recall that for circular wires, this was found to be true only at high frequencies where δ/r_0 was very small. Observe that Eq. (4-48) is identical to Eq. (4-102) if the circumference $2\pi r_0$ is replaced by W, the strip-line width. It may be concluded that as long as δ/r_0 is very small, the plane conductor surface impedance can be used for circular wires: This seems reasonable, since for such cases the curvature of the circular wire will be quite small compared to δ, and hence it appears as a plane conductor.

Thus, δ represents the effective penetration depth. However, as before, it is apparent that current penetrates farther than δ: in fact, δ is just the point where the current has fallen to $1/e$ or 36.8 percent of its amplitude at the surface. The surface resistance appears as if the current were uniformly distributed over a thickness δ.

It is often useful to know the actual value of δ as a function of frequency for a good conductor. Figure 4-7(a) shows such a curve for copper from very low to very high frequencies. The value of δ for other conductors

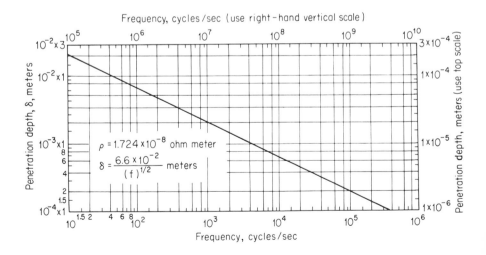

Fig. 4-7(a). Penetration depth of a plane conductor of pure copper at 20°C.

can easily be obtained by multiplication of these values by the square root of the conductivity ratio.

Even though the penetration depth decreases with increasing σ, it should be noted from Eq. (4-103) that the surface impedance is minimized by the use of a good conductor, to make σ as large as possible. At room temperature, pure silver has a slightly larger σ than copper but the high cost of silver limits its use to special, essential applications. At lower temperatures, copper surpasses silver as well as other good conductors, as can be seen from $1/\sigma$ versus temperature in Fig. 4-10.

The surface impedance of a plane, copper conductor as a function of frequency is given in Fig. 4-7(b). The curve is, of course, linear on a log-scale and can be easily extrapolated or calibrated for other frequencies and other conductors.

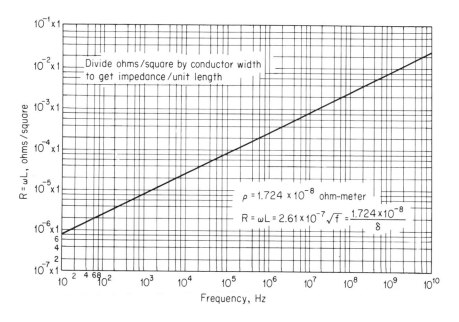

Fig. 4-7(b). Impedance of plane conductor of pure copper at 20°C.

4.8 SURFACE IMPEDANCE OF A PLANE CONDUCTOR OF ANY THICKNESS

If the parallel plane strip conductors of Fig. 4-6 have a finite thickness as shown in Fig. 4-8, analysis similar to that previously used may still be employed. Both conductors are assumed to be of thickness T, width W, and separation S. Propagation in the x direction is still of the form $e^{-\gamma x}$, where γ is to be determined.

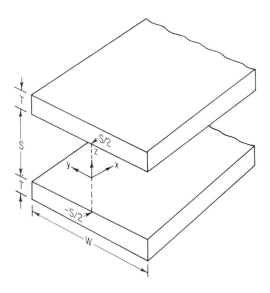

Fig. 4-8. Coordinate system of strip line of finite thickness T.

In the dielectric, the functional variation of the field quantities is unchanged, with E_x, E_z, and H_y being given by Eqs. (4-52), (4-58), and (4-56).

In the conductors, the situation is somewhat different. Previously, it was necessary to use only a negative exponential solution of E_x and H_y. Now, in order to match boundary conditions, it is necessary to use $\pm\Gamma_2$. Equations (4-60) and (4-63) therefore become

$$E_x = e^{-\gamma x}\left(C_2 e^{-\Gamma_2 z} + C_3 e^{\Gamma_2 z}\right) \tag{4-105}$$

$$H_y = \sigma e^{-\gamma x}\left(\frac{C_2}{\Gamma_2} e^{-\Gamma_2 z} - \frac{C_3}{\Gamma_2} e^{\Gamma_2 z}\right) \tag{4-106}$$

In order to evaluate the constants, it is necessary to apply boundary conditions. If it is assumed that W/S is very large, then the H field external to the strip-line conductor must be 0, since the field from currents in the $+x$ and $-x$ direction will cancel at all external points (but will add at internal points). The first boundary condition is thus that at $z = S/2 + T$, $H_y = 0$

and from Eq. (4-106)

$$C_3 = C_2 \exp\left[-2\Gamma_2\left(\frac{S}{2} + T\right)\right] \qquad (4\text{-}107)$$

At the dielectric-conductor boundary, $z = S/2$, and the conditions are as they were before, namely, tangential components of E and H are continuous. Thus, for E_x it is necessary that

$$C_1 \sin\Gamma_1 \frac{S}{2} = \left(C_2 e^{-\Gamma_2 S/2} + C_3 e^{\Gamma_2 S/2}\right) \qquad (4\text{-}108)$$

Substituting Eq. (4-107) into Eq. (4-108) and simplifying gives

$$C_2 = \frac{C_1 \sin\Gamma_1 S/2}{\left(1 + e^{-2\Gamma_2 T}\right) e^{-\Gamma_2 S/2}} \qquad (4\text{-}109)$$

For H_y components at $z = S/2$, it is necessary from Eqs. (4-56) and (4-106) that

$$j\omega\epsilon \frac{C_1}{\Gamma_1} \cos\Gamma_1 \frac{S}{2} = \sigma\left(\frac{C_2}{\Gamma_2} e^{-\Gamma_2 S/2} - \frac{C_3}{\Gamma_2} e^{\Gamma_2 S/2}\right) \qquad (4\text{-}110)$$

Substituting Eqs. (4-107) and (4-109) into the above and simplifying yields

$$\tan\Gamma_1 \frac{S}{2} = \frac{j\omega\epsilon}{\sigma} \frac{\Gamma_2}{\Gamma_1} \coth\Gamma_2 T \qquad (4\text{-}111)$$

This is the general expression relating Γ_1, Γ_2, and γ; it is identical to Eq. (4-65) except for the additional term, which depends on conductor thickness. As T goes to infinity, $\coth\Gamma_2 T$ approaches 1, so that the two equations agree in the limiting case.

The surface impedance can be evaluated as in Sec. 4.7

$$Z_{surf} = \frac{E_y \,(z = S/2)}{W H_y} = \frac{C_2 e^{-\Gamma_2 S/2}\left(1 + e^{-2\Gamma_2 T}\right)}{W\sigma(C_2/\Gamma_2) e^{-\Gamma_2 S/2}\left(1 - e^{-2\Gamma_2 T}\right)} = \frac{\Gamma_2}{W\sigma} \coth\Gamma_2 T$$

$$(4\text{-}112)$$

If we assume that the losses are small, as in Sec. 4.5, then the relationship between Γ_1, Γ_2, and γ is simplified. In Eq. (4-61) $\gamma^2 \ll \omega\mu\sigma$, so that $\Gamma_2 = (1 + j)/\delta$, and thus

$$Z_{\text{surf}} = \frac{1 + j}{W\sigma\delta} \coth \Gamma_2 T \quad \frac{\text{ohms}}{\text{meter}} \qquad (4\text{-}113)$$

In the limit, as T becomes very large, Eq. (4-113) becomes identical to Eq. (4-102).

4.9 PLANE WAVE INCIDENT PERPENDICULARLY ON A THICK, PLANE CONDUCTOR

In Sec. 4.5, a wave propagating along the surface of a strip line was found to also penetrate into the conductor; this phenomenon is nothing more than the usual penetration depth. It is desirable now to show that this penetration depth is identical to that which is experienced by a plane wave incident normally on the surface of a thick-plane conductor. Let us first consider a plane, i.e., a transverse electromagnetic, wave traveling in the $+x$ direction and incident on the conductor, as shown in Fig. 4-9. The origin is taken at the surface of the conductor. The wave was initially traveling in air (for $-x$ values) and it will be shown that E and H propagate with the form of traveling waves given by Eq. (2-48). When the wave

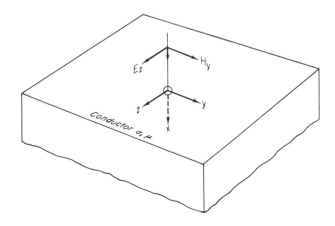

Fig. 4-9. Plane wave incident on a conductor.

strikes the conductor surface, there will be a reflection, propagating in the
$-x$ direction, as well as transmission into the conductor. We shall neglect
the reflection since it is of no consequence here.

H_y and E_z are assumed to be everywhere uniform in the y and z directions, respectively. These vector quantities are governed by Eq. (4-49).
In the dielectric region above the conductor, $\sigma = 0$, so that the differential
equation for E_z becomes

$$\nabla^2 E = \frac{\partial^2 E_z}{\partial x^2} = -\omega^2 \mu\epsilon\, E_z \qquad (4\text{-}114)$$

The solution to this is of the form

$$E_z = E_0 \exp(j\omega t - \gamma x)$$

E_0 is the amplitude, which we can assume to be known. Substitution in
Eq. (4-114) yields

$$\gamma = j\omega \sqrt{\mu\epsilon} = j\beta$$

This is identical in form to Eq. (2-48) for a traveling wave, as might have
been expected.

An identical equation and solution is obtained for H_y, as can be verified
in an anologous manner.

In the conductor, the displacement current is negligible while the conduction current is large, so that Eq. (4-49) becomes

$$\frac{\partial^2 E_z}{\partial x^2} = j\omega\mu\sigma\, E_z \qquad (4\text{-}115)$$

The solution to this is of the form

$$E_z = E_1 \exp(j\omega t - \Gamma x) \qquad (4\text{-}115a)$$

where E_1 must be evaluated. Substitution into Eq. (4-115) gives

$$\Gamma = (j\omega\mu\sigma)^{1/2} = (1 + j)\left(\frac{\omega\mu\sigma}{2}\right)^{1/2} = \frac{1 + j}{\delta}$$

E_1 can be obtained from the boundary conditions at the conductor dielectric surface (the conductor is assumed to be very thick, so that only one boundary condition need be satisfied). The field on the conductor surface at $x = 0$ must equal the amplitude of the traveling wave in the dielectric. Thus $E_1 = E_0$, and in the conductor, Eq. (4-115a) gives

$$E_z = E_0 \exp\left[j\left(\omega t - \frac{x}{\delta}\right)\right] \exp\left(-\frac{x}{\delta}\right)$$

E_0 is the field on the surface of the conductor. An analogous expression is obtained for H_y.

It is apparent that the wave is attenuated as it travels into the conductor with the attenuation constant $\alpha = 1/\delta$. At a distance $x = \delta$, the wave decreases to $1/e = 0.368 E_0$, or is only 36.8 percent of its value on the surface. Also, the wave is phase-shifted as it travels into the conductor with a phase constant

$$\beta = \frac{1}{\delta} \frac{\text{radians}}{\text{meter}}$$

The phase and attenuation constants per unit length have the same amplitude and are identical to those for a strip line with a wave propagating along its surface, as given by Γ_2 of Eq. (4-67).

4.10 INTRODUCTION TO ANOMALOUS SKIN EFFECT

Thus far we have been treating skin effect by assuming that the relationship between current and electric field (or voltage) is governed by Ohm's law, $\mathbf{J} = \sigma \mathbf{E}$. This relationship is valid for most practical cases, but there are cases for which this is no longer true, and it should therefore be understood in order that we may know the limitations of any such analysis. Ohm's law becomes invalid in some cases at very low temperatures and high frequency where the mean free path, i.e., distance between collisions, of the conduction electrons becomes greater than the classical skin depth. Such circumstances give rise to a surface resistance which can be many times larger than that calculated from classical skin effect.

It was known in the late 1800s that the dc resistivity of very thin films was higher than that of bulk material. Thomson [20] suggested that these effects were due to the mean free path constraints on electrons and gave an approximate theory. Lovell [11] made careful measurements on thin films of the alkali metals. However, such work was often hampered by the difficulty of obtaining well-prepared samples. Furthermore, high-frequency, high-power oscillators were not available at this time, so that the effects of frequency and the relationship between the mean free path and the surface impedance of a thick conductor was not known. The latter effects were first observed by London [10] while making resistivity measurements on tin at low temperature and frequency of 1,500 MHz. He noted that at 4°K the resistance of normal tin was several times the expected resistance, and correctly attributed this to the long mean free path of the conduction electrons, although he did not analyze the phenomena.

Anomalous skin effect has been used by physicists to gain further knowledge of the electron porperties of metals since it provides a way to measure the number of free electrons per atom as well as the ratio of the dc conductivity to mean free path σ/Λ. However, since this is a surface phenomenon with very small penetration depths, the resulting measurements are highly dependent on surface preparation, in particular on surface strains. This makes meaningful measurements extremely difficult to obtain, but the technique remains important since there is as yet no better one to replace it.

Although this phenomenon has been used mainly for purposes of studying the electronic properties of conductors, it is becoming increasingly important to the engineers working with high frequencies or high-speed pulse circuits and devices. The trend over the past two decades has been towards higher speed, i.e., high frequency. Also, operation of components at low temperatures is becoming desirable as well as necessary. Under such conditions, it is inevitable that engineers will have to cope with this phenomenon eventually.

As an example, magnetic memory arrays and integrated circuits containing strip lines are greatly miniaturized and hence attenuation is a serious problem. Reduction of the temperature can help this to some extent, but there are limits on the measures that can be taken. It is the purpose of these sections to present the limitations both from a practical as well as a theoretical point of view.

Let us first consider some properties and consequences of anomalous skin effect. It is well known that the conductivity of metals increases or that their resistivity decreases as the temperature decreases. The dc resistivity ρ of pure copper and other conductors as a function of absolute temperature is shown in Fig. 4-10. The penetration depth δ varies as $\rho^{1/2}$, and at 20°C and a frequency of 10^9 Hz, from Fig. 4-7(a), δ is about 2×10^{-6} meters (20 KÅ). At 0°C, the mean free path of conduction electrons in pure copper is about 4×10^{-8} meters (Table 4-1). Thus, near room

Table 4-1. Parameters of Pure Metals

	$*\rho = 1/\sigma\, 10^{-8}$ $(\Omega m)\,(0°C)$	$\dagger\Lambda$, mean free path (m) (0°C)	Theoretical $k = \sigma/\Lambda$ $(\Omega m^2)^{-1}$	$\ddagger$Measured $k = \sigma/\Lambda$ $(\Omega m^2)^{-1}$	T_a, onset of anomalous effects at 1.2 $\times 10^9$ Hz (°K)	R_{ex} (extreme) at 1.2×10^9 Hz (ohms)	$1/R_{ex}$ (ohms)$^{-1}$
Aluminum	2.828(20°C)			20×10^{14}	63	0.00115	870
Copper	1.55	4.2×10^{-8}	15.3×10^{14}	15.4×10^{14}	66	0.00126	790
Gold	2.01	4.1×10^{-8}	12×10^{14}	8.4×10^{14}	50	0.00154	648
Silver	1.47	5.7×10^{-8}	11.9×10^{14}	8.5×10^{14}	63	0.00154^-	653
Tin	0.114(20°K)§ 11.5(20°C)			9.4×10^{14}	25		
Lead	19.3			9.5×10^{14}	14		

* See Fig. 4-10.
† See [7, p. 240], [5, p. 284].
‡ See [3].
§ See [14, p. 389].

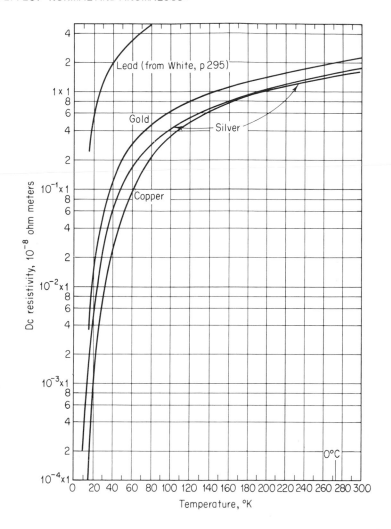

Fig. 4-10. Resistivity as a function of temperature for several good conductors. *(From "American Institute of Physics Handbook," McGraw-Hill Book Company, New York, 1955. By permission of the publishers.)*

temperature, the mean free path is much smaller than δ. Now let us see what happens as the temperature is lowered. First, the temperature-dependence of the mean free path must be determined. This can be done by relating it to the dc conductivity

$$\sigma = \frac{ne^2\Lambda}{m\mathscr{v}_0}$$

where

n = number of conduction electrons per atom
Λ = mean free path
e = electronic charge
m = effective electron mass
ν_0 = velocity of electron at the Fermi surface

Copper is a relatively simple metal with $n = 1$ and m = actual mass of an electron. The parameters n, e, m, and ν_0 are virtually independent of temperature over any temperature range of interest, so that, Λ varies directly with σ. The constant of proportionaltiy can be easily obtained from the known value of Λ and σ at any given temperature. At 0°C, Λ is 4.2×10^{-8} meters, while from Fig. 4-10, $\sigma = 1/\rho = (1.55 \times 10^{-8}\ \Omega m)^{-1}$. Substituting these gives for pure copper

$$\Lambda = 6.5 \times 10^{-16}\, \sigma \text{ meters} \qquad (4\text{-}116)^*$$

where σ = dc conductivity in (ohm-meters)$^{-1}$. More generally

$$\frac{\sigma}{\Lambda} = k \qquad (4\text{-}117)$$

where k is a constant independent of temperature for any metal but varies for each metal, and can also vary from sample to sample of a given metal because of impurities or strain. Values of k for several conductors are given in Table 4-1.

Since σ increases with decreasing temperatures (Fig. 4-10), Λ will do likewise. But the penetration depth decreases with decreasing temperatures, being related to the conductivity by

$$\delta = \left(\frac{1}{\pi f \mu \sigma}\right)^{1/2} \qquad (4\text{-}118)$$

*This equation holds true only as long as the frequency is not so high that extreme anomalous conditions are present.

The onset of anomalous skin effect is expected approximately at the value of σ for which $\Lambda = \delta$ or, from Eqs. (4-117) and (4-118), at

$$\rho_a = \frac{1}{\sigma_a} = \left(\frac{\pi f \mu}{k^2}\right)^{1/3} \tag{4-119}$$

where ρ_a merely represents the dc value of resistivity where anomalous effects set in. For pure copper, this gives, from Eq. (4-116)

$$\rho_a = (1.67 \times 10^{-36} f)^{1/3} \text{ ohm-meter} \tag{4-120}$$

where f = frequency in Hz. At a frequency of 1.2×10^9 Hz, anomalous effects are to be expected at

$$\rho_a = 1.26 \times 10^{-9} \text{ ohm-meter} \tag{4-121}$$

From Fig. 4-10, this corresponds to a temperature of about $66°K$, which is in the range of liquid nitrogen temperature ($77°K$).

It is desirable to apply the above theory to actual measurements. In order to relate to measurable parameters and also show where anomalous effects appear, it is expedient to make use of Eq. (4-104) to relate surface resistance to σ

$$\omega L = R = \left(\frac{\pi f \mu}{\sigma}\right)^{1/2}$$

or

$$\frac{1}{R} = \left(\frac{\sigma}{\pi f \mu}\right)^{1/2} = 5 \times 10^2 \left(\frac{\sigma}{f}\right)^{1/2} \left(\frac{\text{ohms}}{\text{square}}\right)^{-1} \tag{4-122}$$

with σ in mhos/meter and f in Hz. R and the dc conductivity are separately measurable quantities. If the temperature is varied, e.g., lowered, then a curve of $1/R$ versus $\sigma^{1/2}$ for a given frequency should follow a straight line, according to classical theory, up to a value of σ where anomalous effects appear. As the temperature is further decreased, R should be larger than that given by Eq. (4-122).

Figure 4-11 is such a curve for copper measured at 1.2×10^9 Hz. The straight-line portion represents Eq. (4-122) or $1/R = 1.45 \times 10^{-4} (\sigma)^{1/2}$. From Eq. (4-121), anomalous effects are expected in the vicinity of $\sigma^{1/2} = 2.8 \times 10^4$ $(\text{ohm-m})^{-1/2}$ which is actually the case, since the experimental curve begins to deviate from the straight line in this vicinity.

Thus, for pure copper with smooth, strain-free surfaces, anomalous effects should be considered near liquid nitrogen temperatures for frequencies of about 10^9 Hz. For cases where the surface properties differ greatly from bulk properties, σ/Λ will generally be much smaller than above, due to the high surface resistance layer. From Eq. (4-119), a smaller value of k implies that anomalous effects are expected at larger values of ρ or at higher temperatures for a given metal. Thus, for practical work with copper near 77°K and 10^9 Hz, it is desirable to keep k large, and special precautions are needed to obtain the desired surface properties.* Furthermore, silver and gold have smaller values of k than copper (Table 4-1), so that anomalous effects are expected at slightly higher values of ρ for these two metals. The temperature at which anomalous effects begin depends on the value of k and on the ρ-vs.-T curve. If Chambers' [3] values in Table 4-1 are used for k and if Fig. 4-10 is used for ρ versus T, then the onset of anomalous effects for silver and gold at 1.2×10^9 Hz is expected at approximately 63°K and 50°K, respectively. It should be noted that the measured values of k are smaller than the theoretical values: The measurements are very sensitive to surface imperfections and strains, so that accurate results are difficult to obtain.

The value of resistivity at which anomalous effects are expected in tin and lead, which are poorer conductors than silver and gold, can be found by using the value of k from Table 4-1. At 1.2×10^9 Hz, Eq. (4-119) yields $\rho_a = 1.76 \times 10^{-9}$ ohm-meter. This value agrees with that obtained experimentally for tin ([14, p. 388]). From Fig. 4-10, this resistivity corresponds to a temperature of about 14°K for lead. Lead becomes a superconductor below 7.175°K (see Table 6-1), so that anomalous effects set in before superconducting behavior occurs. However, the effect of anomalous behavior on surface resistance will be somewhat small, since the onset of anomalous and superconducting properties is separated by only about 7°K. that is, lead becomes superconducting shortly after anomalous effects occur. Tin is also a superconductor, but with a critical temperature of 3.74°K.

---

*Lending [9] has shown that surface roughness can have a very large effect on the losses in ordinary conductors at microwave frequencies between 3 and 35×10^9 Hz. Furthermore, the aging and corrosive properties of the conductor affect the surface roughness quite substantially at these frequencies and thus can cause the losses to increase with time.

Determination of the temperature where anomalous effects should occur is hampered by lack of accurate data on ρ versus T. If we use the value for tin at 20.4°K of $\sigma = 8.8 \times 10^8$ (ohm-m)$^{-1}$ or $\rho = 1.14 \times 10^{-9}$ ohm-m, given in [14, p. 389], it can be seen that anomalous effects occur above this temperature, in the vicinity of 25°K. Thus, anomalous effects occur before superconducting properties set in at this frequency.*

It should be pointed out that the above results are highly dependent on the values of k and on the ρ-versus-T curve which one uses. Accurate values are often difficult to specify, and the problem becomes more acute at low temperature, which is where anomalous effects usually occur. This difficulty can be understood in view of Matthiessen's rule, which states that the total bulk resistivity varies as $\rho = \rho_i + \rho_0$, where ρ_i is the temperature-dependent, ideal resistivity due to scattering (collisions) of conduction electrons by lattice vibrations, and ρ_0 is the residual resistivity caused by imperfections and impurities and is independent of temperature. At high temperatures ρ_0 is usually negligible compared to ρ_i, but at low temperatures, ρ_0 can be comparable to or larger than ρ_i. Since ρ_0 will vary from sample to sample and cannot be made equal to zero, the low temperature resistance becomes difficult to specify meaningfully. For a detailed account of resistance at low temperature, see [12]. An analysis of the effects of surface roughness and additional information is given in [4].

Chambers [1,2,3] has made measurements which show that silver and gold have anomalous behavior which is similar to that of copper. In fact, if all resistance curves are normalized and scaled to agree with classical theory for small σ, the experimental points fall on the curve in Fig. 4-11. If actual conductance values in mhos are plotted, the curves will obviously be slightly different because of the different values of σ for the materials at a given temperature.

It should be noted that in Fig. 4-11, as σ becomes large, i.e., low temperature, $1/R$ approaches a constant value, independent of temperature. This represents the so-called exterme anomalous range and for copper, silver, and gold, it occurs roughly near 20°K for 1.2×10^9 Hz. The onset of extreme anomalous conditions depends on both frequency and temperature (or σ), the latter being a measure of the mean free path.

Now that the general behavior of anomalous skin effect has been described, let us look into the fundamental concepts of resistivity to try to explain this behavior. A precise, fundamental investigation would require a burdensome digression into the physics of metallic conduction, which is

---

*The surface impedance and transmission-line characteristics of superconductors are considered in Chap. 6.

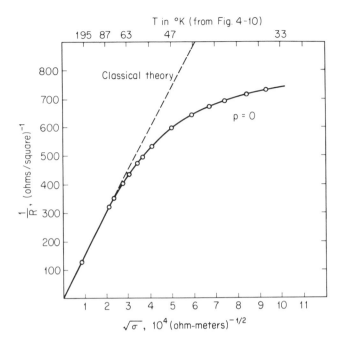

Fig. 4-11. Experimental surface resistance of copper at $f = 1.2 \times 10^9$ Hz and $\sigma/\Lambda = 15.4 \times 10^{14}$ (ohm-meters^2)$^{-1}$. (*After Chambers* [1]).

beyond the scope of this book. Instead, we shall present a somewhat simplified view of the subject and will concentrate on its main features. Extensive references are provided for additional information. We shall first consider Ohm's law in some detail in order to show its limitations.

4.11   LIMITATIONS OF OHM'S LAW—LOCAL VERSUS NONLOCAL THEORY

There are actually two problems associated with the application of Ohm's law, both of which generally occur at low temperatures. In one problem, that of anomalous skin effect, the electric field acting on the electrons varies only with position. The other, extreme anomalous, occurs at higher frequencies or lower temperatures and results from the fact that the electric field acting to accelerate the conduction electrons, besides varying with position, also varies with time due to the high frequency. Obviously, at the frequencies below the extreme anomalous range, the electric field is also varying with time, but the time between collisions of the electrons is short compared to the excitation frequency; thus, the electric field can be considered to be constant with respect to time. In order to clarify these effects, we will proceed with a derivation of Ohm's law, which will show that in order to $J = \sigma E$ to be valid, it is necessary for the electric field to be constant in space and time between collisions of the conduction electrons. Then a physical interpretation of the relation between mean free path and classical skin depth will be presented to show why anomalous conditions arise.

We will now derive Ohm's law from a very simplified point of view based on the free electron theory of metals. A more complete discussion of electrical conductivity can be found in [5] or [7].

Consider the case of free electrons (initially at rest), when a constant electric field E is suddenly applied. Since E is force per unit charge, the electrons will be accelerated and will experience an acceleration or time-changing velocity represented by

$$\frac{d\nu_x}{dt} = -\frac{e}{m} E_x \qquad (4\text{-}123)$$

where ν_x is the average velocity in the x direction (called drift velocity), and m and e are the electron mass and charge, respectively. It is apparent that the acceleration would continue indefinitely (as long as the electron remains in the E field) if there were not some other mechanisms to prevent this. We know from observations that the current does, in fact, reach steady state (Ohm's law) and it is the collision of the electrons with the atomic lattice that brings this about. These collisions and the resulting relaxation time can be analyzed in numerous ways. In the most simple, phenomenological approach, it is expedient to assume that the probability of the collision of an individual electron with the lattice during a time interval dt is dt/τ, where τ is an averaging probability constant, independent of the velocity and energy of the electrons. We also assume that after a collision, the electron is initially at rest, having given up all its energy to the lattice. It will, of course, be accelerated once again by the electric field, but the new velocity is random, independent of previous collisions. Using this picture, it is apparent that the acceleration or change in drift velocity due to E alone is given by Eq. (4-123). In a time interval dt, the change in average velocity $d\nu_x$ due to collisions alone is the average velocity multiplied by the probability for a collision

$$d\nu_x = -\nu_x \frac{dt}{\tau}$$

which gives

$$\frac{d\varkappa_x}{dt}\bigg|_{\text{collision}} = -\frac{\varkappa_x}{\tau} \qquad (4\text{-}124)$$

(negative sign because a collision causes a reduction in velocity). This equation is valid since $\varkappa_x$ is the average velocity and represents the average time between collisions. It is apparent that the total average drift velocity in the direction of E is the sum of Eqs. (4-123) and (4-124) and in steady state must be constant

$$\frac{d\varkappa_x}{dt} = 0 = -\frac{\varkappa_x}{\tau} - \frac{e}{m}\,E \qquad (4\text{-}125)$$

From the above it is seen that

$$\varkappa_x = -\frac{e\tau}{m}\,E \qquad (4\text{-}126)$$

Now the conductivity of an isotropic medium can be considered to consist of a number of electrons, moving with an average velocity, so

$$\sigma = -Ne\varkappa_x \qquad (4\text{-}127)$$

where N is the number of electrons per unit volume and $\varkappa_x$ is the drift velocity given by Eq. (4-126). Thus, it is apparent that

$$\sigma = \frac{Ne^2\tau}{m} \qquad (4\text{-}128)$$

Suppose that after an average steady state has been reached in the metal, the E field is switched off at time T_0. It is apparent that the drift velocity must decay to zero as a result of lattice collisions only; thus, the drift velocity will be governed by Eq. (4-124). This equation can be solved for $\varkappa_x$ for time greater than T_0

$$\varkappa_x = \varkappa_x(T_0)\,\exp\left(-\frac{t - T_0}{\tau}\right) \qquad \text{for} \quad t \geq T_0 \qquad (4\text{-}129)$$

where $\varkappa_x(T_0)$ is the average drift velocity at the instant the electric field is removed. Thus τ, the probability constant, is also the relaxation time and represents the mean free time between collisions for cases where the velocity after collision is random. The conductivity given by Eq. (4-128) is thus a constant independent of E but different for different conductors for which the assumptions are valid.

The simple and obvious conclusion is that in the application of Ohm's law, it is necessary that the conduction electrons see a constant E field between collisions. It has already been shown (Secs. 4.3, 4.7, and 4.8) that for ordinary circumstances, E decreases exponentially from the surface of a plane conductor inward, as shown approximately in Fig. 4-12. If δ is much larger than the mean free path Λ (as was assumed), it is apparent that for any incremental distance into the conductor Δz, which is comparable to Λ, the E field is essentially constant over that distance. This case then clearly satisfies Ohm's law, as was originally assumed without qualification.

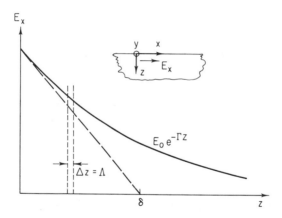

Fig. 4-12. Field penetration of a conductor for $\delta \gg \Lambda$.

It is apparent that as Δz, that is, Λ, approaches δ in Fig. 4-12, the electric field is far from being a constant over the distance Λ, and this variation in space must be included. If the functional variation of E with z remained the same as Λ approached and surpassed δ, it would be a simple matter to include these spatial variations. Unfortunately, E must change functional form (it can no longer be exponential), and its evaluation is somewhat complicated. Based on this simple theory, we have already seen that the onset of anomalous effects does, in fact, occur in the vicinity of $\Lambda \approx \delta$. From Fig. 4-10, for pure copper at 66°K, the anomalous effects due to spatial variation of E begin at $f = 1.2 \times 10^9$ Hz. Based on this simple theory once again, the extreme anomalous effects due to relaxation time should not appear until the applied frequency is at least comparable with the relaxation time

$$2\pi f = \frac{1}{\tau} \quad *$$

τ is found from $\tau = \Lambda / \vartheta_0$ where $\vartheta_0 = 1.6 \times 10^6$ m/sec (velocity at Fermi surface) and is essentially independent of temperature. Thus, for copper at $66°$K

$$\sigma = \frac{1}{\rho} = \frac{1}{1.26 \times 10^{-9}} \qquad \tau = 3.3 \times 10^{-13} \text{ sec}$$

Extremely anomalous effects in this case should then occur roughly at

$$f \geq \frac{1}{2\pi\tau} \quad \text{or} \quad f \geq 500 \times 10^9 \text{ Hz}$$

If the temperature were lowered to about $35°$K, the conductivity increases, and hence τ increases by an order of magnitude over the values at $66°$K. Thus extreme anomalous conditions would then occur in copper at about 50×10^9 Hz at $35°$K.

A simple test of this crude theory can be obtained from Fig. 4-11. The extreme anomalous effects begin very roughly in the region of $20°$K for the measuring frequency of 1.2×10^9 Hz. From Fig. 4-10, at $20°$K, $1/\sigma = 7 \times 10^{-12}$ ohm-meter, so that $\tau = 6 \times 10^{-11}$ sec. Thus, the frequency at which extreme anomalous effects are expected from these simple notions is roughly 2.6×10^9 Hz, which is about twice the actual experimental frequency. Thus, the analysis gives "ballpark" values but cannot be taken literally.

The general relationship between E and J was first derived by Reuter and Sondheimer [18]. Sondheimer [19] has given an excellent detailed account and Pippard [17] has shown that this relation can be expressed

$$J = \frac{3\sigma}{4\pi\Lambda} \int \frac{r(r \cdot [E]) e^{-r/\Lambda}}{r^4} dV \qquad (4\text{-}130)$$

in RMKS units where

σ = dc conductivity in mhos/meter

---

*In the more accurate theory, the frequency must actually be somewhat higher because $1/\tau$ must be replaced by $1/\xi$, where ξ is the time required by an electron to transverse the skin depth— which is actually smaller than τ ([19, p. 38]).

r = distance from origin in three-dimensional space in meters

Λ = mean free path in meters

[E] = electric field at volume dV at time $t - r/\jmath_0$ in volts/meter, that is, retarded field

dV = volume element = $r^2 \, dr \, d\theta$

It is apparent that the current density depends on the E field at all points within the conductor, i.e., nonlocal problem. Equation (4-130) represents the general form of Ohm's law where E can vary in space and time. If E = constant, this reduces to the usual form of Ohm's law, $J = \sigma E$. This latter form represents the so-called local theory where J depends only on E at that point, while the more general form is represented by a nonlocal theory. We shall encounter a somewhat similar situation for superconductors in Sec. 6.2(g). If the frequency is sufficiently low, then [E] can be replaced by E in Eq. (4-130).

4.12 SURFACE IMPEDANCE OF A SEMI-INFINITE HALF-PLANE WITH ANOMALOUS SKIN EFFECT

Surface impedance is generally a very small number (for example, see Fig. 4-11) which can be obtained by measuring the Q (quality factor) of a cavity resonator having one or all parts constructed from the desired material. R is inversely proportional to Q the constant of proportionality being, at least in principle, calculated from the dimensions of the resonator. However, the calculation is usually greatly complicated by the geometry (such as end effects). A simpler method is to measure Q at a temperature high enough so that the classical theory is obeyed; the dc conductivity, classical skin depth, and classical formula for R may be used to obtain the constant of proportionality. The reactive part of the surface impedance has negligible effect on Q but has a very slight effect on, i.e., it lowers, the resonant frequency. We shall not be concerned with methods of measurement but will summarize the general theory and its validity. The reader is referred to [14,15] for details of experimental procedure.

Equation (4-130) relates J to E for the general case from which the surface impedance can be obtained. As was already seen, extreme anomalous effects are generally to be expected at only very high frequencies; thus, relaxation time effects are usually negligible, with [E] = E in Eq. (4-130). The mathematical procedure for obtaining the surface impedance of a half-plane is carried out in [19] and will not be included here. One important point, however, is that in the solution it is necessary to make some assumptions concerning the nature of the reflection of those electrons which strike the surface of the conductor and bounce back into its interior. Reuter and Sondheimer [18] have carried out the details for two cases, namely, completely random (diffuse) scattering and mirror-type (specular) reflection, for which they define the reflection parameters $p = 0$ and $p = 1$, respectively. Although the theoretical solution for $p = 1$ is not much different from that for $p = 0$, all experimental evidence appears consistent with $p = 0$, indicating diffuse scattering. In fact, the solid curve in Fig. 4-11 represents the theoretical curve for $p = 0$ (with appropriate scaling factors to fit classical theory at high temperatures) and fits the experimental points quite accurately.

The general expression for impedance in RMKS units with $\mu = \mu_0$ is

$$Z = \frac{\pi \mu_0 j\omega \Lambda}{\int_0^\infty \ln\left(1 + j\frac{\alpha K(t)}{t^2}\right) dt} \tag{4-131}$$

for $p = 0$, Λ in meters, where $\alpha = 1.5 (\Lambda/\delta)^2$ and

$$K(t) = \frac{2}{t^3} \{(1 + t^2) \tan^{-1} t - t\} \tag{4-132}*$$

Normalized curves for the resistive and inductive part of this surface impedance of a plane conductor as a function of $\sigma^{1/2}$ are given in Fig. 4-13 (A is a constant chosen such that classical theory is valid for small σ). For the classical theory or small σ, $R = X = \omega L$, as was derived in Sec. 4.7. However, in the anomalous region, the resistive and reactive parts are no longer equal and, in fact, in the extreme anomalous region, $X = \sqrt{3}R$ or, more generally[†]

$$Z_{ex} = 3.8 \times 10^{-5} \left(\omega^2 \mu_r^2 \frac{\Lambda}{\sigma}\right)^{1/3} (1 + j\sqrt{3}) \frac{\text{ohms}}{\text{square}} \tag{4-133}$$

σ in mhos/meter, Λ in meters; μ_r is relative permeability, and is usually unity for metals.

From the above, it can be seen that for a given temperature, that is, σ, the extreme anomalous impedance varies with $\omega^{2/3}$, as contrasted with

*See [19, p. 32].
†From [17] but in RMKS units.

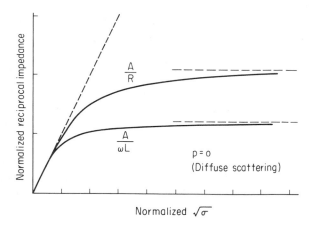

Fig. 4-13. Theoretical curve of real and reactive part of surface impedance as a function of $\sigma^{1/2}$, that is, temperature, covering classical, anomalous, and extreme anomalous regions. (*After Sondheimer.*)

variation of $\omega^{1/2}$ for the classical theory. In between the classical and anomalous range, the frequency-dependence of the impedance is more complicated. The general expression is given by Eq. (4-131). For pure copper at $f = 1.2 \times 10^9$ Hz, Eq. (4-133) gives $R_{ex} = 0.00126$ ohms for the real part, which is in good agreement with the asymptotic value in Fig. 4-11, that is, $1/800$.

From Eq. (4-133), the extreme anomalous impedance varies as $(1/k)^{1/3}$, so that smaller values of k give larger values of the ultimate (lowest possible) impedance at any given frequency. Since gold and silver have smaller values of k, they will have larger values of Z_{ex} and thus are not as effective as copper in this respect. The last two columns of Table 4-1 show a tabulation of some extreme anomalous conditions for several conductors obtained from Eq. (4-133) at 1.2×10^9 Hz. It can be seen that aluminum is a better conductor than the other common conductors in the extreme anomalous range.

Measurements on silver, gold, and lead give normalized resistance curves identical to Fig. 4-13 and are in very close agreement with the Reuter-Sondheimer theory for $p = 0$ at both 1.2×10^9 and 3.6×10^9 Hz [4].

An additional test of the validity of this theory would be to determine whether the extreme anomalous impedance actually varies with frequency as $\omega^{2/3}$. These effects occur only at extremely high frequencies or low temperatures or both, and very little work has been done in regard to this. Pippard ([15, p. 166]) has shown that normal tin just above its critical temperature* follows $\omega^{2/3}$ between 1.2 and 9.4 $\times$ 10^9 Hz, but because of many uncertainties, the results are not generally conclusive.

4.13 EXTREME ANOMALOUS SKIN EFFECT— INEFFECTIVENESS CONCEPT

A somewhat oversimplifed, but nevertheless valuable, way of presenting a physical interpretation of the extreme anomalous effects is in terms of the "ineffectiveness" concept proposed by Pippard [14], which is as follows. When the penetration depth is much smaller than the mean free path, the conduction electrons can be considered to be composed of two types. The first type consists of those electrons which at any given time are moving essentially parallel to or at small angles to the surface. The second type consists of all other electrons. The latter are constrained by the surface and penetration depth; since this distance is small compared to the mean free path, these electrons have little change for a collision and hence are "ineffective" in contributing to the resistance. The former type, moving nearly parallel to the surface, do not experience such constraints and therefore have a high probability of collision (which contributes to the resistance). These electrons will thus move in a constant field during a free path interval, since E is constant in a direction parallel to the surface. Thus, the effective conductivity can be written

$$\sigma_e = \frac{\beta \delta_e}{\Lambda} \sigma \qquad (4\text{-}134)$$

where δ_e is the effective penetration depth (unknown), β is a numerical constant, and σ is the dc conductivity. The ratio $\beta\delta_e/\Lambda$ represents the fraction of effective electrons. The value of β can be deduced from experiments as follows: δ_e can be represented in terms of σ_e by an expression similar to the classical form of Eq. (4-29)

$$\delta_e = \left(\frac{1}{\pi f \mu \sigma_e}\right)^{1/2} \qquad (4\text{-}135)$$

---

*These samples were in the extreme anomalous range at this temperature and frequency.

Substituting Eq. (4-134) into Eq. (4-135) yields

$$\delta_e = \left(\frac{\Lambda}{\pi f \mu \beta \sigma}\right)^{1/3} \tag{4-136}$$

The classical surface impedance should still be valid, provided only that the effective electrons and effective penetration depth are considered to contribute to the resistivity. The classical surface impedance is given by Eq. (4-103) and the extreme anomalous impedance should be of the same form with Eqs. (4-134) and (4-136) used in place of σ_e and δ_e. Thus, making these substitutions

$$R_{ex} = \left(\frac{\pi^2 f^2 \mu^2 \Lambda}{\beta \sigma}\right)^{1/3} \tag{4-137}$$

The imaginary part of the surface impedance can be obtained from the fundamental fact that complex conjugate parts of an analytic function are related to each other, i.e., the real and imaginary parts are not independent. When applied to an electrical network with a well-behaved impedance, i.e., analytic in upper half of complex plane, of the form $Z = R(\omega) + jX(\omega)$, the resistive or reactive part in terms of the other is

$$R(\omega_1) = \frac{2}{\pi} \int_0^\infty \frac{\omega X(\omega) - \omega_1 X(\omega_1)}{\omega^2 - \omega_1^2} \, d\omega$$

$$X(\omega_1) = \frac{2\omega_1}{\pi} \int_0^\infty \frac{R(\omega) - R(\omega_1)}{\omega^2 - \omega_1^2} \, d\omega \tag{4-138}$$

ω_1 is the specific frequency at which the impedance is evaluated and $R(\omega)$ and $X(\omega)$ are the functional variation with respect to frequency, i.e., the integrals must be evaluated from $\omega = 0$ to infinity for each value of ω_1. Rather than evaluating the reactive part at any given frequency, let us simply determine the ratio of reactive to resistive part in the extreme anomalous range where the two values approach a constant, i.e., find

$$\lim_{\omega_1 \to \infty} \frac{X(\omega_1)}{R(\omega_1)}$$

If the ineffectiveness concept has any merit, we should get the same result as that of Eq. (4-133). If we assume that the conductor is in the extreme anomalous condition for all frequencies, e.g., very low temperature, then the dc resistance can be neglected, and from Eq. (4-137) the extreme anomalous resistance should vary as $K\omega^{2/3}$ (K is a constant).*

Using this with Eqs. (4-138) and (4-139) yields

$$\lim_{\omega_1 \to \infty} \frac{X(\omega_1)}{R(\omega_1)} = \frac{2\omega_1}{\pi\omega_1^{2/3}} \left\{ \int_0^\infty \frac{\omega^{2/3}}{\omega^2 - \omega_1^2}\, d\omega - \int_0^\infty \frac{\omega_1^{2/3}}{\omega^2 - \omega_1^2}\, d\omega \right\}$$

The second integral in the above gives no contribution, while the first integral reduces to

$$\frac{2}{\pi} \int_0^\infty \frac{x^{2/3}}{x^2 - 1}\, dx \qquad x = \frac{\omega}{\omega_1}$$

The above integral can be evaluated by the technique of partial fraction expansion (similar to that often used to obtain inverse Laplace transforms) which yields

$$\lim_{\omega_1 \to \infty} \frac{X}{R} = \frac{2}{\pi} \left[\frac{\pi}{2} \sqrt{3} \right] = \sqrt{3}$$

which is in complete agreement with Eq. (4-133).

The value of the constant β can be found by equating Eq. (4-133) to Eq. (4-137) to find†

$$\beta = 7.2$$

The results obtained from the ineffectiveness concept are in good agreement with the more fundamental theory of Sec. 4.12. However, it should be emphasized again that the ineffectiveness concept provides a gross physical interpretation and should not be taken too literally. It does provide some physical notions which are not easily pictured from the detailed theory and is thus satisfying in this respect.

---

*If we allow the resistance to vary as $\omega^{1/2}$ at low frequencies, the results should not change because the low end of the spectrum contributes very little resistance.

†This value of β differs from that given by other authors, e.g., Pippard [17], because of the different system of units often used in the literature, although β has the same physical interpretation in all cases.

4.14 THIN CONDUCTORS

We have not considered the case of very thin conductors operated such that δ is larger than the thickness T, but with the thickness approaching the mean free path of the bulk material. For cases where δ is about equal to or smaller than the thickness, the previous analyses in terms of thick planes are good approximations. If δ is very much larger than T, then the current can be assumed to be uniformly distributed over the conductor. However, if T is very small, the mean free path is now affected by T and the effective σ of such a thin film will be smaller than that of bulk material. This effect, although similar in some respects to anomalous skin effect in a thick conductor, is actually different. For a thick conductor, Sondheimer [19] points out that the anomalous skin-effect theory indicates that electrons from deep down in the metal arriving at the surface are not greatly affected by the surface scattering properties since solutions for diffuse and specular reflection, that is, $p = 0$ and 1, are very similar. This is quite different from the dc properties of very thick conductors where the surface scattering is important. This difference may become important in practical applications where extremely thin conductors are used for transmission lines.

We shall not discuss these differences; for further discussion and review of size effects in metallic conduction, see [2, 12, 19].

APPENDIX 4A

A simplified equation for obtaining the surface impedance of a semi-infinite half-plane conductor can be obtained as follows. From Eqs. (4-97) and (4-99), the general expression for the desired impedance is

$$ Z_{surf} = \frac{1}{W} \frac{E}{H} \Big|_{surf} $$

Now, from Maxwell's equation

$$ \nabla \times \mathbf{E} = -\frac{\partial \mathbf{B}}{\partial t} = -j\omega\mu \mathbf{H} $$

Using the coordinates of Fig. 4-6, there will be only a y component of H

$$ \nabla \times \mathbf{E} = \frac{\partial E_x}{\partial z} = -j\omega\mu \mathbf{H}_y $$

(all other E terms are 0). Thus

$$H_y = -\frac{1}{j\omega\mu} \left.\frac{\partial E_x}{\partial z}\right|_{surf}$$

or substituting into the above equation

$$Z_{surf} = -\frac{j\omega\mu}{W} \left.\frac{E_x}{\partial E_x/\partial z}\right|_{surf} \quad \frac{ohms}{meter}$$

Thus, knowledge of the electric field component in the direction of propagation at the surface of the conductor is sufficient to determine the surface impedance.

PROBLEMS

4-1. Demonstrate why pure copper at ordinary temperatures is easily penetrated by short x-rays (10^{20} Hz).

   Answer: $\sigma/\omega\epsilon$ is very small, so that copper behaves as a dielectric.

4-2. Show that average rural ground with $\epsilon_r = 14$ and $\sigma = 10^{-2}$ mhos/meter behaves as a conductor at low frequencies and as a dielectric at microwave frequencies. Assume that parameters are independent of frequency, which is a gross approximation.

4-3. Determine the penetration depth of ordinary sea (salt) water with $\mu_r = 1$, $\epsilon_r = 80$, and $\sigma = 4\,(\Omega m)^{-1}$ (assume independent of frequency) at (a) 10^3 Hz and (b) 10^7 Hz.

   Answer: (1) 8 meters; (2) 0.08 meter.

4-4. A strip transmission line is being built of pure copper to run high-speed pulse circuit tests as the temperature is lowered. An applied pulse rise time of 35×10^{-12} sec (35 ps) is used. Determine the approximate temperature where anomalous skin effect may be expected to appear.

   Answer: about 87°K.

4-5. The properties of an unidentified metal are known to behave much as is shown in Fig. 4-11; at a frequency of 10^{10} Hz, anomalous effects begin at a temperature of 80°K, and the extreme anomalous resistance is 0.0055 ohms/square. Assume that the material has very good surface properties; identify the metal.

   Answer: Silver.

REFERENCES

1. Chambers, R. G.: Anomalous Skin Effect in Metals, *Nature (London)*, vol. 165, p. 239, 1950.
2. *Ibid.*, The Conductivity of Thin Wires in a Magnetic Field, *Proc. Roy. Soc. (London)*, vol. A202, p. 378, 1950.
3. *Ibid.*, The Anomalous Skin Effect, vol. A215, p. 481, 1952.
4. Chambers, R. G., and A. B. Pippard: *Inst. Metals Monograph,* 13, p. 372, 1952.
5. Decker, A. J.: "Solid State Physics," Prentice Hall, Inc., Englewood Cliffs, N. J., 1957.
6. Dwight, H. B.: "Tables of Integers and other Mathematical Data," 3rd ed., The Macmillan Company, New York, 1957.
7. Kittel, C.: "Introduction to Solid State Physics," 2nd ed., John Wiley & Sons, Inc., New York, 1956.
8. Kraus, J. D.: "Electromagnetics," McGraw-Hill Book Company, New York, 1953.
9. Lending, R.: New Criteria for Microwave Component Surfaces, *Proc. Natl. Electron. Conf., Chicago,* vol. 11, p. 391, October 1955.
10. London, H.: "The High Frequency Resistance of Superconducting Tin, *Proc. Roy. Soc. (London),* vol. A176, p. 522, 1940.
11. Lovell, A. C.: The Electrical Conductivity of Thin Metallic Films I. Rubidium on Pyrex Glass Surfaces, *ibid.,* vol. A157, p. 311, 1936.
12. MacDonald, D. K. C.: Electrical Conductivity of Metals and Alloys at Low Temperatures, in S. Flügge (ed.), "Handbuch der Physik," vol. 14, p. 145, Springer-Verlag, Berlin, 1956.
13. Morese, P. M., and H. Feshback: "Methods of Theoretical Physics," pt. I, p. 372, McGraw-Hill Book Company, New York, 1953.
14. Pippard, A. B.: II. The Anomalous Skin Effect in Normal Metals, *Proc. Roy. Soc. (London),* vol. A191, p. 385, 1947.
15. *Ibid.:* IV. Impedance at 9400 Mc./sec. of Single Crystals of Normal and Superconducting Tin, vol. A203, p. 98, 1950.
16. *Ibid.:* V. Analysis of Experimental Results for Superconducting Tin, vol. A203, p. 195, 1950.
17. *Ibid.:* "Advances in Electronics and Electron Physics," p. 1, Academic Press, New York, 1954.
18. Reuter, G. E., and E. H. Sondheimer: The Theory of the Anomalous Skin Effect in Metals, *Proc. Roy. Soc. (London),* vol. A195, p. 336, 1948.
19. Sondheimer, E. H.: The Mean Free Path of Electrons in Metals, *Advan. Phys.,* vol. 1, p. 1, 1952 (suppl. to *Phil. Mag.*).

20. Thompson, J. J.: Electric Conductivity of Thin Metallic Films, *Proc. Camb. Phil. Soc.*, vol. 11, p. 120, 1901.
21. Watson, G. N.: "Theory of Bessel Functions," 2nd ed., Cambridge University Press, New York, 1952.
22. White, G. K.: "Experimental Techniques in Low Temperature Physics," Clarendon Press, Oxford, 1961.

5 PULSES ON TRANSMISSION LINES

5.1 INTRODUCTION

We have studied the various characteristics of transmission lines with sinusoidal excitation and have shown that traveling waves can propagate on such lines. A similar condition can exist when step functions or pulses are applied to a line. For a long, ideal line, a pulse will propagate with no distortion, arriving at the load with a waveform identical to that at the input end but delayed in time by an amount equal to the length of the line divided by the velocity of propagation, i.e., the delay time of the line. The fact that this is true can be easily demonstrated if we recall that sine waves on a uniform, lossless, dispersionless line undergo no change as they propagate along such a line and all frequencies will propagate at the same velocity. It is well known that a periodic, nonsinusoidal excitation such as a square wave is composed of a set of discrete frequency components, i.e., fundamental plus harmonics, while a single pulse is composed of an infinite Fourier frequency spectrum with the amplitude of the spectrum decreasing for higher frequencies. These frequencies all add together at the input end to form the proper waveform. If all these frequency components travel along the line at the same speed and experience no change, then it is obvious that they must all arrive at the same time at the load and will add together to give the same waveform as at the input end. The phenomenon of pulse distortion arises when the various frequency components travel at different speeds due to dispersion and/or suffer different amounts of attenuation due to frequency-dependent losses. In other words, the various sinusoidal components arriving at the load may have a different amplitude and/or phase relation with respect to one another than they had at the input end. Thus, when they are added together, they must give a different waveform, i.e., the pulse is distorted.

In most practical applications, the transmission lines are chosen, whenever possible, such that the distortion and attenuation are small. Thus, we

shall first consider cases where this is true in order to understand the phenomena of pulses propagating on ideal lines. Then we shall consider pulse reflections from various terminations, and the effects of rise time and losses on
the waveform.

5.2 PULSE WIDTH AND REFLECTION WAVEFORMS

A uniform, lossless, dispersionless, i.e., ideal, transmission line terminated in its characteristic impedance represents a trivial situation since the
pulse anywhere along the line is identical to the applied pulse, though delayed in time by the appropriate electrical length of line. The interesting
aspects of behavior appear when there are reflections on the line because
of improper termination or discontinuities along the line. For the general
case of improper termination at both the load end $(Z_\ell \neq Z_0)$ and generator
end (source impedance $Z_s \neq Z_0$), a single pulse applied to the line will give
rise to multiple reflections traveling back and forth between these two ends
of the line. While in principle it does not matter whether the applied voltage
is a step function, i.e., pulse width equals infinity, or a narrow pulse, i.e.,
pulse width is much less than the electrical length of the line, in the actual
analysis, differently shaped waveforms can be obtained because of the way
in which the reflected voltages add or subtract to give the total (observable)
voltage waveform. It is thus necessary to be aware of the differences that
might occur merely as a result of the choice of pulse width.

Basically, the step function is of prime importance because of the fact
that any pulse can be thought of and analyzed in terms of two step functions,
one positive and one negative, delayed in time with respect to each other
by an amount equal to the pulse width, as shown in Fig. 5-1. It is obvious
that by proper superposition, the pulse reflection waveforms on a given
ideal transmission line can easily be obtained from knowledge of the step-
function behavior of the line. Thus, we shall first study various conditions
with an applied step function and will then present the same cases with
narrow pulses. All cases in between are easily obtained by superposition
and will not be considered here.

5.3 STEP FUNCTION RESPONSE OF IDEAL LINE
WITH IMPROPER TERMINATION

We will first consider the fundamental behavior of an ideal line with
various terminations with an applied step-function voltage. In all cases, it

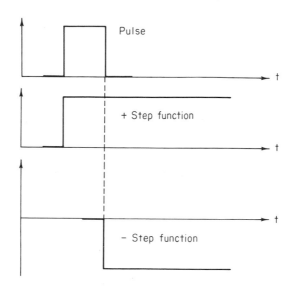

Fig. 5-1. Pulse obtained from superposition of two step functions.

is assumed that the pulse generator internal impedance is equal to Z_0, the characteristic impedance of the given line, as shown in Fig. 5-2. It is also assumed that the rise time of the step function is very small compared to the electric length of the line, giving the appearance of being very sharp. We shall examine the cases for which this condition is not valid later.

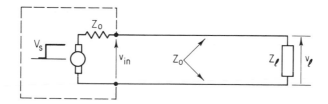

Fig. 5-2. Step function applied to ideal line with arbitrary termination.

Some fundamental concepts which are necessary for determination of pulse behavior can be obtained from Eqs. (2-14) and (2-15)

$$v_x = V_A \exp[j\omega t - \gamma x] + V_B \exp[j\omega t + \gamma x] \qquad (5\text{-}1)$$

$$i_x = I_A \exp[j\omega t - \gamma x] + I_B \exp[j\omega t + \gamma x] \qquad (5\text{-}2)$$

where

$$I_A = \frac{V_A}{Z_0} \qquad I_B = -\frac{V_B}{Z_0} \tag{5-3}$$

V_A and I_A are waves traveling in the $+x$ direction, while V_B and I_B are waves traveling in the $-x$ direction. When pulses are present, a continuous frequency spectrum of sinusoids will be present and it is often simpler to consider only the amplitude of the applied pulse. Thus, the above equations for pulse excitation can be shortened to

$$V_x = V_A (+x) + V_B (-x) \tag{5-4}$$

$$I_x = I_A (+x) + I_B (-x) = \frac{V_A (+x)}{Z_0} - \frac{V_B (-x)}{Z_0} \tag{5-5}$$

The polarity of I_A and I_B are chosen as positive so that they both represent a current flowing in the $+x$ direction in the top conductor and in the $-x$ direction in the bottom conductor, as in Fig. 5-3. Since both currents can be either positive or negative, the direction of current flow in the wire can be different from the direction of propagation of the wave.

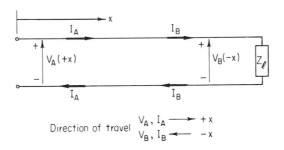

Fig. 5-3. Polarities and direction of travel of voltage and current waves.

In Eqs. (5-1) and (5-2) the phase relationships between the oppositely traveling wave are provided for by the exponential functions. In the short-hand expressions for pulses in Eqs. (5-4) and (5-5), the phase relationships are not present and one must provide the "bookkeeping" to combine various reflections at the proper time. This is easily done since a pulse travels with a constant velocity on an ideal or low-loss line and the points of reflection or time delay between reflected pulses are generally known.

Thus, the rules to keep in mind are that at any given point along a line, the voltage or current at any given instant of time is the algebraic sum of the positively and negatively traveling waves. For instance, two oppositely traveling waves of the same polarity and equal amplitude at a given point and time will add to give twice the individual wave amplitude. The same principle can be applied to points of termination or discontinuity on a line and simply states that the total voltage or current at the point of reflection is the algebraic sum of the incident and reflected wave. The polarities of the reflection must be observed since a positive voltage reflection yields a negative current reflection. If the reflection coefficients are known, then from Eqs. (2-61) and (2-62) the total voltage and current amplitudes are

$$V = V_A (1 + \rho_v) \qquad (5\text{-}6)$$

$$I = I_A (1 + \rho_i) = I_A (1 - \rho_v) \qquad (5\text{-}7)$$

where the incident amplitudes are assumed to be V_A and I_A.

Various waveforms appearing at the input and output ends of the line of Fig. 5-2 for a delay time T_0 (one way) are shown in Table 5-1 for various terminations.

5.4 PULSE RESPONSES OF IDEAL LINE WITH IMPROPER TERMINATIONS

In Sec. 5.3, we considered the various waveforms obtained with a step function (pulse of infinite width). If the pulse is now assumed to have a width which is narrow compared to the electrical length of the line, the individual reflections can be seen separately at the input end. The waveforms for various cases can be obtained in the manner outlined in Sec. 5.2 and are shown in Table 5-2 for cases analogous to those of Table 5-1. As before, the source internal impedance is assumed to equal the line characteristic impedance. For complex terminations, the voltage across each component is not shown but can be obtained from the corresponding cases of Table 5-1. It should be noted that ρ' is not the total reflection coefficient at the load but rather involves only the real part of the load, for reasons that become obvious by observing its use in Tables 5-1 and 5-2.

Table 5-1. Step Function Response of Fig. 5-2 for Various Terminations

$$V_0 = V_s/2, \quad I_0 = V_0/Z_0, \quad T = \ell\sqrt{LC}$$

Termination	Input waveforms v_{in}, i_{in}	Output waveforms v_ℓ, i_ℓ
(a) Short circuit $\quad Z_\ell = 0$	V_0 ; $2I_0$; $2T_0$	always $= 0$; $2I_0$; T_0
(b) Open circuit $\quad Z_\ell = \infty$	$2V_0$, V_0 ; I_0 ; $2T_0$	T_0 ; always $= 0$
(c) Small resistor $\quad R_\ell < Z_0$	V_0 ; $2T_0$; $V_0\dfrac{2R_\ell}{R_\ell+Z_0} = V_s\dfrac{R_\ell}{R_\ell+Z_0}$; $\dfrac{V_s}{R_\ell+Z_0}$; I_0	T_0
(d) Large resistor $\quad R_\ell > Z_0$	V_0 ; $2T_0$; $V_s\dfrac{R_\ell}{R_\ell+Z_0} = (1+\rho_v')V_0$; $\dfrac{V_s}{R_\ell+Z_0}$; I_0	T_0

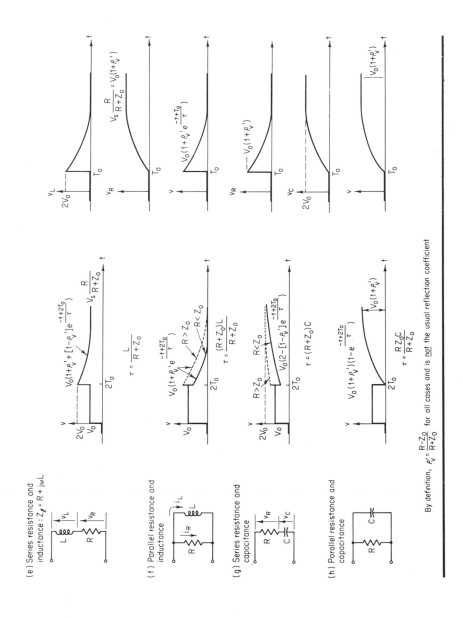

$$\rho_V' = \frac{R - Z_0}{R + Z_0}$$ for all cases and is $\underline{not}$ the usual reflection coefficient

By definition, $\rho_V' = \dfrac{R - Z_0}{R + Z_0}$ for all cases and is $\underline{not}$ the usual reflection coefficient

Table 5-2. Pulse Response of Fig. 5-2 for Various Terminations $V_0 = V_s/2$, $T = \ell\sqrt{LC}$, $\rho'_v = (R - Z_0)/(R + Z_0)$ **(Not Usual Reflection Coefficient for Complex Termination)**

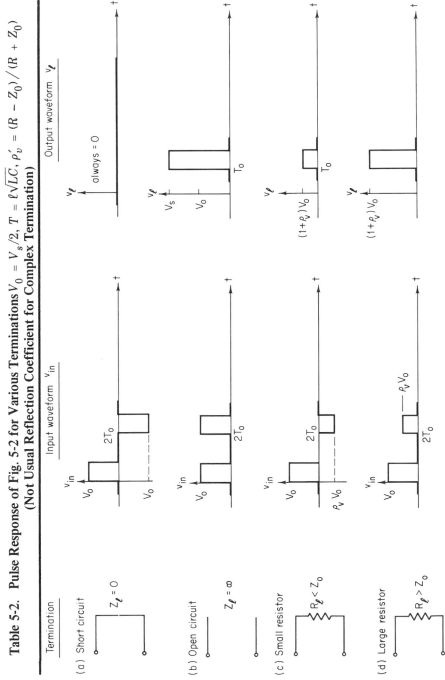

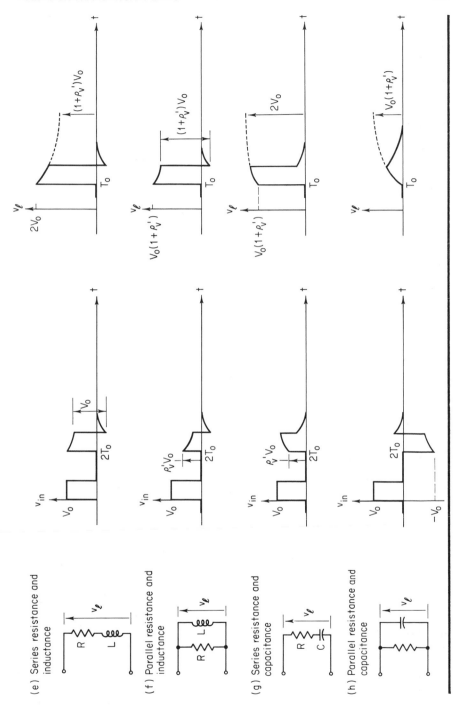

(e) Series resistance and inductance

(f) Parallel resistance and inductance

(g) Series resistance and capacitance

(h) Parallel resistance and capacitance

5.5 DISCHARGING OF AN IDEAL LINE

In Sec. 5.3, we saw that it was possible to charge an open-circuited line to the supply voltage with a waveform as shown in Table 5-1(b). If the power supply is now disconnected, the line will remain charged at V_s. This line can now be discharged by closing the switch in Fig. 5-4, and can yield a very fast rise-time pulse with a pulse width determined by the length of the line. Such a technique has found very widespread application as a pulse generator and can supply the fastest-known rise times obtainable from any device which can simultaneously yield large current and voltage.* We shall now consider this case of a line charged to V_s, with the switch closed at time $t = 0$ to discharge the line through R_ℓ, which can have any (resistive) value. We wish to determine the voltage and current waveforms at the load.

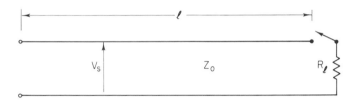

Fig. 5-4. Line initially charged to V_s.

There are several ways of approaching this problem. When $R_\ell = Z_0$, the problem is very simply understood through the construction of a simplified circuit. The charged line can be considered a dc source of value V_s with an internal impedance Z_0 in series with the load of value Z_0, as shown in Fig. 5.5. The voltage source is maintained only for a period while charge exists on the line. This period can be determined from simple energy considerations, since all the energy (capacitive) initially stored in the line must eventually be dissipated in R_ℓ. It should be kept in mind that the source impedance Z_0 is not a true resistance and cannot dissipate any energy. Thus, when the switch is initially closed, the voltage appearing across the load is $V_s/2$ and will remain at this level until all the charge is removed from the line. The

---

*The switching mechanism necessary for a fast rise time is a special coaxially constructed mercury wetted reed relay, since ordinary relays will not do; the rise time of commercial equipment is about 0.1 to 0.2 ns. Elliott [6] has shown that by careful implementation, the reed relay can give a pulse of 18 amperes into 50 ohms with a rise time of 0.070 ± 0.01 ns, that is, 70 ps. Faster devices, such as diodes, are known, but they can deliver only very small voltage and current, for example, 0.5 V at 0.015 ns rise time.

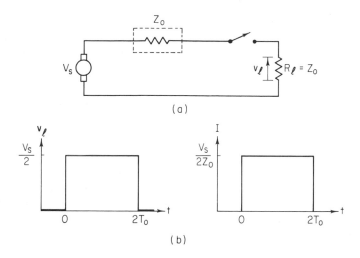

(a)

(b)

Fig. 5-5. Simplified circuit representation and waveforms for a line discharged into its characteristic impedance.

initially stored capacitive energy is

$$\mathcal{E}_c = \tfrac{1}{2} \ell C V_s^2 \tag{5-8}$$

where C is capacitance per unit length and ℓ is length of line; the energy dissipated in the load is

$$\mathcal{E}_\ell = \frac{V_s}{2} I \Delta T = \frac{V_s}{2} \frac{V_s}{2Z_0} \Delta T \tag{5-9}$$

where ΔT is the time period we wish to evaluate. Equating Eqs. (5-8) and (5-9) and canceling terms gives

$$\Delta T = 2 C \ell Z_0 = 2\ell \sqrt{LC} = 2 T_0 \tag{5-10}$$

where T_0 is the delay (one way) of the cable. This voltage and corresponding current waveform are shown in Fig. 5-5(b). The conclusion reached is that a line discharged into its characteristic impedance will give a square voltage pulse of amplitude equal to one-half the charging voltage and a pulse width twice the delay time of the line.

If this same line is discharged into a resistor which does not equal Z_0, then the situation is a little more complicated because of the presence of multiple reflections. The amplitude of the voltage source in Fig. 5-5 will change as the stored charge is dissipated from the line. In principle, it is still possible to analyze this case as was done above, but it is simpler and more descriptive to do it in terms of waves propagating and reflecting on the line. We will now consider the general case.

When the switch is closed, any reflected wave must obviously travel in the $-x$ direction, away from the point of disturbance, so that initially there will be only waves I_B and V_B in Eqs. (5-4) and (5-5). The current must obviously flow in the $+x$ direction in the top conductor, as in Fig. 5-6(a), while the wave travels in the $-x$ direction and hence has a positive polarity by definition; V_B must thus actually be of negative polarity, as given by Eq. (5-3). The voltage across R_ℓ must be the sum of the incident voltage V_s plus the reflected voltage V_B or

$$v_\ell = V_s + V_B \qquad (5\text{-}11)$$

where $V_B = -I_B Z_0$. I_B can be determined from the continuity of current flow, which requires the load current to equal the line current

$$I_B = \frac{v_\ell}{R_\ell} = -\frac{V_B}{Z_0} \qquad (5\text{-}12)$$

Solving the above for V_B and substituting into Eq. (5-11)

$$v_\ell = V_s + \left(-\frac{v_\ell Z_0}{R_\ell}\right)$$

or

$$v_\ell = V_s \frac{R_\ell}{R_\ell + Z_0} \qquad (5\text{-}13)$$

In a similar manner, solving Eq. (5-12) for v_ℓ and substituting into Eq. (5-11) gives for the negatively traveling wave

$$V_B = -V_s \frac{Z_0}{R_\ell + Z_0} \qquad (5\text{-}14)$$

This voltage with negative polarity will propagate down the line in the $-x$ direction and will reduce the total (observable) voltage on the line by this amount. This is illustrated in Fig. 5-6(c) by the voltage observed at the center of the line between the time $t = T_0/2$ to $3T_0/2$. When this wave reaches the open end, it will experience a positive reflection, so that a negative polarity wave will propagate in the $+x$ direction, that is, $V_A = V_B$, again reducing the voltage on the line, as in Fig. 5-6(c), between $3T_0/2$ and $5T_0/2$. When this wave reaches the load, it will experience another reflection, the polarity of which depends on whether R_ℓ is greater or less than Z_0. When R_ℓ is greater than Z_0, a positive reflection of value $\rho_v V_B$ occurs. This reflected wave, which still has negative polarity, propagates in the $-x$ direction and again decreases the total voltage on the line, as shown between $5T_0/2$ and $7T_0/2$ in Fig. 5-6(c). The process continues as before and can be carried as far as is desired. The voltage at the load is easily deduced for this case.

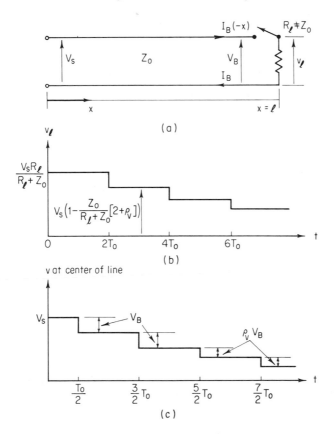

Fig. 5-6. Line discharged into a resistor not equal to the characteristic impedance.

Initially, the voltage is given by Eq. (5-13) and remains at this value until V_A, which equals V_B, reaches the load for the first time at time $2T_0$. From Fig. 5-6(c), it is easily seen that at this time, v_ℓ equals the algebraic sum of $V_s + 2V_B + \rho_v V_B$. Substituting in the value for V_B from Eq. (5-14)

$$v_\ell \bigg|_{2T_0} = V_s \left[1 - \frac{Z_0}{R_\ell + Z_0} (2 + \rho_v) \right] \tag{5-15}$$

where ρ_v is the voltage reflection coefficient. The equation is valid for the general case, but Fig. 5-6(b) shows the decreasing steps obtained when R_ℓ is greater than Z_0. When R_ℓ is less than Z_0, the load voltage starts at a smaller value and oscillates about zero. Figure 5-7 shows two such cases, for R_ℓ equal to $3Z_0$ and $0.5Z_0$. It is easily seen that if $R_\ell = Z_0$, this analysis will yield a pulse identical to that of Fig. 5-5.

The waveforms obtained above for a charged line discharged through a resistor are somewhat disturbing when first encountered, especially the amplitude of voltage pulse obtained. The student is usually familiar with the

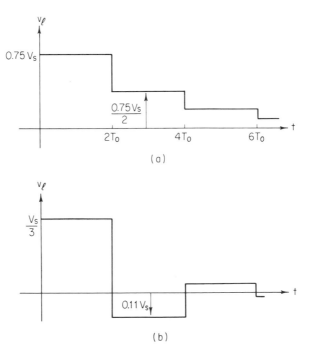

Fig. 5-7. Load voltage for a line discharged into (a) $R_\ell = 3Z_0$, $\rho_v = 0.5$; (b) $R_\ell = 0.5Z_0$, $\rho_v = -1/3$.

discharging of a capacitor through a resistor where the voltage starts at the charge voltage and decreases exponentially to zero with a time-constant RC. When a transmission line is discharged into $R_\ell = Z_0$, the voltage is one-half the charge voltage. In a manner of speaking, this results from the fact that the open end of the line does not know what is happening at the terminated end, so that a wave must propagate in between in order to carry the necessary information and obtain the proper condition.

A useful way of looking at this case is in terms of energy. Initially, all the energy is stored in a static electric field which arises from the charge on the conductors. When the line is suddenly terminated in $R_\ell = Z_0$, half of the static electric field is converted into a dynamic electric field by means of a time-varying magnetic field.* Thus, half of the energy is converted into magnetic energy via a current and the other half remains as a static electric field. This can easily be proved by consideration of the dynamics of converting capacitive energy to magnetic (inductive) energy. In order to do this, it is necessary to consider one aspect of the problem which was completely neglected above, namely, the rise time of the propagating wave. It is during this rise time that the static capacitive energy is converted into magnetic energy, but we did not have to consider it previously, since the problem was handled as a circuit, which can be done independently of the field problem (actually, they are one and the same since circuit equations are derived from field equations).

In order to gain further insight, let us consider a short section of line of length $\Delta\ell$ which is exactly equal to the rise-time equivalent length $T_R/\sqrt{LC}$, where T_R is the rise time. As the wave travels, say in the $-x$ direction, in Fig. 5-8, there is a time-changing flux within the rise-time portion of the line. Thus, the two voltages at the ends of the small section $\Delta\ell$ must be unequal with the difference equal to the time rate of change of flux within this area. Thus

$$V_s - V_2 = \frac{d\phi}{dt} = L\frac{di}{dt}\Delta\ell \qquad (5\text{-}16)$$

If we assume that the rise-time portion is a straight line, as in Fig. 5-8, then the inductive voltage is

$$\Delta\ell L\frac{di}{dt} = \Delta\ell\mu K\frac{V_2/R_\ell}{T_R} \qquad (5\text{-}17)$$

*Total induced electric field is $-E_T = \nabla\Phi + (\partial A/\partial t)$; Φ is the electrostatic potential and represents the capacitance term while A is the vector potential and represents the inductive term. See [10, p. 211] for further discussion of E_T.

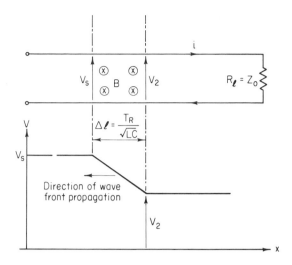

Fig. 5-8. Conversion of static electric field into magnetic field during the discharging of a charged line.

where $di = V_2/R_\ell$, $dt = T_R$, $u\Delta\ell K$ is the total inductance of the section of length $\Delta\ell$, μ is permeability, and K is the geometry coefficient (see Chap. 8). It is also easily seen that $R_\ell = Z_0 = K\sqrt{\mu/\epsilon}$ where ϵ = permittivity. Substituting this and Eq. (5-17) into Eq. (5-16) and solving for V_2

$$V_2 = \frac{V_s}{1 + (\Delta\ell\mu K/K\sqrt{\mu/\epsilon}\, T_R)} = \frac{V_s}{1 + (\Delta\ell\sqrt{\mu\epsilon}/T_R)} \tag{5-18}$$

But since $1/\Delta\ell = \sqrt{\mu\epsilon}/T_R$, then the above becomes $V_2 = V_s/2$. Thus, one-half of the initial charge voltage is converted to a traveling wave within the pulse rise time and one-half remains as a static electric field because of charges on the line. When the pulse is reflected from the open end, the traveling-wave voltage remains at $V_s/2$, but the static electric field and charge are reduced to zero.

5.6 REFLECTIONS FROM DISCONTINUITIES

When a discontinuity occurs at some point along a transmission line rather than at the ends, a reflection will occur which directs some of the energy back to the source and the remainder continues on to the load as determined by the reflection and transmission coefficients of Sec. 2.9.

Discontinuities are usually small, so that only a small amount of energy is reflected, while the major portion is transmitted to the load. The manner in which the reflections occur and the resulting waveforms are no different from those already presented in Secs. 5.3 and 5.4, except that the time at which they occur will depend on the physical location along the line. Examples of reflections expected from several types of discontinuities on an otherwise ideal line with an applied step function are shown in Table 5-3. It is assumed that the source impedance equals Z_0. The results for resistive discontinuities are no different from those of Table 5-1 with various resistance terminations, since the resistive discontinuity can be combined with the line characteristic impedance to give an equivalent resistor.

Table 5-3.   Reflections from Discontinuities with an Applied Step Function

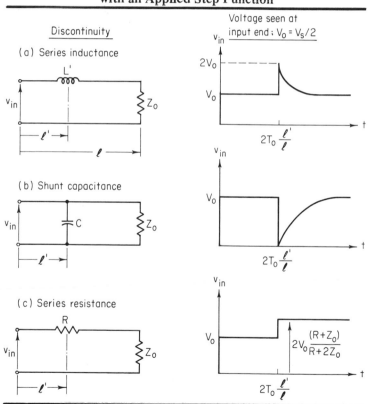

Other cases such as shunt inductance, series capacitance, shunt resistance tance, and various combinations of these are easily deduced from the results of Tables 5-1 and 5-3 and are therefore omitted.

If the applied pulse is truly a step function with zero rise time, then the initial amplitude of the reflected wave will always be equal to the incident wave, that is, V_0, no matter that the values of L and C might be. However, all pulses have finite rise times and therefore for cases where L and C are small, the reflected waves will be quite different. If the applied pulse approximates a linearly rising function with rounded edges, then the reflection will be more nearly that shown in Table 5-4. For cases where the magnitude of the discontinuity is small, it is possible to get a quantitative estimate of its value. For instance, let us consider the series inductance case. The voltage

Table 5-4.   Reflections from Small Discontinuities with Finite Rise Time Pulse

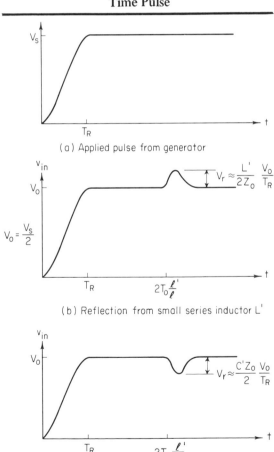

(a) Applied pulse from generator

(b) Reflection from small series inductor L'

(c) Reflection from small shunt capacitor C'

reflection coefficient at the point along the line $x = \ell'$ is

$$\rho = \frac{Z_{\ell'} - Z_0}{Z_{\ell'} + Z_0}$$

But

$$Z_{\ell'} = Z_0 + j\omega L' \qquad (5\text{-}19)$$

since the line beyond the point $x = \ell'$ looks like an inductor in series with a resistor of value Z_0 (line was assumed to be properly terminated). Thus

$$\rho = \frac{j\omega L'}{2Z_0 + j\omega L'} = \frac{V_r}{V_0} \qquad (5\text{-}20)$$

where V_r = reflected voltage. We assumed that the applied pulse had a finite rise time, so that the maximum frequency component will be limited by this rise time. If the rise time is sufficiently large and/or L' is sufficiently small, then $\omega L' \ll Z_0$ throughout the entire frequency range. Thus, Eq. (5-20) simplifies to

$$\rho = \frac{j\omega L'}{2Z_0} = \frac{V_r}{V_0} \qquad (5\text{-}21)$$

or

$$V_r = V_0 \frac{j\omega L'}{2Z_0} \qquad (5\text{-}22)$$

This is written explicitly in terms of a single frequency, whereas we are interested in the response to a pulse containing many frequencies. Equation (5-22) can be used for the pulse response if we recognize that

$$j\omega V_0 = \frac{d}{dt} v_{in} \qquad (5\text{-}23)$$

where

$$v_{in} = V_0 e^{j\omega t} \qquad (5\text{-}24)$$

This merely states that the derivative of each frequency component must satisfy Eq. (5-22). For the pulse response, it is desirable to represent the derivative of each frequency, but not as separate terms. One term is adequate if the rise time of the applied pulse is sufficiently linear. The peak of the reflected pulse V_r will be proportional to the maximum slope of the rise-time portion, as is easily seen upon substitution of Eq. (5-23) into Eq. (5-22)

$$V_r = \frac{L'}{2Z_0}\left(\frac{d}{dt}v_{in}\right)_{max} \tag{5-25}$$

But

$$\left(\frac{d}{dt}v_{in}\right)_{max} \approx \frac{V_0}{T_R} \tag{5-26}$$

so

$$V_r = \frac{L'}{2Z_0}\frac{V_0}{T_R} \qquad L' = 2Z_0 T_R \frac{V_r}{V_0} \tag{5-27}$$

By measuring the value of V_r and V_0, the value of L' can be determined.

A similar analysis can be performed in order to determine the value of a small shunt capacitor.

5.7 EFFECT OF RISE TIME ON WAVEFORMS

In most of the previous cases, we had assumed that the pulses had very steep rise times so that they would not be of any concern in the waveforms. In reality, all pulses and step functions have a finite measurable rise time and in practice, this time is often a substantial portion of the length of the line. When the rise time approaches the delay time of the line, it is often necessary to consider the rise-time portion of the pulse, since it can have considerable influence on the waveforms. We will now consider these effects.

For simplicity's sake, we shall assume that the applied voltage is a step function with a linear, i.e., straight-line, rise time, as shown in Fig. 5-9(a). If this pulse is applied to an ideal line terminated in a resistance of value less

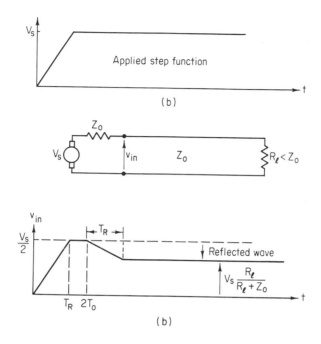

Fig. 5-9. Effect of rise time on step response of mismatched line with $R_\ell < Z_0$.

than its characteristic impedance, the resulting waveforms are similar to those of Table 5-1(c), though modified to include the rise times as shown in Fig. 5-9(b). The reflected wave decreases the amplitude of the final voltage at the input terminals.

As the delay time of the line is decreased for a fixed-step rise time, the input waveform can easily be obtained from a geometrical construction of the applied and reflected voltages superimposed with the proper time sequence. For instance, when the rise time T_R equals just twice the time delay of line (down and back), the waveform is as shown in Fig. 5-10(a). The reflected voltage arrives at the input end the instant the applied voltage has reached peak value. The result is a sharp peak of value $V_s/2$. If the delay time becomes less than the rise time, the reflected voltage arrives at the input end before the applied voltage has reached peak value. The reflected voltage attempts to decrease while the applied voltage attempts to increase the total voltage at the input terminals. The result is a waveform with three distinct slopes, as illustrated in Fig. 5-10(b). It is apparent that the peak of this waveform must be less than $V_s/2$, whereas the final value must be the same as before.

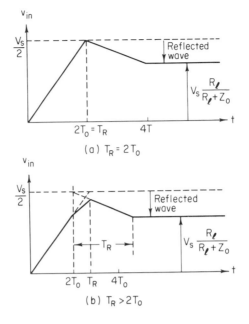

Fig. 5-10. Effects of rise time on step response for $R_\ell < Z_0$: (a) $T_R = 2\,T_0$; (b) $T_R > 2\,T_0$.

5.8 MULTIPLE REFLECTIONS AND EFFECTIVE TIME CONSTANT

We wish to consider an ideal transmission line with multiple reflections which are caused by improper termination at both ends of the line. In particular, we wish to consider the current and voltage buildup in the line in Fig. 5-11(a) where both R_s and R_ℓ are different from each other and from the line characteristic impedance Z_0. We will first consider the voltage build-up on the line in terms of the voltage reflection coefficients at the input and load ends of the line which, from Eq. (2-64), are respectively

$$\rho_0 = \frac{R_s - Z_0}{R_s + Z_0} \qquad \rho_\ell = \frac{R_\ell - Z_0}{R_\ell + Z_0} \qquad (5\text{-}28)$$

and can be positive or negative. When the switch in Fig. 5-11 is initially closed, a step function of value

$$V_0 = \frac{V_s Z_0}{R_s + Z_0} \qquad (5\text{-}29)$$

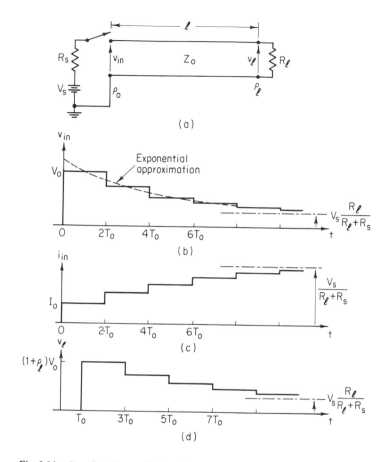

Fig. 5-11. Step function applied to line mismatched on both ends; waveforms shown for negative values of ρ_0 and ρ_ℓ.

appears on the line and travels toward the load. When it reaches the load at time T_0 (delay time one way), the first load reflection occurs, with a reflected wave value $\rho_\ell V_0$. This reflected wave travels toward the generator (battery) and when it reaches the input end at time $2T_0$, the first reflection at the generator occurs, with a value of $\rho_0(\rho_\ell V_0)$. This reflected wave, in turn, travels toward the load, where it is reflected, travels back to the generator where it is once again reflected, etc., until the reflected wave finally approaches zero. These various reflections and the times at which they occur are summarized in Table 5-5 and are shown in Fig. 5-11. It should be noted that we are concerned with the individually reflected waves, and not with the actual (observable) total voltages on the line. The total voltage at any point on the line is a sum of the initial voltage V_0, plus all the reflections, i.e., incident plus reflected waves. Thus the voltage at the input end, after a

Table 5-5. Reflected Wave Amplitude and Time of Occurrence in Fig. 5-11

Reflection number	Reflection at load		Reflection at input	
	Time	Amplitude	Time	Amplitude
1st	T_0	$\rho_\ell V_0$	$2T_0$	$\rho_0 \rho_\ell V_0$
2nd	$3T_0$	$\rho_\ell (\rho_0 \rho_\ell V_0)$	$4T_0$	$\rho_0 (\rho_0 \rho_\ell^2 V_0)$
3rd	$5T_0$	$\rho_\ell (\rho_0^2 \rho_\ell^2) V_0$	$6T_0$	$\rho_0 (\rho_0^2 \rho_\ell^3 V_0)$

etc. until terms become negligible

specified number of round trips of the wave, i.e., at a time equal to an even multiple of T_0, is

$$\frac{v_{in}}{V_0} = 1 + \sum_{n=1}^{m} \left[\rho_\ell^n \rho_0^{n-1} + \rho_\ell^n \rho_0^n \right] \tag{5-30}$$

where m equals the number of round trips, and is equal to the reflection number in Table 5-5. At the output end, the voltage after a specified number of half round trips to the load, i.e., at a time equal to an odd multiple of T_0, is

$$\frac{v_\ell}{V_0} = 1 + \sum_{p=1}^{q} \rho_\ell^p \rho_0^{p-1} + \overset{(q-1) \text{ for } q \geq 2}{\underset{p=1}{\sum}} \rho_\ell^p \rho_0^p \tag{5-31}$$

q is the number of times the wave has reached the load and is equivalent to the reflection number in Table 5-5. The integer q is really equivalent to m in Eq. (5-30), since both represent the number of reflections which have occured at the point in question, but different symbols are used to distinguish between the input and output end. The second summation term in Eq. (5-31) is understood to equal 0 for $q = 1$ and is included only when $q \geq 2$.

In Eqs. (5-30) and (5-31), the first term within the summation sign represents reflections originating at the load while the second term represents reflections occuring at the generator end. The only difference between these two equations is that Eq. (5-30) has one more term to account for the additional reflection which occurs at the input end. These summation expressions are quite simple and obvious, as can be seen by careful examination of Table 5-5. They merely represent a shorthand method of expressing many terms.

In a similar manner, an expression for the current can be derived which will be identical in form to that for the voltage except that the initial step V_0 is replaced by

$$I_0 = \frac{V_0}{Z_0} = \frac{V_s}{R_s + Z_0} \tag{5-32}$$

and the voltage reflection coefficients are replaced by the current reflection coefficients. Recall from Sec. 2.9 that the current reflection coefficient is the negative of the voltage coefficient. Thus, as the total current builds up in amplitude, the total voltage will decrease and vice versa.

As time approaches infinity, the reflections on the line must approach zero and the circuit of Fig. 5-11(a) is essentially a dc circuit after the transient has died away. From simple dc theory it is obvious that at this time

$$v_{in} = v_\ell = V_s \frac{R_\ell}{R_\ell + R_s} \tag{5-33}$$

The same expression can be obtained from Eq. (5-30) by letting m go to infinity. In order to do this, it is helpful to simplify by factoring out ρ_0

$$\frac{v_{in}}{V_0} = 1 + \left(1 + \frac{1}{\rho_0}\right) \sum_{n=1}^{m} (\rho_0 \rho_\ell)^n \tag{5-34}$$

We now let m approach infinity and use the identity

$$\sum_{n=1}^{\infty} x^n = \frac{x}{1 - x}$$

to get, after simplification

$$\frac{v_{in}}{V_0} = \frac{(1 + \rho_\ell) V_0}{1 - \rho_0 \rho_\ell} \tag{5-35}$$

After Eqs. (5-28) and (5-29) are substituted for the reflection coefficients and V_0, this equation reduces to that of Eq. (5-33), as can be checked by carrying out the algebraic operations (left to the reader as an exercise).

A similar operation can be carried out for the current.

Effective time constant. It seems reasonable from an examination of Fig. 5-11 that if the voltage change from maximum to minimum occurs in small increments of short duration, then the waveform will approximate an exponential function as indicated by the dotted line. The smaller and narrower the steps become, the more closely the waveform will approach the exponential. We will now prove that this is, in fact, true.

In order to simplify the problem, we will derive the exponential function representing the slope of the desired curve. The time constant obtained from this must of necessity be identical to the time constant of the actual exponential function, i.e.

$$\frac{d}{dt} e^{t/\tau} \propto e^{t/\tau}$$

After a given number of round trips n, the incremental change in voltage at the input end of the line is

$$\Delta V = V_n - V_{n-1} \tag{5-36}$$

Making use of Eq. (5-34), the above becomes

$$\frac{\Delta v_{in}}{V_0} = \left(1 + \rho_0^{-1}\right)\left(\rho_0 \rho_\ell\right)^{t/2T_0} \tag{5-37}$$

where $t = n2T_0$ (this is done in order to involve time explicitly). In order to get this into an exponential form, it is necessary to make use of an identity

$$a^x = e^{x \ln a}$$

This identity can be proved if a^x and e^z are expanded in an infinite series, whereupon the identity becomes obvious. For our case, $a = \rho_0 \rho_\ell$ and $x = t/2T_0$, and Eq. (5-37) becomes

$$\frac{\Delta v_{in}}{V_0} = \left(1 + \rho_0^{-1}\right) \exp\left(t \frac{\ln \rho_0 \rho_\ell}{2T_0}\right) \tag{5-38}$$

From this, it is evident that the time constant must be

$$\tau = \frac{2T_0}{\ln \rho_0 \rho_\ell} \tag{5-39}$$

This expression can be simplified by expanding the ln term in a series

$$\ln \rho_0 \rho_\ell = (\rho_0 \rho_\ell - 1) - \frac{1}{2}(\rho_0 \rho_\ell - 1)^2 + \frac{1}{3}(\rho_0 \rho_\ell - 1)^3 \cdots \text{ etc. } \tag{5-40}$$

The higher-order terms can be neglected since the absolute value of $(\rho_0 \rho_\ell - 1)$ must always be less than unity and, furthermore, in order for the exponential function to be an accurate representation, ρ_0 and ρ_ℓ must be reasonably large (approaching ± 1) so that the incremental steps are small.* Making use of this, Eq. (5-39) becomes

$$\tau = \frac{2T_0}{\rho_0 \rho_\ell - 1} = -\frac{2T_0}{1 - \rho_0 \rho_\ell} \tag{5-41}$$

The product $\rho_0 \rho_\ell$ is a positive number, less than 1, so that the time constant is a negative number. The negative sign indicates that the exponential function decreases with time, which is the case usually encountered with transient circuits.

We will make use of the above equivalent time constant expression in the next section to find τ for specific cases.

5.9 TRANSITION FROM TRANSMISSION-LINE TO CIRCUIT ANALYSIS

In Sec. 1.1, it was stated that any two conductors can be considered as a transmission line. However, it is well known that in many cases it is more convenient and expedient to use ordinary circuit analysis rather than transmission-line equations. In this section, we will develop some guidelines for determining where each approach is more suitable and will also consider the "gray" region in between.

---

*ρ_0 and ρ_ℓ must also have the same sign in order to yield a continually increasing (or decreasing) waveform; opposite signs will give oscillatory behavior which cannot be represented by a exponential function as above.

An ideal transmission line will always have a time constant equal to zero such that a step function of any rise time, no matter how small, will remain undistorted* (the reason for this was discussed in Sec. 1.5). However, we saw in Sec. 5.8 that if an ideal line is badly mismatched on both ends, the voltage or current will build up in value by a series of incremental steps. These steps can be approximated by an exponential function with a time constant given by Eq. (5-41). Even though the transmission line inherently has a zero time constant, the total circuit, which must include terminating impedances on both ends, produces a time constant which is quite different from zero. Thus, the serious problems and differences arise when a transmission line is not properly terminated. This is equivalent to a circuit whose physical layout has become substantially long compared to the wavelength range of interest and the segments connecting the various elements cannot be implemented as uniform transmission lines of the proper impedance.

When an ideal line is properly terminated, it behaves much as a pure resistor and it usually does not matter if transmission-line or circuit analysis is used, provided that for the latter approach, one realizes the delay inherent in the transmission line. In the final analysis of any electrical network, it does not matter which approach one uses provided that it is done correctly; it is simply a question of expediency in arriving at the final answer.[†] In order to demonstrate this fact, as well as to provide further insights, let us take the example of a long transmission line, improperly terminated on both ends, and derive the time constant of this circuit based on the transmission-line approach and show the equivalence to that obtained from a circuit approach.

More specifically, let us consider the case of a short-circuited transmission line excited by a step function with source impedance not equal to the line characteristic impedance. The general case is shown in Fig. 5-11(a). The voltage reflection coefficients for $R_\ell = 0$ are

$$\rho_0 = \frac{Z_s - Z_0}{Z_s + Z_0} \qquad \rho_\ell = -1 \qquad (5\text{-}42)$$

The approximate time constant for this circuit, from Eq. (5-41), is

---

*In practice, lines are never ideal, so that zero rise time is not possible, but rise times in the picosecond range (10^{-12} sec) can be obtained.

†This is of fundamental importance; a lack of understanding of the relationship between ordinary circuit and transmission lines can lead to much confusion even in some rather simple cases. See Appendix 5A for further discussion.

$$-\tau = \frac{2 T_0}{1 - \rho_0 \rho_\ell} = \frac{2 T_0}{1 + \rho_0} = \frac{T_0 (Z_s + Z_0)}{Z_s} \qquad (5\text{-}43)$$

or

$$-\tau = T_0 + \frac{T_0 Z_0}{Z_s} \qquad (5\text{-}44)$$

We know that $T_0 = \ell\sqrt{LC}$ (delay time one way) and $Z_0 = \sqrt{L/C}$, where ℓ is the physical length of the line and L and C are the per-unit-length parameters. Substitution of these into Eq. (5-44) yields

$$-\tau = T_0 + \ell \frac{L}{Z_s} \qquad (5\text{-}45)$$

It is necessary to have Z_s smaller than Z_0; thus, the reflection coefficients have the same sign to give exponential behavior. Opposite signs give oscillatory behavior. If $Z_s \ll Z_0$, the exponential approximation becomes more accurate, as was discussed in the previous section. If Z_s is very small compared to Z_0, then T_0 is negligible compared to $\ell L/Z_0$, so that Eq. (5-45) becomes

$$\tau \approx -\ell \frac{L}{Z_s} \qquad (5\text{-}46)$$

But ℓL is the total loop inductance L' of the short-circuited transmission line and Z_s is the total series impedance in the circuit. The time constant is then

$$|\tau| = \frac{L'}{R_s} \qquad (5\text{-}47)$$

This is precisely the same time constant that would have been obtained by a circuit-analysis approach if the line were considered an inductance loop in series with a resistor.

In a similar manner, if the transmission line is open-circuited on the far end, then $\rho_\ell = +1$ and

$$-\tau = T_0 \frac{Z_s + Z_0}{Z_0} = T_0 + \frac{T_0 Z_s}{Z_0} \qquad (5\text{-}48)$$

Substituting for T_0 and Z_0

$$-\tau = T_0 + \ell C Z_s \qquad (5\text{-}49)$$

If Z_s is now much larger than Z_0, there will be a voltage buildup on the line, very similar to the current buildup of Fig. 5-11(c). The T_0 term in Eq. (5-49) can be neglected to give

$$|\tau| = \ell C Z_s \qquad (5\text{-}50)$$

Since ℓC is just the total capacitance of the line, this is identical to the RC time constant which would have been obtained from a simple circuit analysis.

In both of the above examples, if $Z_s = Z_0$, there is no exponential type of current or voltage buildup, but instead the input waveform is as shown in Table 5-1(a) and (b). An attempt to use simple circuit analysis would have given completely erroneous results, since there is no longer any time constant associated with the line (assuming an ideal line). A circuit approach would require the transient solution of the problem, which would ultimately reduce to the transmission-line approach. Thus, the latter approach is much more convienent for this case.

An ideal transmission line is composed of distributed inductive and capacitive components, both of which are important. However, for the first example above, the parameters were chosen such that only the line inductance was important and the time constant was determined by L/R, as given by Eq. (5-47). For the second example, the parameters were chosen such that only the capacitance of the line was important; the time constant was determined by RC, as given by Eq. (5-50). By a different choice of parameters, it is possible to obtain all the forms normally associated with circuits, namely, overdamped, critically damped, and underdamped (oscillatory) systems. The two previous examples illustrate overdamped systems. A critically damped transmission line is one that is properly terminated. An underdamped system would, for example, give a waveform as in Fig. 5-12. If $Z_s = 0$, then the oscillations continue indefinitely, although in reality, there are always some losses present to eventually damp out the oscillations.

Thus, the phenomena normally associated with circuit response are all present in transmission-line problems, although the waveforms are somewhat different due to the finite delay time of a transmission line. As the length of the transmission line is reduced, the step-function response becomes more nearly like that of ordinary circuits. This can best be understood by allowing the length of the line to become very small and then determine the effect of this on the input waveform of the line. We shall now do this.

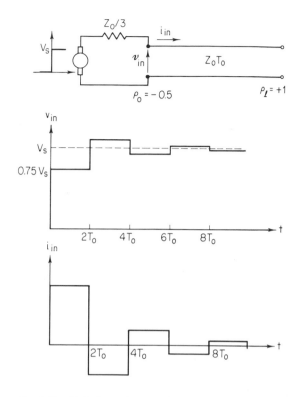

Fig. 5-12. Underdamped transmission line with step rise time $\ll T_0$.

There are actually three general cases which should be considered; they are classified according to the ratio of the electrical length of the line to the rise time of the applied pulse: (1) a long line such that the rise time is essentially zero, (2) a very short line so that the rise time is much larger than the electrical length of the line, and (3) the intermediate case of rise time comparable to the electrical length. The previous examples represent the first case of a very long line. For the other two cases, it is necessary to understand and consider the effect of the rise time on the waveform as was shown in Sec. 5.7.

For the extreme case of a very short line, it is to be expected that to a first approximation, the total L and C of the transmission line can be neglected and the circuit behavior is determined primarily by the source and load resistors. Figure 5-13(b) shows the input waveform with $2T_0 = T_R/8$. It can be seen that v_{in} has very nearly the same rise time as the applied voltage, with these two values becoming more nearly equal as T_0 approaches zero. The rise time of the input voltage is slightly rounded at the initial and final portion due to the finite delay between the various reflected components.

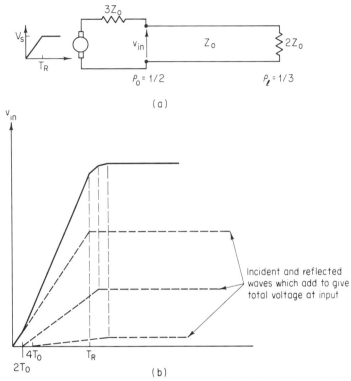

Fig. 5-13. Input waveform for a very short transmission line; $2 T_0 = T_R/8$.

For the region between these two extremes, it is possible to get various waveforms, depending on the ratio of the length of the line to the applied rise time and also on the values of source and load resistance. The value of source and load resistor will determine whether the line is overdamped, critically damped, or underdamped. Since critical damping is a trivial case, we shall consider the other two. An arbitrary value is chosen for the length of line such that the pulse rise time is twice the "down-and-back" time of the line, that is, $T_R = 2(2 T_0)$. The input voltage waveforms for overdamped and an underdamped condition are shown in Figs. 5-14 and 5-15 respectively. The latter merely represents a special case of Fig. 5-12, as can easily be deduced if the rise time is allowed to increase.

In all the figures, the dotted curves represent the incident plus reflected wave components occurring at the specified times; these dotted curves are added together to give the total waveform.

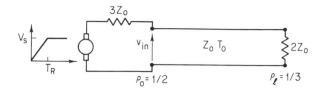

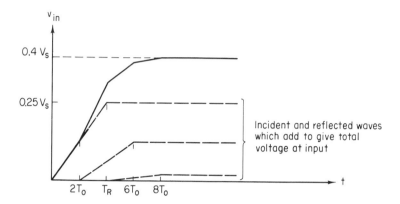

Fig. 5-14. Overdamped transmission line with step function rise time equal to $4\,T_0$.

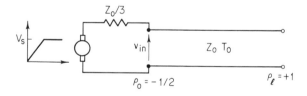

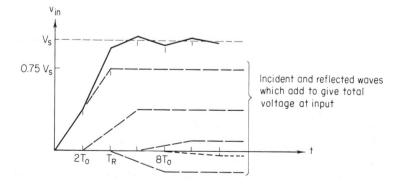

Fig. 5-15. Underdamped transmission line with step function rise time equal to $4\,T_0$.

5.10 INPUT IMPEDANCE AND RESPONSE OF A LONG LINE WITH SMALL SERIES RESISTANCE LOSS

Whenever possible in practice, the series resistance loss is made negligibly small in order to reduce attenuation and distortion. However, it is not always possible to do this; for example, a very long ordinary transmission line must have some finite series resistance which has to be considered in many applications. We wish to consider the input impedance of such a line with small but nonnegligible losses and the response which results when a step function voltage is applied.

If we assume that the conductance is zero, then Eq. (2-33) gives

$$Z_0 = \sqrt{\frac{R + j\omega L}{j\omega C}} = \sqrt{\frac{L}{C}\left(1 + \frac{R}{j\omega L}\right)} \qquad (5\text{-}51)$$

By making use of the identity $(1 \pm \nu)^{1/2} \approx 1 + \nu/2$, this can be expanded in a power series as in Sec. 2.6

$$Z_0 = \sqrt{\frac{L}{C}}\left(1 - j\frac{R}{2\omega L}\right) + \text{higher-order terms} \qquad (5\text{-}52)$$

Since $R/\omega L$ has been assumed to be small, all higher-order terms can be neglected. It is apparent that this impedance is composed of a real part equal to the characteristic impedance with no losses, and a negative imaginary part which yields an equivalent capacitance of value

$$C_e = \frac{1}{(R/2L)\sqrt{L/C}} = \frac{2L}{R}\sqrt{\frac{C}{L}}$$

The equivalent circuit is shown in Fig. 5-16(a). If we assume that the series resistance losses are independent of frequency, then when a step function is applied to the line, the input voltage waveform will be similar to the output voltage waveform shown in Table 5-1(g). The voltage initially jumps to $V_s R_0/(Z_s + R_0)$, as it would if there were no losses, but because of the series resistance, the input voltage must eventually approach the source voltage V_s. It is apparent that the latter will occur only if the line is very long so that the actual, total series resistance of the line is much larger than Z_0, but the entire equivalent circuit is valid only for an infinitely long line or a short line terminated in a series $R_0 C_e$ network.

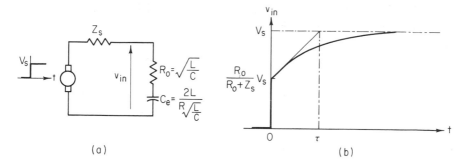

Fig. 5-16. Equivalent circuit and step function response of an infinite transmission line with small series losses.

It is apparent that since the equivalent circuit is a simple series resistance and capacitance, the time constant is just the product of the equivalent capacitance and the total equivalent resistance

$$\tau = C_e (Z_s + R_0)$$

If the source impedance equals R_0, then the input voltage initially jumps to $V_s/2$ and the time constant is

$$\tau = C_e 2 R_0 = 4 \frac{L}{R}$$

Unfortunately, as described in detail in Sec. 5.13, the series losses are determined by skin effect and therefore are frequency-dependent. Thus, the above expressions are an oversimplification but nevertheless give a qualitative understanding of the actual situation.

5.11 DETERMINATION OF LINE PARAMETERS FROM REFLECTION WAVEFORMS

From the preceding sections, it is apparent that the waveforms at the input terminals to a transmission line are determined by the length and characteristic impedance of a line as well as the source and load terminating impedances. It is often desirable to determine accurately the characteristic impedance of a line. This can be done in a number of ways, the most simple one being that which is illustrated in Fig. 5.17. Here, a line with known characteristic impedance Z_0 is fed from a matched source and

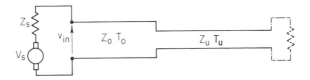

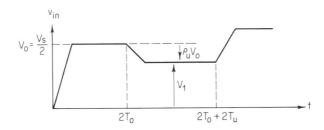

Fig. 5-17. Determination of unknown characteristic impedance using a line and source of known impedance.

terminated in a line of unknown impedance Z_u. When a step function is applied, a reflection from Z_u will appear at the input end at time $2T_0$. The reflected wave will have an amplitude $\rho_u V_0$, where

$$\rho_u = \frac{Z_u - Z_0}{Z_u + Z_0} \qquad (5\text{-}53)$$

Since we do not know Z_u, we cannot terminate the unknown line properly, so that a reflection will occur at the end of the line which will appear at the input terminals at time $2(T_0 + T_u)$. If T_u is sufficiently long to give distinguishable reflections, Z_u can easily be obtained. Referring to Fig. 5-17, it is obvious that $\rho_u V_0 = V_0 - V_1$ or

$$\rho_u = 1 - \frac{V_1}{V_0} \qquad (5\text{-}54)$$

From Eq. (5-53)

$$Z_u = \frac{Z_0 (1 + \rho_u)}{1 - \rho_u} \qquad (5\text{-}55)$$

Substitution of Eq. (5-54) into Eq. (5-55) gives

$$Z_u = Z_0 \frac{[2 - (V_1/V_0)]}{V_1/V_0} = Z_0 \frac{[1 - (V_1/V_s)]}{V_1/V_s} \qquad (5\text{-}56)$$

where $V_0 = V_s/2$. Thus, by measuring the amplitude ratio of the input voltage steps, it is possible to easily determine the unknown impedance, assuming that losses are negligible.

If Z_u is very close in value to Z_0, it is possible to read the unknown value directly from the amplitude of the reflection. This can be seen by rewriting Eq. (5-53)

$$\rho_u = \left(\frac{Z_u}{Z_0} - 1\right)\left(\frac{Z_u}{Z_0} + 1\right)^{-1} \approx \frac{Z_u + Z_0}{2 Z_0} \quad \text{for} \quad Z_u \approx Z_0 \qquad (5\text{-}57)$$

If Z_u/Z_0 is in the vicinity of unity, then Z_u is directly proportional to ρ_u. By proper calibration with known impedances, the reflected wave in Fig. 5-17 can be made to read directly in ohms. As Z_u becomes significantly different from Z_0, the accuracy of this calibration decreases and it is apparent that for the general case, Eq. (5-56) must be used for correct results.

In the above example, it was necessary to have available a transmission line of known characteristic impedance. If such a line is not available, or if the accuracy of the impedance of the available line is not sufficient, it is necessary to use a different, although similar, technique than that of Fig. 5-17. In order to accomplish this, it is necessary to have either a terminating resistor of known value or a source of known internal impedance. If only a terminating resistor is available, then the circuit of Fig. 5-17 is modified to that of Fig. 5-18. Since Z_u is the unknown, the line cannot be properly terminated at either end. Therefore, a reflection will occur at both the source and the load end which gives rise to voltage reflection coefficients ρ_0 and ρ_ℓ, respectively. A voltage measurement across the load will yield

$$v_\ell = V_0 (1 + \rho_\ell) \qquad (5\text{-}58)$$

and a measurement at the input terminals will give V_0, as illustrated. From these two measurements, ρ_ℓ can be determined, and since R_ℓ is known, Z_u can be obtained from Eq. (5-53).

If only a generator of known internal impedance is available, then the line can be left open-circuited on the load end to give $\rho_\ell = +1$. A voltage measurement at the input terminals will give a waveform similar to that in Fig. 5-18(b). Since V_0 is given by the initial value of the step and ρ_ℓ is already known, ρ_0 is easily found from

$$V_1 - V_0 = V_0 \rho_\ell (1 + \rho_0)$$

which gives

$$\rho_0 = \frac{V_1}{V_0} - 2 \qquad (5\text{-}59)$$

Once again, the unknown impedance is obtained from Eq. (5-53).

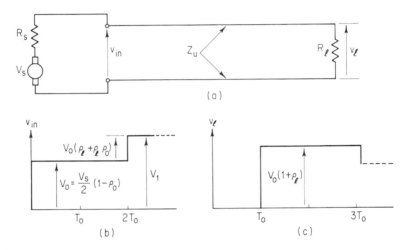

Fig. 5-18. Determination of unknown characteristic impedance for a general case.

It is apparent that this method of pulse reflection can also be used to obtain the delay time of a given length of line simply by measurement of the time delay between reflections. If a certain delay time is required, then the length of line necessary to provide this delay is easily found from the velocity of propagation, which gives

$$T_0 = \frac{\ell}{\nu} = \ell\sqrt{\mu\epsilon} \qquad (5\text{-}60)$$

For all good dielectrics, the relative permeability is unity, so that the above becomes

$$\frac{T_0}{\ell} = \frac{\sqrt{\epsilon_r}}{c} = 0.333 \times 10^{-8} \sqrt{\epsilon_r} \quad \text{sec/meter} \qquad (5\text{-}61)$$

If the above is expressed in English units, a very convenient, easily remembered number results

$$\frac{T_0}{\ell} = 1.018 \sqrt{\epsilon_r} \times 10^{-9} \text{ sec/foot} \qquad (5\text{-}62)$$

$$\approx \sqrt{\epsilon_r} \text{ nsec/foot} \qquad (5\text{-}63)$$

It is easily seen from Eq. (5-62) that an accurate measurement of the dielectric constant of a line can be obtained by measuring the distance between reflections for a known length of line. For instance, in Fig. 5-18, if the line is left open at the far end, then a voltage measurement at the input will give the delay time $2T_0$. If the length of line ℓ is known, the dielectric constant is easily found from Eq. (5-62)

$$\epsilon_r = \left(\frac{T_0}{1.018\,\ell}\right)^2 \approx \left(\frac{T_0}{\ell}\right)^2 \qquad (5\text{-}64)$$

with ℓ in feet. If ℓ is in meters, then Eq. (5-61) should be used.

5.12 RELATIONSHIP BETWEEN RISE TIME AND FREQUENCY RESPONSE

The rise time capability of an amplifier, e.g., oscilloscope, is related to the upper-frequency 3-dB point by the well-known expression

$$T_R = \frac{k}{f(3 \text{ dB})} \text{ sec} \qquad (5\text{-}65)$$

where T_R is the rise time in seconds, $f(3 \text{ dB})$ is the frequency of the upper-half power point in cycles per second, and k is a constant usually between 0.35 and 0.45 cycles, depending on the shape of the frequency-gain curve. If the overshoot on the pulse rise time is less than 5 percent, then the value of 0.35 for k is reasonably accurate.* Figure 5-19 shows a curve of rise time versus frequency for this case as a handy reference.

*See [13, p. 327].

The above expression gives a reasonable working relationship for a single-stage amplifier only; for cascaded stages, the total effective rise time is

$$T_R = \left(T_1{}^2 + T_2{}^2 + T_3{}^2 + \cdots \right)^{\frac{1}{2}} \qquad (5\text{-}66)$$

where T_1, T_2, etc., are the rise times of each stage.

5.13 PULSE DISTORTION ON TRANSMISSION LINES

(a) General concepts. In all the previous sections, we have considered the pulses or step functions to propagate along the given line without changing shape. In other words, we assumed that the lines were ideal. Unfortunately, a pulse propagating on a real transmission line will change shape as a result of two major sources of distortion, namely, attenuation and dispersion. It should be pointed out here once again that as was detailed in Chap. 3, dispersion simply indicates the dependence of phase velocity on the applied frequency, regardless of what mechanisms are responsible for this dependence.

There are numerous sources of both attenuation and dispersion, but for ordinary lines with applied pulses, skin effect is usually the most significant. We shall discuss the subject at length, but first some general comments are in order on just how distortion comes about.

As we know, a single pulse is composed of a continuous frequency spectrum, i.e., a Fourier spectrum. The various frequency components add at the input end to give the applied pulse waveform. If these components all propagate at the same velocity with no attenuation, they will add at any given point to produce the identical waveform. Distortion comes about when attenuation is present and/or when the velocity of propagation is different for the various frequencies, i.e., dispersion. If the attenuation and velocity are the same for all frequencies, then the waveform at any point on an infinitely long line will be identical in shape, but will decrease in amplitude with distance from the input end. This case is not usually encountered in practice since the mechanisms which give rise to attenuation also introduce a small but finite dependence of velocity on frequency, thus causing a distortion of waveform in addition to a reduction in amplitude. However, an extraordinary case where this does, in fact, happen is a superconducting transmission line which is dispersionless for all frequency components below about 1000 megacycles per second, i.e., sub-nanosecond rise time pulse, but small series losses are present to attenuate the wave. This case shall be considered more fully in Chap. 6.

The case usually encountered is that in which losses are present (thus both attenuation and dispersion occur.) In some cases, the losses are independent of frequency, although in general they are dependent on frequency, which only further distorts an applied pulse. In order to further explain the origins of distortion, we shall consider several idealized cases. Afterwards, we will consider the more common skin-effect distortion.

(b) *Series- and shunt-resistance losses.* The finite conductivity of the conductors gives rise to a series resistance while the dielectric insulator gives rise to a shunt resistance as a result of absorption bands (Chap. 3), electrostriction, or conduction. Both of these loss terms will generally be frequency-dependent, but we wish to investigate the idealized case when they are constant and independent of frequency. The losses are usually small, so that the results of Sec. 2.6 are applicable, thus giving the attenuation, phase constant, and phase velocity as

$$\alpha = \frac{R}{2\sqrt{L/C}} + \frac{G}{2}\sqrt{\frac{L}{C}} \tag{5-67}$$

$$\beta = \omega\sqrt{LC}\left(1 - \frac{RG}{4\omega^2 LC} + \frac{G^2}{8\omega^2 C^2} + \frac{R^2}{8\omega^2 L^2}\right) \tag{5-68}$$

$$\nu = \frac{\omega}{\beta} \tag{5-69}$$

From these equations, it is easily seen that frequency-independent series and shunt losses give rise to attenuation and dispersion. The attenuation term is independent of frequency but the phase shift includes the frequency in the higher-order terms. Thus, with a single applied pulse, all the various frequency components will be attenuated the same amount, but each will be shifted in phase by a different amount, so that the waveform will change shape as it propagates.

If the losses are frequency-dependent, then the attenuation will be different for the various frequency components and there will also be a further variation in the phase shift.

(c) *Frequency-dependent reactive components.* In Chap. 3, we saw that the dielectric constant of an insulator varies with frequency, which changes the capacitance in the equivalent circuit. This effect gives rise to dispersion and hence to pulse distortion. Even though there are absorption losses associated with the dispersion frequency range, these are usually quite small compared to the series resistance losses, so that G can usually be neglected. If the series resistance is small but not negligible, then Eqs. (5-67) through (5-69) show that the dispersive character of the dielectric

gives rise to both a frequency-dependent attenuation and, of course, additional phase shift. However, for usual applied pulses, the dielectric constant is essentially constant over the frequency spectrum contained within the pulse. For example, a teflon insulator has a relative dielectric constant which is essentially constant at 2.1 for frequencies well beyond 10 gigahertz (see Fig. 3-8). This would correspond to pulses with rise times well into the picosecond region, as determined from Fig. 5-19. Thus, this type of dispersion effect can usually be neglected.

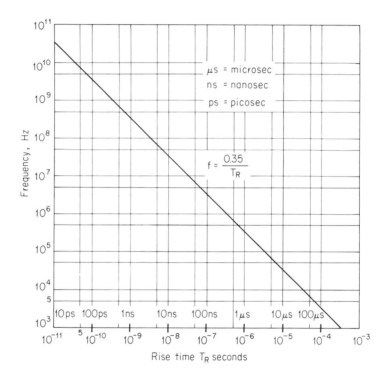

Fig. 5-19. Rise time vs. upper frequency 3 dB point.

(d) Skin-effect distortion. In Chap. 4, the additional series impedance of a plane conductor as a result of skin effect for a sinusoidal excitation was found to have a series resistance and a reactive component, both of which are equal in magnitude and proportional to the square root of the frequency. We wish now to determine the skin-effect distortion of a pulse propagating on a line, assuming that the skin depth obeys Eq. (4-29). The applied pulse is assumed to have a linear rise time and a flat top lasting indefinitely, so that the fall time can be neglected. In the analysis we will first obtain the

Laplace transform response of the line to an impulse, i.e., the transfer function. The response to the pulse is then simply obtained by taking the Faltung integral (commonly known as convolution) of the time-dependent pulse with the impulse response (in time domain) of the line. Once this is obtained, it is possible to plot normalized curves of the response of a general line (parameters specified separately) to pulses with various rise times. Such plots can also be used to represent actual pulses which do not consist simply of a linear rise time, provided that the pulse can be approximated by several short, straight-line segments.

We will assume that the line is terminated in its characteristic impedance, or alternatively that it is infinitely long, so that we may avoid the problem of proper termination and possible reflections. The effects of improper termination are considered briefly in Sec. 5.14.

The impulse response or transfer function of a given length of transmission line can easily be obtained from a fundamental principle of Laplace transform theory, namely, that the transfer function is simply the steady-state response of the line to a sinusoidal excitation. Since it is assumed that there are no reflections and hence no negatively traveling waves on the line, then from Eq. (2-12) the transfer function of a length of line ℓ is given by

$$\frac{v\,(x = \ell)}{v_{in}} = e^{-\gamma} \tag{5-70}$$

where

$$\gamma = \sqrt{(R + j\omega L)\,(G + j\omega C)} \tag{5-71}$$

(L is external (circuit) inductance). We will assume that $G = 0$, since this is very nearly the case. The loss term R is the skin-effect impedance. This skin-effect impedance (containing a resistance and reactance) per unit length for a single plane conductor is given by Eq. (4-102)

$$Z_{sk} = \frac{1 + j}{W\sigma\delta} \tag{5-72}$$

$$\delta = (\pi f \mu \sigma)^{-\frac{1}{2}} \tag{5-73}$$

where W is the width of the conductor (single conductor). For a single round wire at high frequency, W is the circumference of the wire, as shown by Eq. (4-35). Substitution of Eq. (5-73) into Eq. (5-72) gives a general form for the skin impedances

$$Z_{sk} = R_{sk}\sqrt{s} \tag{5-74}$$

where

$$R_{sk} = \frac{1}{W} \sqrt{\frac{\mu}{\sigma}} \quad \text{for a plane conductor} \qquad (5\text{-}75)$$

$$= \frac{1}{2\pi r_0} \sqrt{\frac{\mu}{\sigma}} \quad \text{for circular wire at high frequency} \qquad (5\text{-}76)$$

$$s = j\omega$$

Substituting Eq. (5-74) for R in Eq. (5-71) and then this into Eq. (5-70) gives the transfer function

$$\frac{v(x = \ell)}{v_{in}} = \exp\left[-\ell \left(s^2 LC + s^{3/2} CR_{sk}\right)^{1/2}\right] \qquad (5\text{-}77)$$

In order to obtain the response of this line to a pulse, it will be necessary to have the inverse Laplace transform of Eq. (5-77), which will be the response of the line to an impulse function, i.e., pulse of amplitude approaching infinity, width approaching zero, and area remaining constant at unity. It is not possible to get an analytic solution in closed form for the general case because of the complications introduced by the radical in the exponent. Fortunately, this can be circumvented with the aid of the identity of Eq. (1-25) as follows. The radical of Eq. (5-77) can be written

$$\gamma = s\sqrt{LC} \left(1 + \frac{R_{sk}}{L\sqrt{s}}\right)^{1/2} = s\sqrt{LC}\,(1 + \nu)^{1/2}$$

If it is assumed that $Z_{sk} \ll \omega L$ where L is the external (circuit) inductance, then $\nu \ll 1$; thus, using Eq. (1-25) gives

$$\gamma = s\sqrt{LC} + \frac{1}{2}\frac{R_{sk}}{\sqrt{L/C}}\sqrt{s}$$

$$= j\omega\sqrt{LC} + \frac{1}{2}\frac{R_{sk}}{\sqrt{L/C}}\sqrt{\frac{\omega}{2}}\,(1 + j) \qquad (5\text{-}78)$$

$$= \alpha + j\beta$$

where

$$\alpha = \frac{1}{2}\frac{R_{sk}}{\sqrt{L/C}}\sqrt{\frac{\omega}{2}} \qquad \beta = \omega\sqrt{LC} + \frac{R_{sk}}{2\sqrt{L/C}}\sqrt{\frac{\omega}{2}} \qquad (5\text{-}79)$$

The second term of the above is very similar to the phase constant as given by Eq. (2-46), while the first is very similar to the attenuation term for small losses as given by Eq. (2-45). Equation (5-77) therefore simplifies to

$$\frac{v(x = \ell)}{v_{in}} = \exp(-s\ell\sqrt{LC}) \, \exp\left(-\ell\frac{\sqrt{s}\,R_{sk}}{2\sqrt{L/C}}\right) \tag{5-80}$$

Since the first exponential term merely represents the time delay of the line, the inverse transform of Eq. (5-80) is the inverse transform of the second exponential term, delayed by the appropriate amount. This can be obtained from standard tables*

$$\mathcal{L}^{-1}\frac{v(x = \ell)}{v_{in}} = g(t) = A u^{-3/2} e^{-B/u} \quad \text{for} \quad u \geq 0 \tag{5-81}$$
$$= 0 \quad \text{for} \quad u \leq 0$$

where $u = t - \ell\sqrt{LC}$

$$A = \frac{\ell R_{sk}}{4\,(\pi L/C)^{1/2}} \tag{5-82}$$

$$B = \left(\frac{\ell R_{sk}}{4\sqrt{L/C}}\right)^{2} = \pi A^{2} \tag{5-83}$$

Note that A and B depend only on the physical parameters of the line.

We will make use of Eq. (5-81) to obtain the pulse response of the line. The pulse is assumed to have a linear rise time and is defined as

$$p(t) = 0 \quad \text{for} \quad t < 0$$
$$= \frac{t}{T_R} \quad \text{for} \quad 0 \leq t \leq T_R \tag{5-84}$$
$$= 1 \quad \text{for} \quad t > T_R$$

where T_R is the rise time. The time domain response of the line to this pulse will be denoted by $f(t)$ and is given by the convolution of $p(t)$ with

*See [3, p. 299, no. 82], [4, p. 328, no. 82], or [7, p. 423, no. 33].

$g(t)$ of Eq. (5-81).* Thus

$$f(t) = \int_0^t p(t - \tau)\, g(\tau)\, d\tau \tag{5-85}$$

It is apparent that no voltage can appear at the output end of the line until a time equal to the delay time of the line. Thus, if we denote the delay time as $T_0 = \sqrt{LC}\,\ell$, the output response, $f(t)$ is zero for $t < T_0$. For the time interval corresponding to the rise time of the applied pulse, after it has arrived at the output end, the response is

$$f(t) = \int_0^u \frac{u - \tau}{T_R}\, \tau^{-3/2}\, e^{-B/\tau}\, d\tau \tag{5-86}$$

$$\text{for } T_0 \le t \le T_0 + T_R$$

For the remaining time beyond this, the response is

$$f(t) = \int_0^{u-T_R} \tau^{-3/2}\, e^{-B/\tau}\, d\tau + \int_{u-T_R}^u \frac{u - \tau}{T_R}\, \tau^{-3/2}\, e^{-B/\tau}\, d\tau \tag{5-87}$$

for $t > T_0 + T_R$ where, as before, $u = t - T_0$. Equation (5-87) contains Eq. (5-86) if only positive values of τ are used, so that we will evaluate only the former. After making use of several mathematical manipulations,[†] this reduces to a simple function

$$f(t) = \frac{1}{T_R} \int_{u-T_R}^u \operatorname{cerf} \sqrt{\frac{B}{\tau}}\, d\tau \tag{5-88}$$

for $u \ge 0$, that is, $t \ge T_0$, and with the lower limit equal to zero for $u < T_R$. This equation allows us to evaluate the skin-effect deterioration of a given transmission line with a very wide applied pulse of rise time T_R. In order to obtain general curves of $f(t)$ for different applied pulse rise times, it is desirable to make Eq. (5-88) independent of the physical parameters of the line, i.e., independent of B. This can easily be done by making the

---

*See [3, p. 36] and [12, p. 316] for proof and details.
†See [14].

substitution $u = Bt'$, which yields

$$f(t') = \frac{1}{T'_R} \int_{t'-T'_R}^{t'} \text{cerf} \sqrt{\frac{1}{t'}} \ dt' \quad \text{for} \quad t' \geq 0 \qquad (5\text{-}89)$$

where

$$t' = \frac{t - T_0}{B} \qquad (5\text{-}90)$$

$$T'_R = \frac{T_R}{B} \qquad (5\text{-}91)$$

Equation (5-89) defines a family of normalized pulse responses $f(t')$ as a function of normalized time t' for various values of normalized rise time $T'_R = T_R/B$. The family of curves is plotted in Fig. 5-20. For $T'_R > 2000$, the pulse distortion becomes insignificant. Thus, for a given line, the physical parameters can be used to evaluate B from Eq. (5-83) with R_{sk} given by Eq. (5-75) or Eq. (5-76). Once this is done, the skin-effect distortion of a pulse of given rise time can be obtained, provided that the assumptions used for the analysis are valid. The assumptions are that the surface resistance or attenuation follows the $\omega^{1/2}$ law for the bandwidth of interest and that $Z_{sk} \ll \omega L$. If the bandwidth extends to such low frequencies that the skin depth is greater than r_0 (or conductor thickness for a plane), then the resistance is constant for these lower frequencies (see Fig. 4-5) and the above analysis is incorrect. Similarly, if the frequency becomes too high, anomalous skin effect could appear, although this is not likely for copper at room temperature. If the series impedance becomes too large, then the above analysis becomes less accurate, since higher-order terms are required in Eq. (5-78).

In order to determine B, it is necessary to know physical parameters of the line such as r_0 and σ. These are often not known or not easily obtained. Fortunately, B can be determined directly by measurement of the attenuation of a given line for a specified frequency. Any line which obeys the $\omega^{1/2}$ law for skin effect will have an attenuation constant given by α of Eq. (5-79). But since B is given by Eq. (5-83), it follows that

$$B = \left(\frac{\ell\alpha}{2\sqrt{\omega/a}}\right)^2 = \frac{(\ell\alpha)^2}{4\pi f} \text{ sec} \qquad (5\text{-}92)$$

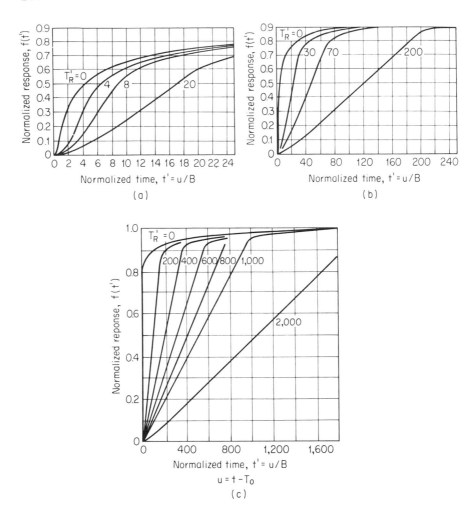

Fig. 5-20. Calculated, normalized response of a perfectly terminated line to a step function of normalized rise time $T'_R = T_R/B$, assuming skin effect distortion. *(After Wigington and Nahman.)*

where α is in radians per unit length* and is measured (or calculated) at the frequency f; ℓ is total line length.

Before we proceed, it should be mentioned that some indication of the accuracy of assuming skin-effect distortion as a function of the square root of the frequency to be the major factor can be obtained from Eq. (5-79). Any line which obeys this relationship will have a straight-line relationship

---

*Since α is usually given in dB, the conversion factor is α (radians per unit length) = 0.115 α (dB per unit length).

of attenuation versus frequency plotted on a log-log scale. If such a plot deviates severely from a straight line, additional factors are significant and will have to be included for a more exact analysis. For most commercially available coaxial cables, the straight-line relationship is very nearly obeyed.*

Below are some comparisons of calculated pulse response which make use of the above theory with actual measurements on coaxial cables[†] and strip lines.[‡]

For the commercial coaxial lines, Fig. 5-21 shows the applied pulse and the straight-line segments used to represent it for calculations. The normalized curves of Fig. 5-20 are used for each straight-line segment and are superimposed to obtain the total waveform. The measured and calculated pulses at the ends of various lines are shown in Fig. 5-22: all have 50

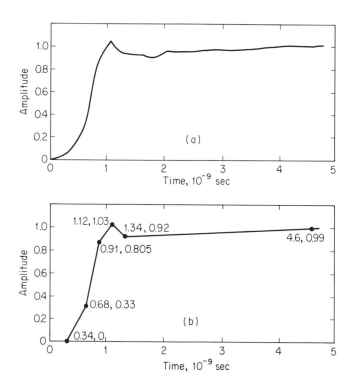

Fig. 5-21. Pulse applied to coaxial cables. (b) Actual pulse (normalized); (b) straight line segment approximation. (*After Wigington and Nahman.*)

*[11, p. 614].
†See [14].
‡See [5].

ohms impedance though each has a different value of B. Two conclusions are immediately apparent. First, the measured and calculated curves are generally in good agreement; second, significant pulse distortion is obtained, the amount varying for different cables. It should be pointed out that B is a measure of the distortion to be expected, with larger values of B giving larger distortion. This is seen to be true from Fig. 5-22, with the smallest distortion occurring for cable (d) (which has a large diameter and hence the smallest value of B).

Similar results have been obtained by Eastman [5] for a 32-inch-long strip line fabricated from a 0.002-inch-thick copper ground plane, a copper

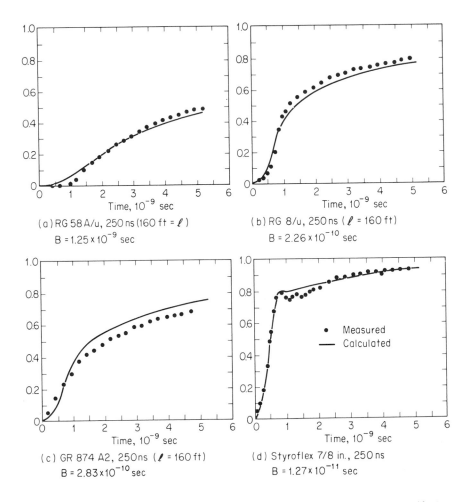

Fig. 5-22. Comparison of calculated and measured response of several common coaxial cables to the pulse of Fig. 5-21. (*After Wigington and Nahman.*)

strip conductor of 0.0007-inch thickness, and mylar sheet insulation as the
dielectric (see, e.g., Fig. 8-9). The insulation thickness was varied to obtain
various values of impedance and therefore different values of B. The value

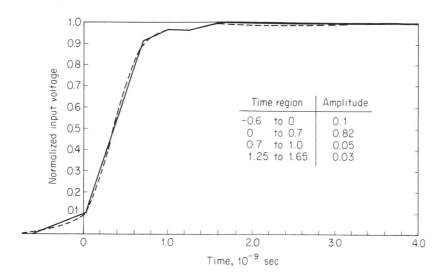

Time region	Amplitude
-0.6 to 0	0.1
0 to 0.7	0.82
0.7 to 1.0	0.05
1.25 to 1.65	0.03

Fig. 5-23. Pulse applied to strip line and straight line segment approximation (solid line).
(*After Eastman.*)

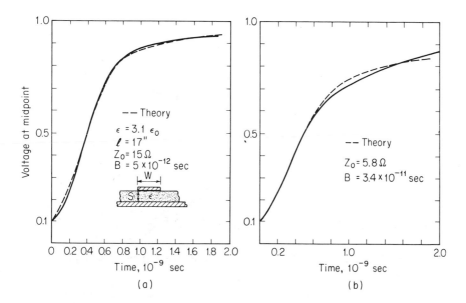

Fig. 5-24. Measured and calculated response of a strip line to pulse of Fig. 5-23 with $W = 0.037$
inch. (a) $S = 0.002$ inch; (b) $S = 0.0005$ inch. (*After Eastman.*)

of B was calculated from Eqs. (5-75) and (5-83) by using the measured value of $\sqrt{L/C}$. Since skin-effect distortion is rather large in this line, measurements were made at the midpoint ($\ell = 17$ inches) to prevent reflections caused by improper termination from interfering with the measurements. Fig. 5-23 shows the applied pulse and straight-line segment approximation. Fig. 5-24 shows measured and calculated response for a 0.037-inch-wide strip for two values of S and therefore of Z_0.

An important point to keep in mind is the fact that skin-effect distortion is not just a function of the series resistance and inductance alone, but rather varies as the ratio of these series parameters to the characteristic impedance. For example, the skin-effect distortion becomes smaller as the impedance $(L/C)^{1/2}$ becomes larger. This is just another way of stating that larger values of B give more distortion, since for a given conductor resistance, the attenuation is larger for smaller values of L/C.

5.14 SKIN-EFFECT DISTORTION WITH IMPROPER TERMINATION

In Sec. 5.13, the skin-effect distortion was analyzed on the assumption that there were no reflections on the line, i.e., that the line is perfectly matched or is infinitely long and observations are made at some intermediate point. It is often difficult as well as impractical to terminate a line in its characteristic impedance; the question then quite naturally arises as to the behavior of an improperly terminated line with fast-rise-time applied pulses, and hence skin-effect distortion. It is possible to analyze such a line, but only for applied pulses which are assumed to be periodic in time, i.e., repetitive with a reasonable repetition rate. This is because each frequency component must be treated separately, since each frequency present on a line will have its own particular velocity of propagation, impedance, and attenuation, so that for a fixed termination, each experiences a different reflection. Each frequency must be applied separately to the line and the final behavior obtained by reassembling the various components. A single applied pulse has a continuous (infinite) frequency spectrum, whereas a periodic wave can be represented by a converging series of discrete harmonics. Even with modern-day, high-powered scientific computers, only the periodic wave can be handled with approximations (although very good approximations in many cases), but transmission lines are usually pulsed at some finite repetition rate, so that a periodic wave is quite realistic.

An analysis of skin-effect distortion of the type presented in Sec. 5.13, though for various terminations, is rather straightforward in principle; it essentially requires nothing more than Eq. (5-79) and a Fourier harmonic decomposition of the applied excitation. However, a scientific computer is necessary to carry through the details of calculation, bookkeeping, and reassembly of the harmonics. Bertin [2] provides a general outline of the technique, while McNichol and English [9] provide a detailed Fortran program and numerous results for strip lines in a magnetic-film memory array. A few of the interesting results from the latter are included here.

A periodic pulse of 25 ns base width and 100 ns period is represented by 40 harmonics, as shown in Fig. 5-25. This is applied to a silver strip line above an infinite silver ground plane all at $0°C$ ($\rho = 1.47 \times 10^{-6}$ ohm-cm); the strip width W is 0.010 inch, thickness T is 8×10^{-5} inch (20 KA), and separation S is 0.0005 inch above the ground plane by a dielectric of $\epsilon_r = 6.8$. The source impedance is 50 ohms, and the line is 12 inches long and has $\sqrt{L/C}$ equal to 7 ohms.

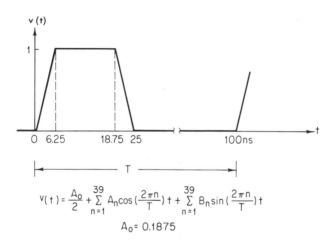

$$v(t) = \frac{A_0}{2} + \sum_{n=1}^{39} A_n \cos\left(\frac{2\pi n}{T}\right)t + \sum_{n=1}^{39} B_n \sin\left(\frac{2\pi n}{T}\right)t$$

$$A_0 = 0.1875$$

Fig. 5-25. Unit amplitude periodic pulse train (source voltage) and its harmonic representation used to obtain Fig. 5-26.

Figure 5-26 shows the calculated current waveform (one pulse shown only) at the end of the line with various load resistances as given by curves $A, B, C,$ and E; in order to show the effect of improper termination on the line, the waveform at a distance of 12 inches from the source for an essentially infinitely long line is shown by curve D, that is, this is the waveform at that point for a properly matched line with skin-effect distortion. It is apparent that the termination has considerable effect on the waveform; also, termination in a resistance of value $(L/C)^{1/2}$ gives reasonable results, although general conclusions cannot be made since the waveform at a point

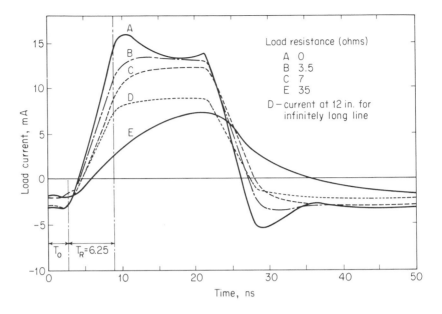

Fig. 5-26. Calculated skin effect response at end of 12 inch strip line for various terminations with excitation of Fig. 5-25; T_0 = 2.6 ns, $\sqrt{L/C}$ = 7 ohms. (*After McNichol and English.*)

can change significantly with any of the variables such as source impedance, line length, etc. Calculations for specific cases should be carried through before conclusions are drawn.

An important point regarding the definition of rise time deserves mention. In pulse work, the rise time is usually taken as the time interval between the 10 and 90 percent points of the amplitude. It can be seen from Fig. 5-26 that skin effect has a significant effect on the high frequencies, so that the final portion of the rise time can be very rounded. Thus, this usual definition is obscured in many cases, especially in memory arrays and other networks where the signal is proportional to the slope of the applied pulse.

Another important point is that skin effect in a strip line conductor will cause the current to crowd near the edges of the conductor, thereby increasing the internal conductor impedance over that obtained from the uniform field calculations of Sec. 4.7 or Sec. 4.8. In a similar manner, the current distribution in the ground plane will change with frequency; in some applications, the effect of the so-called "current spreading in ground plane" must be taken into account.* We will not consider these effects any further. However, it should be emphasized that these effects only change the internal

*See [8].

conductor impedance, since the external (circuit) inductance is a function only of geometry, as detailed in Chap. 8.

APPENDIX 5A

DISCRETE CIRCUIT COMPONENTS

In the past, a great deal of confusion has resulted from a failure to include the transmission-line analysis of a circuit or component, i.e., discrete resistor, inductor, or capacitor, when the physical situation justified such an approach, e.g., fast-rise-time applied pulse. One often hears about the so-called "dilemma of lumped circuits," as discussed, for instance, by Adler *et al.* [1]. They attempt to define two problems: first, a parallel plate strip line and its static solution from a circuit point of view; second, the use of the same value of C (or L) in both static and dynamic analyses. It is important to recognize that the first problem or so-called "dilemma," while present in some situations, is not intrinsic to the subject matter but rather is a consequence of the approach often used. In order to fundamentally understand any device or circuit component it is always necessary to specify in exact detail the means of excitation and it is also necessary to perform the transient as well as the steady-state solution.* If one considers only the steady-state solution, then some uncertainty can arise. It is easily seen in view of the results of Sec. 5.9 that if a step function, i.e., dc voltage, is suddenly applied to a pair of open-circuited parallel plates (see, e.g., Fig. 8-3),[†] a transient voltage will be required to charge the plates, and if the rise time of the applied step function is sufficiently fast, then the parallel plates will charge up in discrete steps similar to what is shown in Fig. 5-11(c). In the limit of very small incremental steps in the waveform, smaller T_0, and slower rise time of the step function, the device gives a waveform which can be described by a time constant RC, where C is the "static" capacitance. A completely analogous situation exists when the parallel plates are shorted and appear as an inductor. In reality, all components or devices must be treated as transmission lines when an ideal step function is applied, but as the devices become smaller and/or the rise time of the step function becomes larger (slower), the transmission-line analysis gives way to ac circuit analysis because one type of field or lumped parameter dominates the behavior. This is not a dilemma, but rather a natural extension or simplification

*For example, see Prob. 5-2.
†See [1, Fig. 2-1(a)].

of the exact solution to such a problem. For example, in Sec. 4.5, the exact solution for propagation between parallel conductors yielded a complicated solution, but the use of certain very realistic approximations gave exactly the usual lumped-circuit representation.

The second problem mentioned above, that of using static parameters for ac analysis, is also a natural consequence of a more fundamental behavior of electromagnetics. For a TEM mode, Appendix 8A shows that the ac and dc parameters, if viewed properly, are always identical.

Thus, the correct conclusion conerning any device or circuit component is that a properly designed device can behave as a lumped element as closely as is desired, but only within a certain range of excitation. In general, as the applied frequency or pulse rise time increases, all devices look more and more like transmission lines and will eventually require analyses as presented in Chap. 5. In many cases, these circuits will be nonuniform transmission lines, which only further complicates the analysis.

PROBLEMS

5-1. Show that a small capacitive discontinuity on a line has an approximate value of $C' = 2 T_R V_r / Z_0 V_0$.

5-2. An ordinary composition resistor of 50 ohms is used to terminate a 50-ohm transmission line. The leads of the resistor are bent to form a square of area 1/2 in. × 1/2 in. Assuming a typical 1/4-watt resistor with wire diameter of approximately 1/32 in., show that the loop formed by the termination can be represented by a series LR circuit. and that the time constant is approximately 0.4 ns. (Thus, rise times comparable to or smaller than 0.4 ns require a smaller loop termination.)

5-3. For an experimental setup as in Fig. 5-17, with $Z_s = Z_0$ and $T_u \gg T_0$, v_{in} is observed to be as below

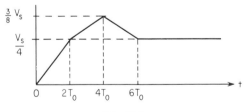

Fig. 5P-1.

Z_0 and T_0 are known; find Z_u in terms of Z_0 if the applied pulse rise time is $4 T_0$.

Answer: $Z_u = 1/3 Z_0$.

5-4. A pulse of 0.5 ns rise and fall time and 150 ns base width is applied to 100 ft of $RG\,8/u$ coaxial cable. Obtain the output pulse at the end of the line, assuming only skin-effect distortion. Evaluate α and therefore B at 0.5×10^9 Hz.

Answers: α = 6 db/100 ft = 0.69 rad/100 ft
$\alpha\ell$ = 0.69 rad, B = 7.6 x 10^{-11}
T_R' = 6.5 (use Fig. 5-20(a) for T_R' = 8)

REFERENCES

1. Adler, R. B., L. J. Chu, and R. M. Fano: "Electromagnetic Energy Transmission and Radiation," chap. 2, John Wiley & Sons, Inc., New York, 1960.
2. Bertin, C. L.: Transmission Line Response Using Frequency Techniques, *IBM J. Res. Develop.,* vol. 8, no. 1, p. 52, January 1964.
3. Churchill, R. V.: "Modern Operational Mathematics in Engineering," McGraw-Hill Book Company, New York, 1944.
4. *Ibid.,* "Operational Mathematics," McGraw-Hill Book Company, New York, 1958.
5. Eastman, D. E.: "High Speed Pulse Transmission in Strip-line Arrays," master's thesis, M.I.T., June 1963 (no other publication).
6. Elliott, B. J.: Picosecond Pulse Measurements of Volt Ampere Characteristics of Materials with Bulk Negative Differential Conductivity, Conf. Precision Electromagnetic Measurements, Boulder, Colo., June 1968.
7. Goldman, S.: "Transformation Calculus and Electrical Transients," Prentice-Hall, Inc., Englewood Cliffs, N. J., 1949.
8. Liniger, W., and S. Schmidt: Transient Magnetic and Electric Fields Above a Conducting Ground Plane of Arbitrary Thickness, *IEEE Trans. Magnetics,* vol. MAG-2, no. 4, p. 727, December 1966.
9. McNichol, J. J., and T. D. English: "Transmission Properties of Strip Lines Used in a Thin Film Memory," *IBM Rept.* RC 1417, June 9, 1965 (summary of this appears in *IEEE Trans. Magnetics,* vol. MAG-1, no. 4, p. 272, December 1965.
10. Ramo, S., and J. R. Whinnery: "Field and Waves in Modern Radio," 2nd ed., John Wiley & Sons, Inc., New York, 1958.
11. International Telephone and Telegraph Corp., "Reference Data for Radio Engineers," 4th ed., American Book-Stratford Press, Inc., New York, 1956.

12. Stratton, J. A.: "Electromagnetic Theory," McGraw-Hill Book Company, New York, 1941.

13. Terman, F. E., and J. M. Pettit: "Electronic Measurements," 2nd ed., McGraw-Hill Book Company, New York, 1952.

14. Wigington, R. L., and N. S. Nahman: Transient Analysis of Coaxial Cables Considering Skin Effect, *Proc. IRE,* vol. 45, p. 166, February 1957.

6 SUPERCONDUCTING TRANSMISSION LINES

6.1 INTRODUCTION

In previous chapters, we considered lines for which the losses were small, and in many cases we assumed the losses to be negligible in order to develop the general concepts of wave propagation. In most cases of practical interest, losses will be present and these will set limits on the range of applicability of a given transmission line. If a line is made too long or the characteristic impedance too small or if the applied frequency is too high, attenuation and dispersion will not be negligible, resulting in a distortion of an applied signal.

As an example, power companies are vitally concerned with the losses (primarily series resistance although other forms occur) on long lines even with a low frequency of 60 Hz. The $i^2 R$ losses are so large that it is economically feasible to greatly increase the voltage and reduce the current in order to transmit the same power with reduced losses. Thus, power transmission voltages of 100,000 to 300,000 or more are quite common. The saving in $i^2 R$ losses more than compensates for the additional expense of step-up and step-down transformers plus a variety of high-voltage equipment.

It is possible to construct a low-power transmission line with virtually no losses over a wide frequency range; it would thus behave as the ideal case considered in Sec. 1.5. Such lines make use of superconductors for the conductors and ordinary dielectric materials for the insulators. The major problem with such lines is that there are no known materials which act as superconductors above 18°K, that is, 18° above absolute zero. Thus, such lines require additional refrigeration equipment, which is complex and expensive; therefore, they have not found widespread use. Nevertheless, such lines are of great academic as well as experimental interest and are beginning to find practical applications. In this chapter, we shall consider some of the properties of superconducting transmission lines and we will determine their limitations.

Before we proceed with an analysis of superconductors, a very brief background in some of the more important concepts concerning superconductivity will be discussed. The fundamentals of superconductivity in conjunction with Maxwell's equations will then be applied to the derivation of the surface impedance of a plane superconductor. This surface impedance, which merely represents the series impedance of a conductor of a superconducting transmission line, can then be used to determine the propagation constant and the characteristic impedance of such a line as was done in Sec. 5.13(d). All of these analyses will be carried out with the assumption of an applied sinusoidal excitation. This analysis will allow us to make a number of predictions concerning the frequency response of such lines.

We shall find that very long lines which are carrying relatively small currents can be made with no attenuation or dispersion even though an electric field exists in the direction of propagation. This is true only for frequencies below about 1,000 m/sec, since above this frequency superconductors exhibit a finite resistance. For many cases of practical interset, however, superconducting lines behave as the ideal case. Some interesting experimental results of the pulse behavior of superconducting coaxial lines will be compared to the calculated response expected from ordinary lines to show the improvement expected. All of the above analyses will be performed for lines which are assumed to be carrying relatively small currents so that the material remains primarily in the superconducting state. In Sec. 6.7 we shall briefly touch upon the subject of power transmission using high-field (hard) superconductors carrying large currents so as to be in the so-called mixed state where the surface impedance analysis is no longer correct.

6.2 PROPERTIES OF SUPERCONDUCTORS

In order to understand the properties of the "ideal" transmission line, it is essential to know at least a few of the fundamental properties of superconductors. Some of the more important of these fundamental properties are:

a. Resistance as a function of temperature
b. Meissner effect
c. Resistance as a function of frequency
d. Resistance as a function of magnetic field; type-I and type-II superconductors
e. London's phenomenological equations and penetration depth

f. Coherence length and relationship to type-I and type-II superconductors

g. Local (London) versus nonlocal (Pippard) theory

h. Ac losses in type-I and type-II superconductors

We shall briefly discuss each of these separately, concentrating on the properties and concepts which are more relevant to transmission-line behavior. Further discussion of the individual topics can be found in the references cited.

(a) Resistance versus temperature. It is an established fact that below a certain so-called "critical" temperature, a number of materials lose all of their dc electrical resistance. It cannot be stated with certainty that the resistance is identically equal to zero, but it is so small that it cannot be detected by any known means. Measurements have shown that the maximum value of resistivity for a superconductor must be smaller than 10^{-23} ohm-cm, a value which is a factor of 1.7×10^{17} smaller than the resistivity of ordinary copper at room temperature. A number of experiments have been performed whereby a circulating current was established in a superconductor and the current continued to flow (in the absence of any emf) for several years without any measurable decrease in amplitude. Thus, for all practical purposes, superconductors can be considered to have zero resistance for dc and, in some cases, even for frequencies below approximately 10^9 Hz. The fact that the resistance of a superconductor is not zero at high frequency is shown by Fig. 6-1 for tin at 1.2×10^9 Hz. The dc resistance R_n of a superconductor at an infinitesimal temperature above the critical temperature is known as the "residual resistance" and is a measure of the impurity content of the material, i.e., the ability of the material to scatter the conduction electrons (see Sec. 4.10).

(b) Meissner effect. The Meissner effect very simply and clearly states that a magnetic field is excluded for a superconductor. In order to understand the significance of this, let us consider a sphere made of pure tin or lead, in the normal conducting state above the critical temperature. If this sphere is brought into a uniform magnetic field, then during the transition period there will be induced eddy currents, but these will be damped out by normal losses so that eventually the magnetic field will be uniform through the sphere, as shown in Fig. 6-2(a). If the temperature of the sphere is now lowered, nothing happens until the critical temperature is reached. At this point, when the sphere becomes superconducting, the magnetic field is essentially "pushed out" or excluded from the interior of the metal because

of the supercurrents which are induced on the surface. These supercurrents create an equal and opposite field to H_0 which gives a net field of 0 in the interior, as shown in Fig. 6-2(b).

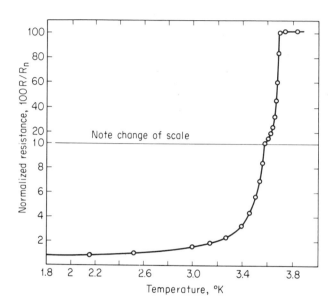

Fig. 6-1. Measured resistance of superconducting tin at 1.2×10^9 Hz; $T_c = 3.74°K$. (*After Pippard* [11].)

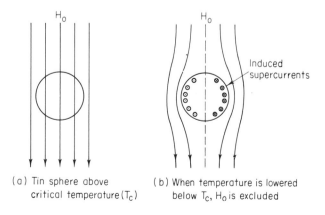

(a) Tin sphere above critical temperature (T_c)

(b) When temperature is lowered below T_c, H_0 is excluded

Fig. 6-2. Meissner effect for solid sphere.

If the sphere were not a solid conductor, but rather, say, a ring with a hole in the center, it is possible for the magnetic field to be "pushed" into the hole or, in other words, it is possible to obtain trapped flux. This is done in a manner similar to the above for the sphere, with the result as shown in Fig. 6-3. This example is important, since often materials are made which are not solid superconductors but rather can be thought of as being composed of many very small rings of superconductors with a nonsuperconducting hole in the center. Such a material can trap flux and on a macroscopic level appears not to exclude the magnetic field, although it actually does so, though only on a microscopic level.

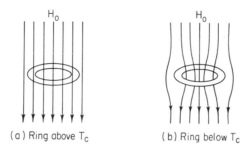

(a) Ring above T_c (b) Ring below T_c

Fig. 6-3. Flux trapped in a superconducting ring.

Superconductors are therefore perfect diamagnets with an induced magnetic moment equal and opposite to the applied magnetic field.

(c) Resistance as a function of frequency. As the excitation frequency is increased, for example, 1.2×10^9 Hz, a superconductor will have a finite resistivity for all temperatures. The resistivity will decrease as the temperature approaches absolute zero, but it does not necessarily decrease to zero. This finite resistance results from the high electric field present in superconductors at high frequencies. This large field will accelerate the normal conduction electrons as well as the superconducting electrons, the former giving rise to the losses (see Sec. 6.2(h) for further details and comparisons). Historically, this high-frequency resistance of superconductors was first observed in 1940 by London, who, in a series of measurements on tin at low temperature, found that at a frequency of 1,500 MHz, there was a finite resistance in the superconducting state. This is, or course, in contrast to dc measurements where the resistance is zero below the transition temperature ($3.74°$K). As the temperature was further lowered, London found that the resistance of tin at this frequency further decreased, apparently tending toward zero at absolute zero. That the resistance is zero only at dc and has a finite value at all other frequencies and temperatures above absolute zero

(though in some cases the latter requirement is not necessary) appears to be completely general, i.e., to apply to all superconductors. However, in most cases the resistance is so small that it is generally negligible for all but the high frequencies above roughly 1,000 MHz.* The effect of frequency on resistance is shown in Fig. 6-4 for tin. Note that at a sufficiently high frequency (2×10^{13} Hz) there is no difference between the normal and superconducting resistance.

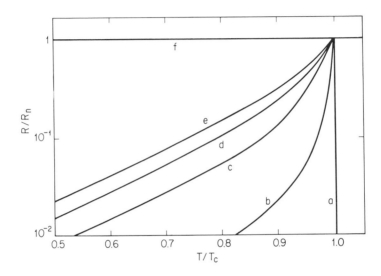

Fig. 6-4. Temperature variation of the resistance of superconducting tin at (a) dc; (b) 1.2×10^9; (c) 10^{10}; (d) 2.4×10^{10}; (e) 3.75×10^{10}; (f) 2.14×10^{13} Hz. R_n is normal resistance at an infinitesimal temperature above T_c. (*After Pippard* [16, p. 41].)

In Secs. 6.4 and 6.5 we will investigate the surface impedance and transmission-line characteristics of superconductors as a function of frequency, where it will be seen that this finite resistance at high frequency imposes a bandwidth limitation on such lines.

(d) Resistance as a function of magnetic field. If a superconductor already below its transition temperature is placed in a magnetic field of some small value, supercurrents will be induced to exclude the external field in a manner analogous to the Meissner effect described previously. Since the material is a perfect diamagnet, then as the external field is increased, the

---

*The exception to this is type-II superconductors carrying high currents such as that used in power lines (Sec. 6.7).

induced currents or diamagnetic moments increase in direct proportion, as shown in Fig. 6-5. However, when the external field reaches some critical value, this relationship is no longer valid. For type-I superconductors, at a field value of H_c, the supercurrents are destroyed or the material is no longer a superconductor. In other words, the material has been switched to the normal state. If the magnetic field is reduced, the M-H curve will retrace itself, becoming a superconductor once again at some infinitesimal field below H_c.

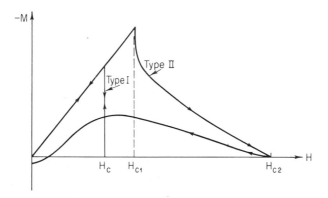

Fig. 6-5. M-H curves for superconductors, Type-I–(Pippard) reversible; Type-II–(London) irreversible for $H > H_{c1}$.

This critical field is dependent on temperature; it is lower for higher temperature, as is obvious since at the critical temperature no magnetic field is required for the transition. The relationship between temperature and critical field is given very closely by

$$H_c = H_0 \left[1 - \left(\frac{T}{T_c} \right)^2 \right]$$ (6-1)

where T_c is the critical temperature in the absence of any magnetic field and H_0 is the critical field at absolute zero temperature. Some values for the parameters are given in Table 6-1. The magnetic field expressed in Eq. (6-1) does not necessarily have to be external, but rather can be the surface magnetic field created by a current flowing in the superconductor. This is known as Silsbee's hypothesis and is necessary for the operation of many superconducting devices. These type-I materials are referred to as

Table 6-1.   Important Parameters of Some Common Superconductors

	T_c (°K)[‡]	H_0 (Oe)[‡]	λ_0[†] (10^{-10} m)	ξ[†] (10^{-10} m)	$\rho = 1/\sigma$[‡] (10^{-6} ohm-cm)(dc)
Aluminum	1.196	99	500	16,000	2.828 (20°C)
Cadmium	0.56	30	1,300		7.54 (18°C)
Indium	3.407	293	640	4,400	8.37 (0°C)
Lead	7.175	802.6	390		19.8 (0°C)
Mercury*	4.153	412	380		95.783 (20°C)
	3.949	339.3	450		
Molybdenum	0.92	98			5.33 (22°C)
Niobium	9.25	1944			14.5 (22°C)
Tin	3.74	305	510	2,300	11.5 (20°C)
Zinc	0.91	53			5.75 (0°C)
Zirconium	0.55				42.4 (22°C)
Nb_3Sn	18.07	$> 10^5$			

*Anisotropic
†See [8].
‡See [2].

"soft" because they are usually soft mechanically (low melting point and high ductility) and have relatively low H_c values (magnetically soft).

In addition to the type-I superconductors described above, there is another group of materials which generally have high values of H_c and are usually hard materials, classified as type-II.* The linear relationship between M and H (Fig. 6-5) exists up to a field of H_{c1}. At this point, M begins to decrease with a further increase in H. This decrease is associated with the occurrence of the so called mixed state, where the material is a conglomeration of superconducting and nonsuperconducting regions. In such a state, the material may be likened to a conglomeration of superconducting rings as in Fig. 6-3(b), with flux being trapped, etc. Further increases in H will cause M to disappear at H_{c2}. The details of this behavior are unimportant for our discussion; the significant point is that if H is now decreased, the MH curve does not retrace itself, but instead shows some hysteresis effect as indicated. This hysteresis represents losses in type-II material and gives rise to significant losses even at low frequencies in power transmission lines (Sec. 6.8).

In type-II superconductors, H_{c1} can be larger than H_c for type-I materials and H_{c2} can be very large, exceeding 10^5 oersteds. While these high-field,

*This distinction as hard and soft is now known to be inaccurate, since "soft" superconductors can be made "hard" by severe straining or addition of impurities. However, this is of little importance for our purposes.

high-current capabilities of type-II materials are due to impurities and imperfections which act as pinning sights to the magnetic flux, these defects are also responsible for the hysteresis losses for time-changing field. These losses can be calculated from the critical state model as outlined in Sec. 6.2(h).

(e)   London's phenomenological equations and penetration depth. There are two concepts associated with superconductors which must be accounted for. The Meissner effect states that a superconductor excludes magnetic fields and this is accomplished by the establishment of surface currents which produce equal and opposite magnetic fields to cancel the applied field, thus giving zero net field within the superconductor itself. However, since these surface currents must flow within a finite thickness along the surface of the superconductor,* and since magnetic fields are associated with currents by means of Ampere's circuital law, it is apparent that the magnetic field must penetrate a small but finite distance into the superconductor. The second concept which must be accounted for is the fact that the relationship between current density and electric field, given by Ohm's law ($J = \sigma E$), is not adequate to describe supercurrents, since there is no resistance but instead a finite current, and for dc conditions, no electric field (no losses).

In order to describe these phenomena, F. and H. London developed a macroscopic theory which gives rise to two phenomenological equations describing the current and field relationships in a superconductor. These two equations are like Maxwell's field equations to the extent that they cannot be proved or disproved but appear to be correct, phenomenologically, and therefore are adequate for purposes of describing macroscopic behavior of superconductors. Historically, the development of the macroscopic theory of superconductivity is very interesting and deserves a few words. Originally, superconductivity was believed to be the limiting case of zero resistance. While it is true that a superconductor has zero resistance, this is not sufficient since this assumption only implies that the magnetic field within a super-conductor must remain a constant, whereas, in fact, the Meissner effect clearly shows that the magnetic field must be zero. This situation was clarified and the relationship between current density and magnetic field derived by F. and H. London [7]. We shall briefly outline the major points in the development of the macroscopic theory of superconductors.

It is first necessary to replace Ohm's law by an expression relating current to electric field, incorporating the fact that the resistance must be

---

*Otherwise infinite current density would result, an impossible situation.

zero. In Sec. 4.11 we developed a simplified version of Ohm's law by considering the electrons to be accelerated by the electric field and decelerated by collisions with the lattice (atoms) of the bulk material. It is these collisions and the resulting deceleration which give rise to resistance, so that in the absence of resistance, the electrons experience no retarding force. If we assume that a superconductor is composed of a density of free electrons n_{sc} which are accelerated only by an electric field, then these electrons are governed by Eq. (4-123)

$$a = \frac{d\mathit{v}}{dt} = \frac{-eE}{m} \qquad (6\text{-}2)$$

The current density is the time rate of flow of the total charge density or it equals the total charge density times the velocity of the charge. Thus

$$J_{sc} = -en_{sc}\mathit{v}$$

or

$$\frac{dJ_{sc}}{dt} = -en_{sc}\frac{d\mathit{v}}{dt} \qquad (6\text{-}3)$$

Substituting the expression for acceleration from Eq. (6-2)

$$\frac{dJ_{sc}}{dt} = \frac{e^2 n_{sc}}{m}E = \frac{E}{\mu\lambda} \qquad (6\text{-}4)$$

This is one of the fundamental London equations describing the relationship between the current density and electric field and can be considered as an alternative to Ohm's law. It obviously describes the situation in a superconductor at dc since there is no time rate of change of current and hence no E field. This equation also states that when the current is changing with time, there is an E field in the direction of J. This E field will tend to accelerate the normal electrons as well as the superconducting electrons, and since normal electrons will experience lattice collisions, there will be normal conduction losses associated with ac excitation of superconductors. The equivalent circuit for a superconductor is shown schematically in Fig. 6-6. The superelectrons can give rise to only a reactive impedance while the normal electrons have reactance and resistance, the latter being the losses.

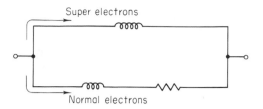

Super electrons

Normal electrons

Fig. 6-6. Equivalent circuit for the impedance of an incremental section of a superconductor.

These losses are generally quite small but become important at very high frequencies, imposing a bandwidth limitation on superconducting transmission lines, as we will see in Sec. 6.5.

Equation (6-4) is not in itself sufficient, as it generally describes a material of zero resistance and does not include the Meissner effect. The second London equation, which includes the Meissner effect, relates current density to the magnetic field and can easily be obtained with the aid of Maxwell's curl equation

$$\nabla \times \mathbf{E} = -\frac{\partial \mathbf{B}}{\partial t} = -\mu \frac{\partial \mathbf{H}}{\partial t} \tag{6-5}$$

where μ is the permeability of the medium, assumed to be a constant. Taking the curl of Eq. (6-4)

$$\nabla \times \frac{\partial \mathbf{J}_{sc}}{\partial t} = \frac{e^2 n_{sc}}{m} \nabla \times \mathbf{E} \tag{6-6}$$

Substitution of Eq. (6-5) yields

$$\nabla \times \frac{\partial \mathbf{J}_{sc}}{\partial t} = \frac{e^2 n_{sc}}{m} \left(-\mu \frac{\partial \mathbf{H}}{\partial t} \right) \tag{6-7}$$

Time integration of Eq. (6-7) gives

$$\nabla \times \mathbf{J}_{sc} = -\frac{e^2 n_{sc}}{m} \mu (H - H_0) \tag{6-8}$$

H_0 is a constant of integration: For a material considered only to be the limiting case of zero resistance, H_0 can have any value, while for a super-conductor, the Meissner effect specifies that $H_0 = 0$; thus Eq. (6-8) becomes

$$\nabla \times J_{sc} = -\frac{e^2 n_{sc} \mu}{m} H \qquad (6\text{-}9)$$

This is the second London equation, which will be used throughout this chapter along with Eq. (6-4) to describe the behavior of superconductors.

The phenomenon of penetration of dc fields into a superconductor is very similar to the ac penetration of fields into a normal conductor. This can easily be seen when the expression for the penetration of current into a superconductor is derived; we will now do this in a manner similar to that of Sec. 4.9.

We assume an infinite half-plane, with a uniform magnetic field at the surface. The one-dimensional differential equation describing this situation can be obtained by taking the curl of Eq. (6-9)

$$\nabla \times \nabla \times J = \nabla(\nabla \cdot J) - \nabla^2 J = \frac{-e^2 n_{sc} \mu}{m} \nabla \times H \qquad (6\text{-}10)$$

But

$$\nabla \cdot J = 0 \qquad (6\text{-}11)$$

and from Maxwell's equation, if displacement current and normal conduction current are neglected

$$\nabla \times H = J_{sc} \qquad (6\text{-}12)$$

Thus Eq. (6-9) becomes

$$\nabla^2 J_{sc} = \frac{e^2 n_{sc} \mu}{m} J_{sc} = \frac{J_{sc}}{\lambda^2} \qquad (6\text{-}13)$$

where

$$\lambda^2 = \frac{m}{e^2 n_{sc} \mu} \qquad (6\text{-}14)$$

It is obvious that the most general solution to Eq. (6-14) is

$$J_{sc} = J_0 \, e^{-z/\lambda} + J_1 \, e^{z/\lambda} \tag{6-15}$$

Since the conductor is assumed to be infinitely thick, it is apparent that J must decrease with z or the positive exponential term must be zero. Thus

$$J_{sc} = J_0 \, e^{-z/\lambda} \tag{6-16}$$

where J_0 is the current density at the surface $z = 0$. This is identical in form to Eq. (4-115a) with λ being equivalent to δ (the normal penetration or skin depth). Thus λ can be considered to be the superconducting penetration depth. This concept is very important and should be thoroughly understood since it will be used extensively in the remaining sections to analyze and describe the surface impedance and transmission-line characteristics of superconductors.

The superconducting penetration depth λ has received considerable investigation and can, in fact, be measured, as we will see in Sec. 6.5. This parameter has only a very slight, and therefore usually negligible, dependance on an applied magnetic field. However, it depends rather heavily on temperature, as shown in Fig. 6-7, where

$$\lambda = \frac{\lambda_0}{[1 - (T/T_c)^4]^{\frac{1}{2}}}$$

At absolute zero, λ_0 has been measured to be about 5×10^{-8} meters (500 Å) for a number of materials. We will make use of this temperature variation of λ in Sec. 6.5 to show the temperature variation of the transmission characteristics of a superconducting line.

A summary of some of the important parameters of the more common superconductors is given in Table 6-1. This list is by no means exhaustive, and the reader is referred to references for more complete listings.

(f) Coherence length and relationship to type-I and type-II superconductors. While it is beyond the scope of this book to engage in any discussion of the fundamental theory of superconductivity, it is necessary to at least be aware of the meaning and significance of the so-called "coherence length" of a superconductor. The concept of coherence length arises from the Bardeen-Cooper-Schrieffer (BCS) theory, which shows that the superconducting state is characterized by the formation of electron pairs which are coupled over relatively long distances, the coherence length ξ.

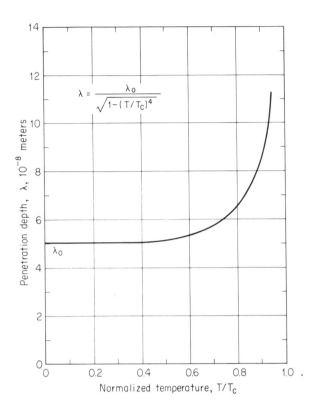

$$\lambda = \frac{\lambda_0}{\sqrt{1-(T/T_c)^4}}$$

Fig. 6-7. Penetration depth vs. temperature for tin (also approximately for A1, Hg, In, and Pb).

This distance, which is on the order of 10^{-4} cm, is great compared to the electron mean free path in a normal metal (about 10^{-6} cm). This phenomenon is important to us here, since the relation that the coherence length ξ bears to the superconducting penetration depth λ is similar to the relation that the electron mean free path bears to the normal skin depth δ. To be more specific, we saw in Sec. 4.10 that when the electron mean free path became greater than the normal skin depth, the local theory (from which Ohm's law is derived) breaks down and the phenomenon of anomalous skin effect appears. In a superconductor, the London Eqs. (6-4) and (6-9), which are based on a local theory, are valid when the coherence length is much smaller than the penetration depth. As with a normal metal, when $\xi > \lambda$, the London theory breaks down and the field relations are then expressed reasonably well by the phenomenological, nonlocal theory of Pippard or the Ginzburg-Landau theory. The latter type of superconductors are known as type-I while the former are known as type-II or London superconductors.

(g) Local (London) versus Nonlocal (Pippard) Theory. The London theory and resulting equations are strictly true only for type-II superconductors, which are the so-called "hard" materials. For type-II materials, the coherence length is smaller than the penetration depth, so that current density at a point may be described by a constant, local field. To be more specific, we can specify that the current density is directly proportional to the magnetic vector potential at that point. This is easily seen from Eq. (6-9) if $\nabla \times \mathbf{A} = \mathbf{B}$ is used

$$\nabla \times \mathbf{J}_{sc} = -\frac{1}{\mu\lambda^2} \mathbf{B} = -\frac{1}{\mu\lambda^2} \nabla \times \mathbf{A}$$

so that

$$\mathbf{J}_{sc} = -\frac{1}{\mu\lambda^2} \mathbf{A} \quad \text{for} \quad \xi \ll \lambda \tag{6-17}$$

When the coherence length becomes larger than the penetration depth, the above point relationship between $\mathbf{J}_{sc}$ and $\mathbf{A}$ is no longer strictly true. For such a case, it is necessary to use the Pippard nonlocal theory, which relates $\mathbf{J}_{sc}$ to a weighted average of all vector potentials in some neighborhood of each point. This is completely analogous to the type of nonlocal theory used to describe anomalous skin effect in Sec. 4.11. We will not engage in a discussion of the nonlocal theory of superconductors but will rather be content to indicate that Eq. (6-17) must be replaced by an integral of $\mathbf{A}$ over space which involves the coherence length (see [15]). It is obvious that the nonlocal theory complicates the mathematical analysis of superconductors. The London equations are much simpler.

The question which requires consideration is which type of superconductor is to be used in a transmission line and what mathematical analysis can be used. Both types can be and are used (Sec. 6.7) and, furthermore, for most of the more common superconductors, e.g., tin, lead, ξ is larger than λ (see Table 6-1), which indicates that these are type-I and are to be described by the Pippard nonlocal theory. However, the London theory can be used for type-I superconductors with very little error. For instance, Pippard has calculated the penetration of a magnetic field into a type-I material for $\xi \approx 23\lambda$ according to the local and nonlocal theories (see Fig. 6-8). This represents a relatively extreme case, since ξ is not usually quite so much larger than λ. Thus the error between the two theories will be reasonably small for many cases. Furthermore, it has been shown that lead and indium

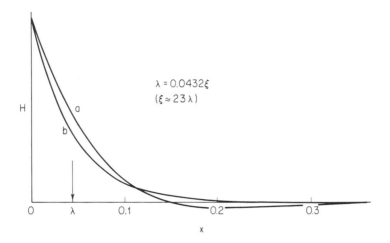

Fig. 6-8. Variation of magnetic field with depth, calculated from (a) Pippard, type-I (nonlocal) theory and (b) London, type-II (local) theory. (*After Pippard* [15, p. 560].)

(type-I) behave approximately as a London superconductor and tin lies in between [6]. Thus we shall use the London local theory throughout this chapter.

(h) Ac losses in type-I and type-II superconductors. Let us return briefly to the losses associated with ac excitation of superconductors. In a type-II (London) superconductor when there are time-changing currents, Eq. (6-4) specifies that an electric field is also present. This electric field can then accelerate the normal conduction electrons and therefore normal losses will be encountered for ac conditions in a superconductor. These losses will be small for all but the very high frequencies (microwave), since the electric field induced by a time-changing current is small in comparison to usual electric fields encountered in normal conductors.

As an example, let us compare the electric field generated by a time-changing current in a superconductor such as tin with the field of a steady current of the same amplitude in the same metal but in the normal state, above the critical temperature.

For tin at low temperature, well below T_c

$$\lambda \approx 500 \text{ Å} = 0.5 \times 10^{-7} \text{ meter}$$

Assume a current density of 1 A/cm^2 = 10^4 A/m^2 and frequency of 10^6 Hz

$$J_{sc} = 10^4 \, e^{j\omega t} \text{ A/m}^2$$

From London's equation

$$E = \lambda^2 \mu_0 \frac{dJ}{dt} = \mu_0 \lambda^2 j\omega J_{sc}$$

Substituting in the above values

$$|E| = 0.6 \times 10^{-9} \text{ V/m}$$

In contract to this, tin at $20°K$ carrying a current of 10^4 A/m^2 would require an electric field of $E = \rho J$, where $\rho = 0.114 \times 10^{-8}$ ohm-meter (see Table 4-1). Thus $|E| = 0.114 \times 10^{-4}$ V/m. It is easily seen that the above field in tin in the normal state, which is required to move the electrons through this lossy medium, is more than four orders of magnitude larger than the field generated by the same current at 10^6 Hz in the superconducting state. Thus, the E-field generated in a superconductor is quite small and requires the frequency to be very high in the microwave region (above 10^9 Hz) in order to become large enough to influence the normal electrons within the superconductor. We shall return to these losses in Sec. 6.5 to investigate their influence and limitations imposed on transmission-line characteristics.

Critical State Model

The above analysis can usually be applied to both type-I and type-II superconductors, provided that the applied current and/or the magnetic field does not drive the material out of the superconducting state for type-I or into the mixed state for type-II. If a type-II superconductor is driven into the mixed state above H_{c1} (see Fig. 6-5), then significant losses will occur even for low frequencies, for example, 60 Hz. Such circumstances are often encountered where superconductors are required to carry large amounts of power, and these low-frequency ac losses must be considered. For such a condition, the losses can be estimated by the so-called "critical state model" (see [4]), which is a reasonable approximation for relatively low frequencies. The critical state model assumes that time-changing fields induce persistent currents up to a limiting or critical current density J_c. The currents are induced by the usual time-changing flux (Lenz' law) so as to minimize the change in total flux linking the material. Initially, the critical current density J_c is induced only in portions of the superconductor, for example, near the surface of a wire. The current density at that point can no longer increase, so that further increases in field or total current causes penetration farther into the conductor until eventually the entire wire has a uniform current density. The electric field and current density are assumed to be related as in Fig. 6-9: The electric field increases at a fixed value of current density J_c as the current penetrates farther into the material. In reality, the slope is finite, not infinite as shown. However, it is sufficiently steep that this approximation is reasonable for many cases. This relationship plus Maxwell's equations are sufficient for purposes of calculating the current and field distributions as well as the power losses. We will not pursue this model but will instead summarize a few results.

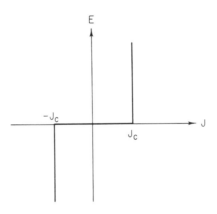

Fig. 6-9. Electric field vs. current density for critical state model: E and J are parallel in the superconductor.

For a plane conductor of thickness a, fields parallel to the surface, and a large applied magnetic field

$$H = H_0 \sin \omega t, \qquad H_0 \gg aJ_c\left(\frac{4\pi}{10}\right)$$

$$\frac{\text{loss}}{\text{cycle vol}} \approx \frac{aJ_c H_0 \times 10^{-7}}{5} \quad \frac{\text{joules}}{\text{cycle cm}^3}$$

where J_c is in A/cm^2 and H_0 is in oersteds.

For a current carrying wire of radius r, carrying a current $I = I_0 \sin \omega t$, the loss is

$$\frac{\text{loss}}{\text{cycle unit length}} = \frac{2 \times 10^{-9} I_0^{\,3}}{3\pi r^2 J_c} \frac{\text{joules}}{\text{cycle cm}} \quad \text{for} \quad I_0 \ll \pi r^2 J_c$$

and

$$2 \times 10^{-9} I_0^{\,2} \frac{\text{joules}}{\text{cycle cm}} \quad \text{for} \quad I_0 = \pi r^2 J_c$$

It is apparent that the power loss varies linearly with frequency. These calculations are good only for power frequencies.

6.3 SURFACE IMPEDANCE OF AN INFINITE HALF-PLANE SUPERCONDUCTOR

The ultimate goal is to determine the propagation properties of a superconducting strip line, and this can be achieved in several ways. The exact general solution can be obtained in a manner identical to that used in Sec. 4.5 with the addition of the London equations to Maxwell's equations and the addition of supercurrent to normal current. However, this boundary-value problem can be greatly simplified by calculating only the surface impedance of an infinitely wide plane superconductor and using this in the usual lumped-parameter transmission-line expressions to find the propagation constant and characteristic impedance.* Both approaches to transmission-line behavior must ultimately give the identical result, as was shown in Secs. 4.5 and 4.6, so that we shall choose the latter. Thus, it is necessary first to calculate the surface (series) impedance of a superconductor, which is the subject of this section. These results will be used in Sec. 6.5.

In the calculation of the series resistance and inductance parameters of a superconducting line, it is possible to proceed in two different ways: we can use either the lumped-circuit or the field-theory approach. The two approaches must, of course, ultimately give equivalent results, but the path by which they are reached can be quite different. The classical engineering approach is to calculate the inductance in terms of the flux linkages per unit current, as in Sec. 4.2, a method usually learned in elementary circuits

---

*If an exact analysis were to be performed on a strip line, we would run into exactly the same problems as in Sec. 4.5, namely, the propagation constant is determined by a transcendental equation which can only be solved approximately. By making certain assumptions which are nearly always valid, the problem is greatly simplified and simple, closed-form expressions are possible.

courses.   However, because of the rather unusual properties of superconductors, this approach cannot be used. This is because there is an electric
field in the direction of propagation, even when there are no losses present.
The flux-linkage-per-unit-current concept can be used only when the line
propagates a transverse electromagnetic, i.e., TEM, wave. Since the lowest-
order mode that can be obtained on a superconducting line is a TE or TM
wave with a nonnegligible axial field, a more complex calculation is involved.
Meyers [10] has shown that the inductance can be obtained for frequencies
up to about 10^9 Hz simply from consideration of the static magnetic field
energy, provided that the kinetic energy of the superelectrons is included.
The latter is necessary because of the penetration effects. We will not concern ourselves with this lumped-circuit type of calculation, but rather will
approach the problem from a field solution point of view.

In order to determine the surface impedance, we will make use of the
London equations (6-4) and (6-9) in addition to Maxwell's field equations
in order to obtain the necessary differential equation. The differential equation describing the problem can be obtained in terms of the current density
or magnetic (or electric) field. The two approaches must, of course, be
equivalent, but we shall use the latter.

In order to derive the necessary differential equations, we will make
use of the so-called London "two-fluid" model of a superconductor. This
model simply states that the total current density in a superconductor is
the sum of the normal current density J_n, plus the supercurrent J_{sc} (see Fig.
6-6). In other words, whenever we apply Maxwell's equations to a superconductor, it is necessary to replace the current density by

$$\mathbf{J} = \mathbf{J}_n + \mathbf{J}_{sc} \tag{6-18}$$

J_n is still determined by Ohm's law

$$\mathbf{J}_n = \sigma \mathbf{E} \tag{6-19}$$

while J_{sc} is governed by Eqs. (6-4) and (6-9)

$$\frac{\partial \mathbf{J}_{sc}}{\partial t} = \frac{\mathbf{E}}{\mu \lambda^2} \tag{6-20}$$

and

$$\nabla \times \mathbf{J}_{sc} = -\frac{\mathbf{H}}{\lambda} \tag{6-21}$$

In order to derive the surface impedance, we will proceed as follows. Using Maxwell's two curl equations for E and H in conjunction with the above expressions for the super and normal current density, it is possible to obtain a differential equation for the penetration of magnetic field into the conductor. Once this differential equation is obtained, the general solution is obvious; this results in the propagation constant relating the propagation of H into the superconductor. In other words, the expression for the penetration depth is obtained. Once this is done, the analysis becomes very similar to that of Sec. 4.7 for a normal metal.

The surface impedance per unit length of conductor for a specified width W is just the surface voltage drop per unit length divided by the total current I contained in the width of conductor W. Since the voltage drop per unit length is just the electric field at the surface, and the total current is just the width times the total current density in the strip, the surface impedance per unit length is

$$Z_{\text{surf}} = \frac{V/\ell}{I} = \frac{E_x\big|_{z=0}}{WJ} \qquad (6\text{-}22)$$

The total surface impedance of a truly infinite half-plane is zero since the impedance of a conductor varies inversely with the width. In other words, if W in Fig. 6-10 goes to infinity and the current density remains constant over this width, then the impedance approaches zero. Thus, we shall be interested in only a section of the infinite half-plane, namely, the section of width W. We will assume that the return path is infinitely far away so that the fields are uniform throughout the dielectric, that the magnetic field H has only a y component which is uniform and of constant amplitude everywhere external to the conductor, and varies as a function of z only within the conductor as shown. We also assume sinusoidal variation with time.

The differential equation for magnetic field can be obtained from Maxwell's curl equations, using Eq. (6-18) for total current

$$\nabla \times H = J_{\text{sc}} + J_n + j\omega\epsilon E \qquad (6\text{-}23)$$

$$\nabla \times E = -\frac{\partial B}{\partial t} = -j\omega\mu H \qquad (6\text{-}24)$$

where it is assumed that

$$B = \mu H \qquad (6\text{-}25)$$

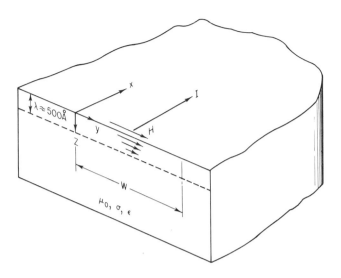

Fig. 6-10. Infinite half-plane superconductor showing H field penetration to a depth λ.

Taking the curl of both sides of Eq. (6-23)

$$\nabla \times \nabla \times \mathbf{H} = \nabla(\nabla \cdot \mathbf{H}) - \nabla^2 \mathbf{H} \tag{6-26}$$

$$= \nabla \times \mathbf{J}_{sc} + \nabla \times \mathbf{J}_n + j\omega\epsilon \nabla \times \mathbf{E} \tag{6-27}$$

Substituting Eqs. (6-21) and (6-24) into Eq. (6-27) for the appropriate terms, and using the well-known fact that

$$\nabla \cdot B = 0 \tag{6-28}$$

gives

$$\nabla^2 \mathbf{H} = \left(\frac{1}{\lambda^2} + j\omega\mu\sigma - \omega^2\mu\epsilon \right) \mathbf{H} \tag{6-29}$$

There is only a y component of H and it will propagate in the x and z direction so

$$\frac{\partial^2 H_y}{\partial x^2} + \frac{\partial^2 H_y}{\partial z^2} = \left(\frac{1}{\lambda^2} + j\omega\mu\sigma - \omega^2\mu\epsilon \right) H \tag{6-30}$$

We assume a solution of the form

$$H_y = H_0 \, e^{\pm \gamma x} \, e^{\pm \Gamma z} \tag{6-31}$$

which upon substitution gives

$$\gamma^2 + \Gamma^2 = \frac{1}{\lambda^2} + j\omega\mu\sigma - \omega^2\mu\epsilon \qquad (6\text{-}32)$$

The displacement current in the conductor is represented by the $\omega^2\mu\epsilon$ term in the above equation. If it is to be neglected, then it is necessary that

$$\frac{1}{\lambda^2} \gg \omega^2\mu\epsilon \qquad (6\text{-}33)$$

Furthermore, for nearly all cases of practical interest, it will be found that $\gamma^2 \ll \omega\mu\sigma$ and also that $\gamma^2 \ll 1/\lambda^2$, so that γ^2 can also be neglected. We will assumed this to be true and will check the region of validity in Sec. 6.5. Using the above two assumptions, Eq. (6-32) becomes

$$\Gamma^2 = \frac{1}{\lambda^2} + j\omega\mu\sigma \qquad (6\text{-}34)$$

The time-dependence has not been included explicitly but is assumed to be present. As before, it is apparent that if Γ is a positive quantity, then only the negative sign of the exponent is permissible, since the magnetic field must decrease with z, whereas the positive sign would given an ever-increasing H_y with z, an impossible situation. Thus, it is obvious that

$$\mathbf{H}_y(z) = H_0 \exp\left[-\left(\frac{1}{\lambda^2} + j\omega\mu\sigma\right)^{1/2} z\right] \qquad (6\text{-}35)$$

H_0 is the amplitude of the field at the conductor surface and is unknown. Its evaluation is not necessary since it does not enter into the expression for the surface impedance as we would expect. The $e^{-\gamma x}$ variation is omitted in the above equation since γ is much smaller than Γ and we are interested in impedance per unit length, i.e., variation of H_y in the x direction is negligible over a small unit length compared to variation in the z direction.

Since we are interested only in the impedance of the section of the plane of width W, it is necessary to determine the total current in that section from the surface $z = 0$ to infinite depth. This is easily done with Ampere's law

$$\oint \mathbf{H} \cdot d\ell = I$$

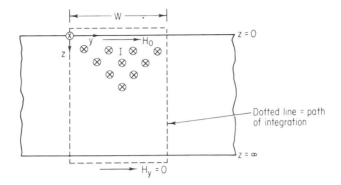

Fig. 6-11. Superconducting half-plane showing path of integration.

The path of integration to be used is as shown in Fig. 6-11. We can break this path into four separate paths as shown, namely, the top and bottom surface and the two sides. We know that there is no z component of $\mathbf{H}$ anywhere, and so the line integral for the vertical sides must be zero. The integral then becomes

$$\int_0^W H_y(z = 0) \cdot dy + \int_W^0 H_y(z = \infty) \cdot dy = I \tag{6-36}$$

We know that H_y at infinite depth must be zero, so that only the first integral term gives a contribution. From Eq. (6-35) H_y evaluated at $z = 0$ is a constant of value H_0; thus

$$I = H_0 W \tag{6-37}$$

An expression for E_x can easily be obtained from Eq. (6-23) with substitution of Eqs. (6-19) and (6-20). $\mathbf{H}$ has only a y component and varies only in the z direction, so that there can be only an x component to the curl in the conductor. Thus

$$\nabla \times \mathbf{H} = -\frac{\partial \mathbf{H}_y}{\partial z} = \left(\sigma + \frac{1}{j\omega\mu\lambda^2} \right) \mathbf{E}_x \tag{6-38}$$

The derivative of H_y with respect to z can be obtained from Eq. (6-35). If this is found and substituted into Eq. (6-38), the result for the field at the surface is

$$E_x\Big|_{z=0} = \frac{[(1/\lambda^2) + j\omega\sigma]^{1/2}\, H_0}{\sigma + (1/j\omega\mu\lambda^2)} = \frac{j\omega\mu\, H_0}{[(1/\lambda^2) + j\omega\mu\sigma]^{1/2}} \tag{6-39}$$

Now the surface impedance is easily obtained from Eqs. (6-37) and (6-39). After simplification, this becomes

$$Z_{surf} = \frac{E_x}{I} = \frac{1}{W}\, \frac{j\omega\mu}{[(1/\lambda^2) + j\omega\mu\sigma]^{1/2}} \tag{6-40}$$

The above equation is a very basic and general expression, valid for both high and low frequencies, subject to certain assumptions. It includes both normal skin effect and superconducting penetration depth. The only assumptions implicit in the above are the following: $\gamma^2 \ll 1/\lambda^2$, $\gamma^2 \ll j\omega\mu\sigma$, the displacement current is negligibly small, the conductor is very thick, and anomalous skin effect (Sec. 4.10) is not present.

Equation (6-40) can be put into another form which gives further insight into its significance. Recall that the normal skin depth is given by

$$\delta^2 = \frac{2}{\omega\mu\sigma} \tag{6-41}$$

Substituting this into Eq. (6-40)

$$Z_{surf} = \frac{1}{W}\, \frac{j\omega\mu}{[(1/\lambda^2) + (2j/\delta^2)]^{1/2}} \tag{6-42}$$

This equation now shows more physically the roles of λ and δ in the impedance of a superconductor. λ is independent of frequency (but dependent on other things such as temperature) while δ varies with the reciprocal of the square root of the frequency.

As long as the superconducting penetration depth is very small compared to δ, it will be the factor limiting the penetration of H_y into the conductor and thus will dominate the surface impedance. This can be expressed more exactly by letting δ become very large in Eq. (6-42) (for example, very low frequency) which gives

$$Z_{surf}\Big|_{\delta=\infty} = \frac{j\omega\mu\lambda}{W} \tag{6-43}$$

It is seen from the above equation that for this limiting case, the super-conductor looks like a pure inductance and dissipates no energy.

If, on the other hand, λ becomes very large compared to δ,* then the latter becomes the limiting factor and the surface impedance reduces to that of a normal conductor. This can be seen by letting λ become very large in Eq. (6-42) to get

$$Z_{surf} = \frac{\omega\mu\delta}{W}\left(\frac{j}{2}\right)^{1/2} = \frac{1}{W}\left(\frac{\pi f \mu}{\sigma}\right)^{1/2}(1 + j) \tag{6-44}$$

which is identical to Eq. (4-102).

Equations (6-43) and (6-44) give the surface impedance for two extreme cases, namely, normal conduction current negligible and normal conduction current predominating. We wish to consider a very important case in which the normal conduction current is small, but not completely negligible. This is anologous to the situation considered in Secs. 1.6 and 2.6 where losses are small but not negligible. In other words, we wish to consider the case where

$$\frac{\lambda^2}{\delta^2} \ll 1 \tag{6-45}$$

but not negligible. The problem then is to expand the radical in the de-nominator of Eq. (6-42) into its real and imaginary parts. In order to do this, let us rewrite the radical as

$$\left(\frac{1}{\lambda^2} + \frac{2j}{\delta^2}\right)^{-1/2} = \lambda\left(1 + 2j\frac{\lambda^2}{\delta^2}\right)^{-1/2} \tag{6-46}$$

Now we can expand this in a series expansion of the form

$$(1 + \nu)^{-1/2} = 1 - \frac{1}{2}\nu + \frac{1}{2}\left(\frac{3}{4}\right)\nu^2 - \frac{1}{2}\left(\frac{3}{4}\right)\frac{5}{6}\nu^2 + \text{etc.} \tag{6-47}$$

with the series converging for all $\nu < 1$. Using this series simplifies Eq. (6-42) for surface impedance to

---

*λ/δ can become very large, for instance, by letting the temperature approach the critical temperature as shown in Fig. 6-7, or by letting the frequency become very high.

$$Z_{surf} = \frac{j\omega\mu}{W} \lambda \left[1 - \frac{1}{2} 2j \frac{\lambda^2}{\delta^2} + \frac{3}{8} \left(2j \frac{\lambda^2}{\delta^2} \right)^2 + \text{etc.} \right] \qquad (6\text{-}48)$$

In view of the restriction imposed by Eq. (6-45), and the ever-decreasing coefficients, only the first two terms of the expansion need be retained*

$$Z_{surf} \bigg|_{\lambda^2/\delta^2 < 1} = \frac{1}{W} \left(\frac{\omega^2 \mu^2 \lambda^3 \sigma}{2} + j\omega\mu\lambda \right) \qquad (6\text{-}49)$$

where use has been made of Eq. (6-41) for δ^2.

From Eq. (6-49) it is apparent that the normal conduction losses vary directly as the square of the frequency and directly proportional to the

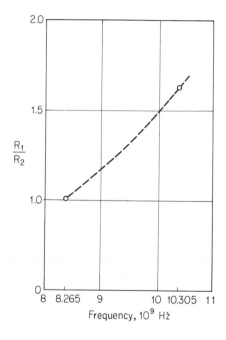

Fig. 6-12. Relative surface resistance vs. frequency for a niobium-lead line. (*After Shizume and Vaher.*)

*Of course, if λ^2/δ^2 is nearly equal to 1, additional terms would have to be included in the series expansion.

conductivity σ. This is in contrast to the normal skin-effect losses which vary as the square root of frequency and inversely with the square root of conductivity as given by Eq. (6-44).

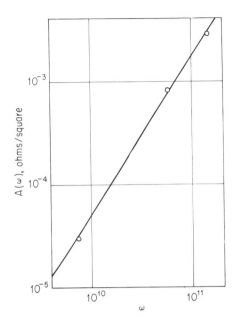

Fig. 6-13. Frequency variation of superconducting tin at low temperature showing $\omega^{1.5}$ dependence. (After Pippard [16, p. 42].)

In the low temperature region where normal conduction losses are small, as above, it is expected theoretically that at low frequencies the surface resistance should vary as ω^2. As the frequency is increased, the exponent of ω would decrease, eventually giving way to an $\omega^{2/3}$ law when extreme anomalous effects are present, as was shown in Sec. 4.12.* Experimental verification of these expectation is far from complete. Shizume and Vaher [18] have shown that a niobium-lead line exhibits a $\omega^{2.1}$ dependence roughly between 8 and 10 $\times$ 10^9 Hz (Fig. 6-12). Within experimental error, this appears to be an ω^2 dependence as predicted. Measurements on tin at some-

---

*When the frequency becomes sufficiently high, δ becomes much smaller than λ_0, so that the superconductor behaves as an ordinary metal. Thus, extreme anomalous effects are expected in superconductors much as they are in normal metals.

what higher frequencies have shown an $\omega^{1.5}$ dependence (Fig. 6-13).* The exponent appears to be decreasing but much more extensive measurements at both low and high frequencies are required before it can be concluded that theoretical and experimental agreement exists.

6.4 SURFACE IMPEDANCE OF A SUPERCONDUCTING PLANE OF ANY THICKNESS

In Sec. 6.3 we derived the expression for the surface impedance of a half-plane which was assumed to be infinitely thick. This single-plane conductor can be considered to be the limiting case of the impedance of a strip transmission line, as shown in Fig. 6-14. The two conductors are shown with finite width, finite separation, and finite thickness. As such, it would be necessary to do an exact analysis of the structure as in Sec. 4.5. However, this can be greatly simplified by letting the width and separation of the

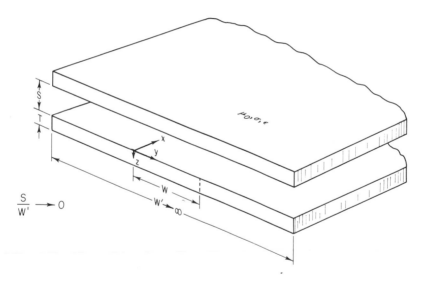

Fig. 6-14. Geometry for determining surface impedance of finite thickness superconductor.

*For temperatures well below T_c, Pippard [16] has shown that superconducting resistance can be represented by

$$R = A(\omega) \; \frac{t^4(1 - t^2)}{(1 - t^4)^2}$$

where $t = T/T_c$. $A(\omega)$ is independent of temperature, varying only with ω and having a different amplitude for different materials. Figure 6-13 depicts $A(\omega)$ versus frequency.

conductors as well as the ratio W/S approach infinity while the thickness T remains fixed. The problem then reduces to the plane conductor. Actually, it is not necessary to have the dimensions go to infinity; only the width of the conductors must be very large compared to the separation. Since the geometrical impedance of such a structure varies as width divided by separation (Sec. 8.4(a)), this impedance will still become 0. For boundary conditions required later, it is helpful to consider the two conductors as being some finite distance apart in order to be able to state that the magnetic field external to the structure is everywhere equal to zero. Thus, we will use the geometry of Fig. 6-14 with W' approaching infinity and $S/W' = 0$ where W' is the total width of the conductors. We will derive the surface impedance of a section of finite width W of the lower conductor in terms of the magnetic field in a manner anaolgous to that employed in Sec. 6.3. It is assumed once again that all fields vary sinusoidally with time, that there is only a y component of H, an x and z component of E, and an x component of J. These three vector quantities E, H, and J vary only with z and only within the conductor itself, i.e., they are assumed to be uniform within the dielectric.

The initial part of the analysis of this problem is identical to that given by Eqs. (6-18) through (6-30). We obtain a differential equation describing the penetration of the magnetic field into a conductor, and this equation must obviously be identical for an infinite or finite conductor, namely, Eq. (6-30). (It is suggested that the reader become familiar with the analysis for an infinite conductor before proceeding.) The point of departure in the analysis comes in the final expression describing H as a function of z as a result of the different boundary conditions. For the physical situation illustrated in Fig. 6-14, since the width-to-separation ratio is assumed to be large, then it is apparent that the magnetic field evaluated externally to the structure must be zero. Also, since the tangential components of magnetic field across any boundary must be continuous, then it is apparent that the magnetic field just at the lower surface of the bottom conductor must be zero, or, in other words

$$H_y \Big|_{z=T} = 0 \tag{6-50}$$

where T is the conductor thickness. This boundary condition must now be applied to the solution of Eq. (6-30). The general solution to this equation is

$$H_y = H_1 e^{\Gamma z} + H_2 e^{-\Gamma z} \tag{6-51}$$

where

$$\Gamma = \left(\frac{1}{\lambda^2} + j\omega\mu\sigma \right)^{\frac{1}{2}} \tag{6-52}$$

and is the propagation constant in the z direction from Eq. (6-34). H_1 and H_2 must be determined from boundary conditions. Applying the boundary condition of Eq. (6-50) to Eq. (6-51)

$$H_2 = -H_1 e^{2\Gamma T} \tag{6-53}$$

Thus Eq. (6-51) becomes

$$H_y = H_1 (e^{\Gamma z} - e^{2\Gamma T - \Gamma z}) \tag{6-54}$$

The second boundary condition is the same as that used in Sec. 6.3, namely, that at the surface of the conductor H is uniform and of value H_0 (assumed to be known)

$$H_y \Big|_{z=0} = H_0 = H_1 (1 - e^{2\Gamma T}) \tag{6-55}$$

This gives us an expression for H_1 in terms of H_0 and when this is substituted into Eq. (6-54) the final expression is

$$H_y(z) = \frac{H_0}{1 - e^{2\Gamma T}} (e^{\Gamma z} - e^{2\Gamma T - \Gamma z})$$

$$= \frac{H_0}{1 - e^{2\Gamma T}} (e^{\Gamma(z - T)} - e^{-\Gamma(z - T)}) \tag{6-56}$$

In order to obtain the surface impedance, we need to know E_x at the surface of the conductor. This can easily be obtained with the aid of Eq. (6-38)

$$\left(\sigma + \frac{1}{j\omega\mu\lambda^2} \right) E_x = -\frac{\partial H_y}{\partial z} = \frac{-H_0}{e^{-\Gamma T} - e^{\Gamma T}} (\Gamma e^{\Gamma(z - T)} + \Gamma e^{-\Gamma(z - T)}) \tag{6-57}$$

At $z = 0$, this reduces to

$$\left(\sigma + \frac{1}{j\omega\mu\lambda^2}\right)E_x = \Gamma H_0 \coth\Gamma T \tag{6-58}$$

To complete the evaluation of the surface impedance, we need an expression for the total current in the conductor section of width W. This can be obtained as in Sec. 6.3 by using Ampere's circuital law and performing the integration along the top and bottom surfaces of the conductor. Since the magnetic field is zero on the bottom side ($z = T$), then the integral must be identical to that previously obtained with the required current given by Eq. (6-37). Using this and Eqs. (6-52) and (6-58) yields

$$Z_{surf} = \frac{j\omega\mu}{W[(1/\lambda^2) + j\omega\mu\sigma]^{1/2}} \coth\left(\frac{1}{\lambda^2} + j\omega\mu\sigma\right)^{1/2} T \tag{6-59}$$

Equation (6-59) is the general expression for the surface impedance of a superconductor of thickness T including both normal skin-effect and superconducting penetration effects as expressed by the London equations. If anomalous skin effect is present, then Eq. (6-59) is no longer valid. If the skin depth becomes much smaller than λ, then Eq. (6-59) reduces to Eq. (4-113) for a normal conductor, as it should

Note that as the thickness T becomes very large, the coth term approaches unity and Eq. (6-59) becomes identical to Eq. (6-40), as it should.

If applied frequency is sufficiently low, then

$$j\omega\mu\sigma \ll \frac{1}{\lambda^2} \tag{6-60}$$

or, in other words, if the ordinary skin depth is much greater than λ

$$\frac{1}{\lambda^2} \gg \frac{2j}{\delta^2} \tag{6-61}$$

Equation (6-59) then simplifies to

$$Z_{surf} = \frac{j\omega\mu\lambda}{W} \coth\left(\frac{T}{\lambda}\right) \tag{6-62}$$

It should be understood in the above that δ can be very large, even though the conductor thickness T itself is small: δ is determined by the electrical parameters ω, μ, and σ and not by geometrical parameters. Thus, Eq. (6-62) can be valid for any thickness $T > \lambda$, as long as the frequency is sufficiently low and as long as the temperature is sufficiently below the critical temperature so that λ is small (Fig. 6-7).

One shortcoming of the surface-impedance calculations in this and the previous section is the complete neglect of anomalous skin effect. If anomalous effects are present to a significant extent in the superconducting state, then a combination of the Reuter-Sondheimer [17] theory with London's equations is necessary, the former relating normal current to electric field in general (anomalous or not) and the latter characterizing the superconductor. Such an analysis with experimental measurements at 24×10^9 Hz is presented by Maxwell, Marcus, and Slater [9], but we shall not consider this problem.

Perhaps the most complete collection of surface impedance measurements of superconductors is given by Bardeen [3], but there are still many discrepancies left unresolved, and reliable engineering information is rather difficult to obtain. One further criticism of the preceding analyses is that there is still some doubt as to the validity of the London two-fluid model ([16, p. 43]), but since a better theory has not been advanced, the above serves as a good engineering approximation.

6.5 ANALYSIS OF A SUPERCONDUCTING TRANSMISSION LINE

In the analysis of a transmission line, it is often expedient to assume that the line is ideal, having no losses and therefore no attenuation or dispersion and a characteristic impedance which is purely resistive. We have used such ideal lines in presenting a number of cases of interest. We have also considered, particularly in Secs. 1.6, 2.6 and 5.12, the effect of non-ideal characteristics on transmission-line behavior. For most ordinary transmission lines of relatively short length, and with a low applied frequency, the losses are often negligible, although as the length increases, the total attenuation of the line becomes significant. We saw in Sec. 5.12(d) that skin effect is very important in high-speed pulse propagation.

In Secs. 6.3 and 6.4, the surface impedance of a superconductor was seen to be reactive and therefore lossless for reasonably high frequencies (up to about 1,000 megahertz). Thus, we might expect superconducting transmission lines to behave nearly as ideal lines up to this frequency, even for very long lines. This is, in fact, true, and we will derive the expression for the propagation constant of a superconducting transmission line and show the range of validity as well as the effect of small losses when the frequency becomes sufficiently high.

We wish to consider the case of a strip transmission line of width W and separation S with W/S very large, as shown in Fig. 6-15. The permeability of the conductors is taken as that of free space, namely, μ_0.

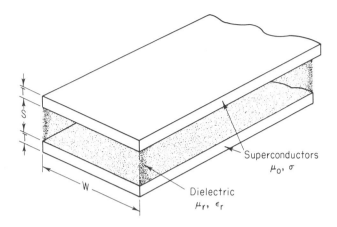

Fig. 6-15. Superconducting strip line.

From Secs. 1.8 and 2.5, we know that for an applied ac, the propagation constant of any transmission line is equal to the square root of the product of the total series impedance per unit length and the total shunt admittance per unit length

$$\gamma = (Z_{\text{series}} Y_{\text{shunt}})^{1/2} \tag{6-63}$$

For the strip line of Fig. 6-15, we know that for the most general case the series impedance is composed of the surface impedance of the conductors plus the geometrical inductance due to magnetic flux between the strip lines. Thus, for the case of different parameters for conductor 1 and 2

$$Z_{\text{series}} = Z_{\text{surf}_1} + Z_{\text{surf}_2} + j\omega L \tag{6-64}$$

We will assume that the two conductors have identical thickness and material, so that T, λ, μ, σ, and H_c are the same. Thus the series impedance becomes

$$Z_{\text{series}} = 2 Z_{\text{surf}} + j\omega L \tag{6-65}$$

(All impedances are per unit length, as always.)

For a line with $W/S \gg 1$, L is given by Eq. (8-20). Using this and Eq. (6-59)

$$Z_{series} = \frac{2\,j\omega\mu_0}{W\Gamma}\, \coth\Gamma T + j\omega\mu_0\,\mu_r\,\frac{S}{W} \qquad (6\text{-}66)$$

where μ_r is the relative permeability of the insulator.

Let us sidetrack for a moment to consider an interesting case where the frequency is sufficiently low so that the normal skin depth is large compared to λ: Then Eq. (6-66) becomes

$$Z_{series} = \frac{j\omega\mu_0}{W}\left[2\lambda\,\coth\left(\frac{T}{\lambda}\right) + \mu_r S\right] \qquad (6\text{-}67)$$

If we further assume that the conductor is thick compared to the super-conducting penetration depth (usually but not generally valid), then Eq. (6-67) becomes

$$Z_{series} = \frac{j\omega\mu_0}{W}\,(2\lambda + \mu_r S) \qquad (6\text{-}68)$$

This impedance is a pure inductance, as can be seen by rewriting it

$$Z_{series} = j\omega L = j\omega\,\frac{\mu_0 S}{W}\left(\mu_r + \frac{2\lambda}{S}\right) \qquad (6\text{-}69)$$

The effect of the superconductor surface impedance on the total inductance of the line is to effectively add an additional insulator of relative permeability $\mu_r = 1$ and of a thickness λ to each side (top and bottom) of the actual insulator. In other words, the superconductor makes the situation appear as if the conductors were separated by an additional distance 2λ with an air (or free space) dielectric, a rather interesting result.

Now let us return once again to determine the propagation contant. The shunt admittance is

$$Y_{shunt} = G + j\omega C \qquad (6\text{-}70)$$

G, which represents the dielectric losses, will be assumed to be negligible.

Thus the propagation constant for this line becomes

$$\gamma = \left(2j\omega C Z_{surf} - \omega^2 LC\right)^{\frac{1}{2}} \tag{6-71}$$

From Sec. 8.4(a) we know that L and C for $W/S \gg 1$ are given by

$$L = \mu_0 \mu_r \frac{S}{W} \qquad C = \epsilon_0 \epsilon_r \frac{W}{S} \tag{6-72}$$

per unit length. Substituting these and Eq. (6-59) for the surface impedance results in

$$\gamma = j\omega \sqrt{\mu_0 \epsilon_0} \left(\frac{2\epsilon_r \coth \Gamma T}{\Gamma S} + \mu_r \epsilon_r\right)^{\frac{1}{2}} \tag{6-73}$$

where

$$\Gamma = \left(\frac{1}{\lambda^2} + \frac{2j}{\delta^2}\right)^{\frac{1}{2}} = \left(\frac{1}{\lambda^2} + j\omega\mu_0\sigma\right)^{\frac{1}{2}} \tag{6-74}$$

Equation (6-73) represents the general case of a superconducting strip line of identical strips of thickness T and separation S, the insulator having relative permeability and dielectric constants μ_r and ϵ_r, respectively. The only assumptions made were, first, that the width-to-separation ratio W/S must be large so that the fields are uniform, and, second, that the normal conduction electrons obey the classical skin-effect equations while the super electrons obey London's equations. For the general case, the propagation constant is very complicated and the real and imaginary parts are not given by simple expressions. However, reasonable simplifying assumptions are often possible which give relatively simple expressions to serve as good engineering approximations. We will now undertake this and, in particular, we will consider the propagation constant as a function of various parameters and whenever possible will derive simplified expressions which are valid for various ranges of parameters.

If we consider the case when conduction current is completely negligible, then a simple expression for γ and the velocity of propagation can be obtained. In other words, if we assume

$$\lambda \ll \delta \tag{6-75}*$$

---

*This inequality is independent of thickness T of the superconductor and merely states that the frequency and temperatures must be sufficiently low.

then $2j/\delta^2$ can be neglected in Eq. (6-74) and the propagation constant becomes

$$\gamma = j\omega \sqrt{\mu_0 \epsilon_0} \left[\frac{2\lambda}{S} \epsilon_r \coth\left(\frac{T}{\lambda}\right) + \mu_r \epsilon_r \right]^{1/2} = j\beta \qquad (6\text{-}76)$$

It is apparent that the term under the radical is real, so that γ is purely imaginary, indicating no attenuation for frequencies for which Eq. (6-75) is valid. Also, none of the parameters in Eq. (6-76) are frequency-dependent and since the phase velocity is

$$\upsilon = \frac{\omega}{\beta} = \frac{c}{[(2\lambda/S)\,\epsilon_r \coth(T/\lambda) + \mu_r \epsilon_r]^{1/2}} \qquad (6\text{-}77)$$

where $c = 1/\sqrt{\mu_0 \epsilon_0}$, there is no dispersion for this case (normal conduction current negligible).

There is an E field in the direction of propagation given by Eq. (6-20) even though there are no losses or dispersion. In a normal conductor, such an axial field can be present only when there are losses (voltage drop) along the line, but such an axial field must exist in a superconductor even though no losses are present.

Equation (6-77) gives the phase velocity for the general case of a lossless superconducting strip line with both conductors having the same thickness and other identical properties, and assuming that the relative permeability of the conductors is unity. The dielectric insulator has relative permeability and dielectric constant of μ_r and ϵ_r, respectively. Most dielectrics have $\mu_r = 1$, so that if we assume this, Eq. (6-77) can be put into a simpler form

$$\upsilon = \frac{\upsilon_0}{[(2\lambda/S)\coth(T/\lambda) + 1]^{1/2}} \qquad (6\text{-}78)$$

where $\upsilon_0 = c/\sqrt{\epsilon_r}$ is velocity of light in the bulk dielectric.

Equation (6-78) relates the effect of the superconductor on the propagation velocity and can be used to investigate penetration depth. It can be seen that for a given value of λ, a traveling wave on such a line will be slowed down more and more as the dielectric and conductor thickness S and T are made smaller and smaller. If all the physical parameters are fixed, then λ varies with temperature, which produces a temperature-dependence of the propagation velocity. This dependence is shown in Fig. 6-16 for several

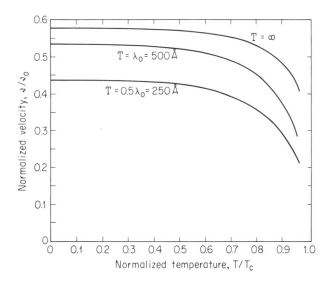

Fig. 6-16. Theoretical phase velocity vs. temperature for a strip line of identical superconductors of same thickness T and no losses; v_0 is the velocity of light in bulk dielectric, $S = \lambda_0 = 500$ Å; λ vs. temperature taken from Fig. 6-7.

values of T and for $S = \lambda_0 = 500$ Å, where Fig. 6-7 is used to obtain the temperature variation of λ. It can be seen that as the superconductor thickness is decreased, the velocity decreases. The dielectric thickness used is rather small in order to exaggerate the effect on propagation velocity. If S is made very large compared to λ, the effect on the velocity is negligible except near the critical temperature where λ also becomes very large.

In many experimental situations, superconducting lines are constructed by using a very thick ground plane with a thin dielectric and superconducting strip deposited on top of this. Under such circumstances, one conductor is essentially an infinitely thick plane (Sec. 6.3) and its surface impedance is independent of T. Thus, Eq. (6-78) must be modified slightly in an obvious manner, and if we assume that the superconductors are identical in all other respects, the propagation velocity for the lossless case becomes

$$\frac{v}{v_0} = \frac{1}{[1 + (\lambda/S) + (\lambda/S)\coth(T/\lambda)]^{1/2}} \tag{6-79}$$

If the penetration depths of the two superconductors are different, then the two values of λ will be different, corresponding to the respective conductor.

Equation (6-79) gives a slightly different temperature-dependence for the velocity, as shown in Fig. 6-17 for cases similar to those of Fig. 6-16.

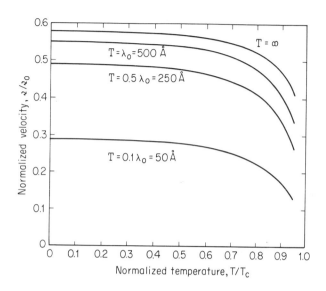

Fig. 6-17. Theoretical phase velocity vs. temperature for a strip superconductor of thickness T, above an infinite superconducting ground plane, assuming no losses. Parameters the same as in Fig. 6-16.

For the parameters chosen in Figs. 6-16 and 6-17, the curves are valid only for a temperature of up to very roughly $0.9\ T/T_c$. As the temperature increases, λ increases, the normal conduction current increases, and losses begin to become important. A more detailed analysis of the range of parameters for which the equations are valid will be given later. For the present, we wish to consider the effect of losses on the superconducting line characteristics. More specifically, we wish to consider Eq. (6-73) for the case when normal conduction losses are small but not negligible. We will do this for two cases: first, when the conductor thickness is very large, and second, when it is small and must be considered.

When the conductor thickness is very large we know that $\coth T/\lambda = 1$, so that Eq. (6-73) becomes

$$\gamma = j\omega\sqrt{\mu_0\,\epsilon_0}\left(\frac{2\epsilon_r}{\Gamma S} + \mu_r\epsilon_r\right)^{1/2} \tag{6-80}$$

Since we are assuming that the normal conduction current is small but not negligible, then from Eqs. (6-46) through (6-48)

$$\frac{1}{\Gamma} = \lambda \left(1 + 2j \frac{\lambda^2}{\delta^2}\right)^{-\frac{1}{2}} \approx \lambda \left(1 - j \frac{\lambda^2}{\delta^2}\right) \tag{6-81}$$

or for large T

$$\gamma = j\omega \sqrt{\mu_0 \epsilon_0} \left[2\epsilon_r \frac{\lambda}{S}\left(1 - j\frac{\lambda^2}{\delta^2} + \mu_r \epsilon_r\right)\right]^{\frac{1}{2}} \tag{6-82}$$

It is necessary to separate the above equation into real and imaginary parts to obtain the attenuation and phase constant, respectively. This can be done in a manner identical to that used in Sec. 2.6 by using a series expansion; as a matter of fact, we will obtain expressions analogous to Eqs. (2-45) and (2-46) for α and β. In general

$$\alpha = \frac{\text{Re}(Z_{series})/2}{[\text{Im} Z_{series}/\omega C]^{\frac{1}{2}}} \tag{6-83}$$

$\text{Im}(Z_{series})$, which is ωL is given by Eq. (6-68); $C = \epsilon_0 W/S$ and Real(Z_{series}) can be obtained from the real part of Eq. (6-49). Substituting these

$$\alpha = \frac{1}{2} \frac{\omega^2 \mu_0 \sqrt{\mu_0 \epsilon_0 \epsilon_r}\, \sigma\lambda^3}{S[\mu_r + 2(\lambda/S)]^{\frac{1}{2}}} \tag{6-84}$$

To a first approximation, the normal conduction losses do not affect the phase constant

$$\beta = \omega \sqrt{LC} = \omega \sqrt{\mu_0 \epsilon_0} \left[\epsilon_r\left(\mu_r + 2\frac{\lambda}{S}\right)\right]^{\frac{1}{2}} \tag{6-85}$$

When the conductor thickness is small, and therefore must be included, the general expression for the propagation constant, neglecting G, is

$$\gamma = (Z_{series}\, j\omega C)^{\frac{1}{2}} \tag{6-86}$$

Since the normal conduction current is small, but not negligible, the series impedance given by Eq. (6-66), using the approximation of Eq. (6-81) for $1/\Gamma$, yields

$$Z_{series} \approx \frac{\omega\mu_0}{W}\left[\frac{2\lambda^3}{\delta^2}\coth\Gamma T + j(2\lambda\coth\Gamma T + \mu_r S)\right]$$

Since the losses are assumed to be small, we can approximate Γ by an expression similar to that of Eq. (6-81) for $1/\Gamma$, namely

$$\Gamma = \frac{1}{\lambda}\left(1 + 2j\frac{\lambda^2}{\delta^2}\right)^{1/2} \approx \frac{1}{\lambda}\left(1 + j\frac{\lambda^2}{\delta^2}\right) \qquad (6\text{-}87)$$

Substituting this approximation into the previous equation yields

$$Z_{series} = \frac{\omega\mu_0}{W}\left[\frac{2\lambda^3}{\delta^2}\coth\left(\frac{1}{\lambda} + j\frac{\lambda}{\delta^2}\right)T + j\left(2\lambda\coth\left(\frac{1}{\lambda} + \frac{j\lambda}{\delta^2}\right)T + \mu_r S\right)\right] \quad (6\text{-}88)$$

The propagation constant for this can now be obtained by substituting this into Eq. (6-86). The attenuation and phase constant would then be found by expanding γ into its real and imaginary parts as in Sec. 2.6. However, since the approximate formula for attenuation is given by Eq. (6-83), it is much simpler to use this formula as was done previously. Identical answers are obtained by either approach (left as an exercise). Before we proceed, it is necessary to separate the coth term in Eq. (6-88) into its real and imaginary parts (otherwise terms will be left out which can be significant). This can be done by using the identity

$$\coth\left(\frac{T}{\lambda} + j\frac{\lambda T}{\delta^2}\right) = \frac{\sinh(2T/\lambda) - j\sin(2\lambda T/\delta^2)}{\cosh(2T/\lambda) - \cos(2\lambda T/\delta^2)} \qquad (6\text{-}89)$$

We will assume that $2\lambda T/\delta^2 \ll 1$, since this is often true except near the critical temperature where λ becomes large. Using this approximation

$$\sin\frac{2\lambda T}{\delta^2} \approx \frac{2\lambda T}{\delta^2} \qquad \cos\frac{2\lambda T}{\delta^2} \approx 1 \qquad (6\text{-}90)$$

By making use of the identities

$$\sinh 2\theta = 2 \sinh\theta \cosh\theta$$

$$\cosh 2\theta = 2 \sinh^2\theta + 1$$

(6-91)

Eq. (6-89) simplifies to

$$\coth\left(\frac{T}{\lambda} + j\frac{\lambda T}{\delta^2}\right) \approx \coth\frac{T}{\lambda} - j\frac{\lambda T/\delta^2}{[\sinh(T/\lambda)]^2}$$

(6-92)

Substituting this into Eq. (6-88) and simplifying yields

$$Z_{series} = \frac{\omega\mu_0}{W}\left\{\frac{2\lambda^3}{\delta^2}\coth\frac{T}{\lambda} + \frac{(2\lambda^2/\delta^2)\,T}{[\sinh(T/\lambda)]^2}\right.$$

$$\left. + j\left[\mu_r S + 2\lambda\coth\frac{T}{\lambda} - \frac{(2\lambda^4/\delta^4)\,T}{[\sinh(T/\lambda)]^2}\right]\right\}$$

(6-93)

Now making use of the approximate expression, Eq. (6-83) for attenuation, we get, after some simplification

$$\alpha = \frac{\omega^2\mu_0\sqrt{\mu_0\,\epsilon_0\,\epsilon_r}\,\,\sigma\lambda^3\left[\coth\dfrac{T}{\lambda} + \dfrac{T}{\lambda}\bigg/\left(\sinh\dfrac{T}{\lambda}\right)^2\right]}{2S\left[1 + 2\dfrac{\lambda}{S}\coth\dfrac{T}{\lambda} - \left(\dfrac{2\lambda^4}{\delta^4}\dfrac{T}{S}\right)\bigg/\left(\sinh\dfrac{T}{\lambda}\right)^2\right]^{1/2}}$$

(6-94)

where μ_r is assumed to equal 1 for the dielectric. This equation represents the general case of a strip line of identical superconductors of thickness T with small losses. The assumptions implicit in this are

$$\frac{2}{\delta^2} = \omega\mu\sigma \ll 1$$

(6-95)

and

$$\frac{2\lambda T}{\delta^2} \ll 1$$

(6-96)

For many cases of practical interest, the entire sinh term under the radical is completely negligible.

The phase constant when the losses are small can be obtained in a similar way, using the approximate formula derived in Sec. 2.6

$$\beta \approx \omega \sqrt{LC} \left(1 + \frac{R^2}{8\omega^2 L^2}\right)$$

For our case here, this becomes

$$\beta \approx \omega \left[C \operatorname{Im}\left(\frac{Z_{series}}{\omega}\right)\right]^{\frac{1}{2}} \left[1 + \frac{(\operatorname{Re} Z_{series})^2}{8\omega^2 (\operatorname{Im} Z_{series}/\omega)^2}\right]$$

We will not evaluate this but rather will point out that the small losses have only a second-order effect on the phase velocity, i.e., varies as square of losses, whereas they have a first-order effect on attenuation. Thus, to a first approximation, small losses give a small attenuation constant but usually negligible dispersion.

We will now investigate α as a function of temperature: λ varies with temperature according to Fig. 6.7. The conductivity of normal electrons in the superconducting state is related to the conductivity σ_n of the normal electrons at an infinitesimal temperature above the critical temperature T_c by the approximate equation

$$\sigma_{sc} = \left(\frac{T}{T_c}\right)^4 \sigma_n \qquad\qquad (6\text{-}97)^*$$

The above expression should be used to determine σ. Figure 6-18 is a plot of α versus temperature using $\sigma_n = 10^9/\text{ohm-meter}^\dagger$ and a frequency of 10^9 Hz. Both the conductor and dielectric thickness were assumed to be small since this tends to exaggerate the influence of the superconductor. It should be noted that values read from the curve must be multiplied by $\sqrt{\epsilon_r}$ to obtain the actual attenuation.

At a temperature of about $T/T_c = 0.94$, $1/\lambda^2 = 80 \times 10^{12}/\text{m}^2$, while $\omega\mu\sigma = 5.2 \times 10^{12}/\text{m}^2$. Thus, the inequality of Eq. (6-95) is satisfied with more

---

*See [19, p. 194].

†This number represents an order of magnitude value for some common superconductor, e.g., lead in Fig. 4-10.

than an order of magnitude difference in these two quantities. At a temperature of $T/T_c = 0.99$, these two quantities are nearly equal, so that the simple expression for attenuation breaks down. Figure 6-18 is a reasonable approximation up to T/T_c of roughly 0.95.

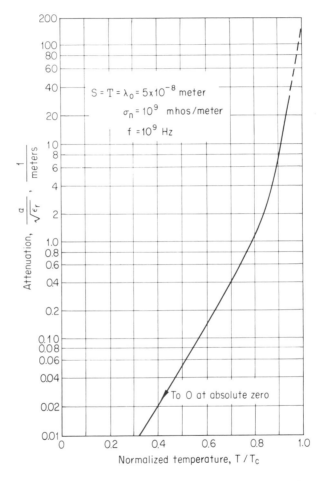

Fig. 6-18. Theoretical attenuation (Eq. (6-94)) vs. temperature for strip line of identical superconductors with small losses.

As the temperature approaches the critical temperature, the losses become too large, and so it is necessary to use the general statement of Eq. (6-73), which can be expressed in its real and imaginary parts by various techniques, all of which involve lengthy algebraic expressions.* We will not

---

*For example, see [20].

perform these evaluations, but will instead make use of the fact that at the critical temperature, λ becomes infinite and normal skin effect dominates. At or above T_c, λ will be much larger than δ (for reasonably high frequencies), so that the $1/\lambda^2$ can be neglected in Eq. (6-74). Upon substitution and the use of identities for coth ΓT and $\sqrt{j}$, Eq. (6-73) becomes

$$\gamma = j\omega\sqrt{\mu_0\epsilon_0\epsilon_r}\left\{1 + \frac{\delta\{\sinh(2T/\delta) - \sin(2T/\delta) - j[\sinh(2T/\delta) + \sin(2T/\delta)]\}}{S[\cosh(2T/\delta) - \cos(2T/\delta)]}\right\}^{1/2}$$

(6-98)

where it is assumed that $\mu_r = 1$, as previously. This equation represents the normal skin-effect equation of strip conductors both of finite thickness, and is a more general expression for γ than are Eqs. (4-72) and (5-78). If both the conductor and dielectric thickness T and S are allowed to become very large compared to δ, the above equation will give an attenuation which is exactly twice that given by Eq. (5-79), as should be the case since the latter was obtained for only one conductor whereas the above equation includes both conductors. (Verification of this is left as an exercise to the reader.) Both expressions are valid only for cases where the losses are small.

At the critical temperature, $T/T_c = 1$, the parameters of Fig. 6-18 give $\delta = 5 \times 10^{-7}$ m, and the propagation constant becomes very nearly

$$\gamma = 148\sqrt{\epsilon_r}\,(1 + j)$$

(6-99)

So the attenuation is

$$\alpha = 148\sqrt{\epsilon_r}\ \text{per meter}$$

(6-100)

This is about an order of magnitude larger than the value at $T/T_c = 0.92$ and about two orders of magnitude larger than that at $T/T_c = 0.82$. These numbers indicate the very significant improvement that can be obtained by using superconducting lines. One significant point which needs mention is that these calculations have been made assuming that the dielectric thickness $S = \lambda_0 = 5 \times 10^{-8}$ meter. This small value of S makes the line characteristic impedance very small and any series losses are then more significant, as shown by Eq. (6-83). If S is made much larger, the attenuation in the normal state will not be nearly so large, but these numbers have been given only as an example.

In deriving all the equations thus far, we have made various assumptions or approximations concerning the applied frequency and the relationship between various parameters. We wish now to consider these restrictions in more detail.

The first restriction imposed on all expressions in the entire chapter is that the displacement current within the superconductor must be small, or, from Eq. (6-33), $1/\lambda^2 \gg \omega^2 \mu_0 \epsilon_0$; thus it is necessary that

$$\omega \ll \frac{1}{\lambda \sqrt{\mu_0 \epsilon_0}} = \frac{3 \times 10^8 \text{ m/sec}}{\lambda} \qquad (6\text{-}101)$$

At low temperatures, $\lambda = \lambda_0 = 5 \times 10^{-8}$ m, and so

$$f \ll \frac{1}{2\pi} 0.6 \times 10^{16} \text{ Hz}$$

This is such a high frequency that it is apparent the displacement current can be neglected, even for much larger values of λ, as long as the frequency is in the microwave region.

A second restriction, which was implicitly assumed, is that anomalous skin effect is not present, so that the normal conduction current is governed by classical skin effect, i.e., Ohm's law in Maxwell's equations for normal electrons. This requires that the classical skin depth δ must be much larger than the electron mean free path Λ (Sec. 4.10). This condition cannot be specified in any general form since the point where anomalous effects begin is highly dependent on temperature and material. For example, at 1.2×10^9 Hz, anomalous skin effect begins in lead at $14°$K (Table 4-2), which is slightly above its critical temperature of $7.17°$K, while in aluminum it begins at $63°$K, which is well above its critical temperature of $1.196°$K. For lead, the anomalous effects are expected to have a very small influence on its superconducting properties, while in aluminum the influence should be rather important.

When anomalous effects are to be considered, it is necessary to replace Ohm's law for the normal electrons by the more general Reuter-Sondheimer theory expressed by Eq. (4-130). It is apparent that the analysis becomes extremely complicated and shall not be considered here. Maxwell, Marcus, and Slater [9] have considered such a case theoretically and experimentally for tin at 24×10^9 Hz.

A third restriction which has been assumed is that concerned with the time required for the traveling wave to penetrate into the superconductor. We have just assumed that fields penetrate the conductor to their (near) maximum extent in a time which is short compared with the time to travel a distance along the surface which is a significant fraction of the wavelength of the applied frequency. This problem is discussed in Sec. 6.3, Eq. (6-34), and is* $y^2 \ll 1/\lambda^2$, where y is the propagation constant. Since we are concerned primarily with cases for which the losses are negligible or are quite small, then y is essentially the wavelength of the applied frequency. The third restriction then becomes that the penetration depth λ must be very much smaller than the applied sinusoidal wavelength λ_w

$$\lambda \ll \lambda_w = \frac{v_0}{f} = \frac{c}{f\sqrt{\epsilon_r}} \tag{6-102}$$

where v_0 is the velocity of the traveling wave and ϵ_r is the relative dielectric constant of the insulator. At low temperatures, $\lambda = 5 \times 10^{-8}$ m, so that the above expression gives

$$f \ll \frac{c}{\lambda_0 \sqrt{\epsilon_r}} = \frac{0.6 \times 10^{16}}{\sqrt{\epsilon_r}} \text{ Hz}$$

It is apparent that this is such a high frequency that we need not concern ourselves with this restriction, even when the temperature increases such that λ(penetration depth) increases by several orders of magnitude.

The fourth and final restriction from Eqs. (6-81) and (6-45) is that the losses, while not negligible, must be small or

$$\omega\sigma\mu = 2/\delta^2 \ll 1/\lambda^2$$

The frequency limit imposed by this restriction is

$$f \ll \frac{1}{2\pi\sigma\mu\lambda^2} \tag{6-103}$$

The severest restriction imposed by conductivity will be when it is at its maximum value or $\sigma = \sigma_n$; we shall thus use this value and assume it to be constant. For temperatures sufficiently below T_c, λ is essentially $\lambda_0 = 5 \times 10^{-8}$ m. Thus, for low temperature and high conductivity, substitution of appropriate values yields

$$f \ll 50 \times 10^9 \text{ Hz}$$

*This is similar to the assumption required in Eq. (4-67) for the exact solution of an ordinary line.

If we specify that the symbol $\ll$ means at least an order of magnitude less than, then the above gives

$$f \leq 5 \times 10^9 \text{ Hz} \tag{6-104}$$

Thus, of the four restrictions, this is the severest. This is still a reasonably high frequency range which is in the lower microwave region. If we allow the temperature to increase, the frequency restriction will become more severe than Eq. (6-104). In particular, if we allow the temperature to increase to, say 0.95 T_c, then, from Fig. 6-7, λ increases by a factor of about 2.24 (or λ^2 increases by a factor of 5), so that the new restriction on frequency is

$$f \leq 1 \times 10^9 \text{ Hz}$$

At a temperature of 0.99 T_c, λ increases by a factor of 5 and it is necessary that

$$f \ll 2 \times 10^9 \quad \text{or} \quad f \leq 0.2 \times 10^9 \text{ Hz}$$

The frequency range has dropped well below the microwave region, into the region where ordinary transmission lines give reasonable performance and therefore begin to become competitive (although superconducting lines are still superior).

The conclusion to be drawn from the above analysis is that Eqs. (6-84), (6-85), and (6-94) are valid for quite high frequencies up to roughly 10^9 Hz as long as the temperature is sufficiently low, less than about 0.95 T_c. As the temperature approaches the critical temperature, the frequency restriction becomes severer and must be taken into account.

Even though the losses are small, the attenuation is finite, especially for T/T_c above about 0.8. The approximate formula plotted in Fig. 6-18 is valid for T/T_c in the neighborhood of 0.95 for the particular parameters chosen. It is advisable to carry through such calculations for specific cases of interest.

6.6 CHARACTERISTIC IMPEDANCE OF A SUPERCONDUCTING LINE

In general, the characteristic impedance of a line is given by

$$Z_0 = \frac{\gamma}{Y} \tag{6-105}$$

where γ is the propagation constant and Y is the shunt admittance. If we assume that the dielectric losses are negligible, then $Y = j\omega C$, where, for the previous line of Sec. 6.5 with W/S very large, $C = \epsilon_0 \epsilon_r W/S$. If the losses are assumed to be negligibly small, γ is given by Eq. (6-76). Substituting these into Eq. (6-105)

$$Z_0 = \frac{S}{W} \sqrt{\frac{\mu_0}{\epsilon_0 \epsilon_r}} \left(\frac{2\lambda}{S} \coth \frac{T}{\lambda} + \mu_r \right)^{\frac{1}{2}} \qquad (6\text{-}106)$$

If the conductor thickness is very large so that $T/\lambda \gg 1$, then $\coth T/\lambda \approx 1$; thus, letting $\mu_r = 1$, Eq. (6-106) yields

$$Z_0 = \frac{S}{W} \sqrt{\frac{\mu_0}{\epsilon_0 \epsilon_r}} \left(1 + \frac{2\lambda}{S} \right)^{\frac{1}{2}}$$

or, more generally

$$Z_0 = Z_n \left(1 + \frac{2\lambda}{S} \right)^{\frac{1}{2}}$$

where Z_n is the characteristic impedance of the same line using normal conductors with no losses. This impedance is a pure resistance with a slightly increased value due to the penetration of the wave into the superconductor. This is similar to the effect that classical skin effect has on impedance except that the latter also introduces a reactive term, whereas the superconductor introduces only a pure resistance term.

It is apparent that when the losses are no longer negligible, the impedance is no longer a pure resistance and the expression becomes very involved. We will not consider this.

6.7 FAST RISE TIME PULSES PROPERTIES OF SUPERCONDUCTING TRANSMISSION LINES

In all of the preceding sections, we have been concerned with the fundamentals of superconducting lines and analysis of plane conductors or strip lines. We saw in Sec. 6.5 that a strip line can be made which has virtually no attenuation or dispersion for frequencies below approximately 1,000

MHz. From Fig. 5-19, this would correspond to a pulse rise time of about 0.35 ns. In other words, it is to be expected that very long superconducting transmission lines should be able to transport pulses of less than 0.5 ns rise time with essentially no distortion of the pulse at the output end. We also know that if the thickness of the conductors as well as their separation is large compared to the penetration depth λ, then the characteristic impedance is not significantly influenced by the superconductor and the pulse travels with the velocity of light in the dielectric insulation, i.e., the superconductor does not slow down the pulse. Such transmission lines can, in fact, be obtained. Allen and Nahman [1] describe several such lines which were constructed as coaxial cables. Even though all of our previous analyses were done for plane conductors, the conclusions remain the same for co-axial cables, with the only necessary changes occurring in the multiplying coefficients, i.e., geometrical constants, as was the case in Sec. 5.13, for instance.

One such cable is shown in Fig. 6-19. The external diameter is quite small (0.090 in.), while the length is quite long, 1,360 ft (415 meters). We would expect that such a cable constructed from ordinary wire would significantly distort and attenuate a pulse of 0.5 ns rise time, while the superconducting line should be much better. The response of the line of Fig. 6-19 to a pulse of about 0.5 ns rise time and 7 ns base width is shown in Fig. 6-20. It can be seen that the rise time is only very slightly affected and there is insignificant attenuation. It is important to point out that for such lines, the end connections or small nonuniformities on the line itself become very important and may be the dominant factor in any observed distortion.

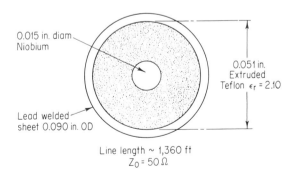

0.015 in. diam. Niobium

0.051 in. Extruded Teflon $\epsilon_r = 2.10$

Lead welded sheet 0.090 in. OD

Line length ~ 1,360 ft
$Z_0 = 50\,\Omega$

Fig. 6-19. Experimental superconducting coaxial line.

The delay time for this line was measured to be 2.014 sec, giving a relative dielectric constant of 2.10 for the Teflon, assuming that the

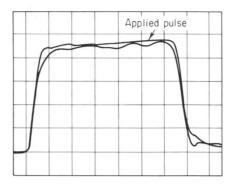

Fig. 6-20. Pulse response of superconducting co-axial cable—delay not shown: 1 ns/div. (*After Allen and Nahman.*)

superconductor does not affect the velocity of propagation. This value agrees quite well with handbook values, as was expected (see Fig. 3-8).

A rough idea of the improvement in performance over ordinary lines can be obtained by comparing the results of Fig. 6-20 with those of Fig. 5-22. The latter were obtained for some common cables at room temperature with an applied pulse of nearly the same rise time (0.5 ns) but with cables only 160 ft. long. Thus, even though these ordinary cables are a factor of 8.5 shorter in length, the pulse distortion and attenuation is significantly greater than that of the longer superconducting line.

In order to have a more meaningful comparison of the improvement expected with superconducting lines, let us determine the pulse response to be expected from the same length of a common cable, using the normalized curves of Fig. 5-20 to obtain the output pulse. Let us pick a coaxial cable such as RG 58A/U and apply a very wide pulse with a linear rise time equal to that applied to the superconducting line, namely, $T_R = 0.5$ ns. In order to use the normalized curves of Fig. 5-20, it is necessary only to determine the value of B from which the normalized parameters t' and T_R' are found, with the latter indicating the particular curve of the family of curves to be used and t' indicating the time scale from which actual time can be found. B can be calculated from Eq. (5-83), provided that R_{sk} can be calculated. R_{sk} for the center conductor can be obtained from Eq. (5-76); for the outer conductor, the same formula can be used as an approximation with r taken as the inner radius of the outside conductor. However, a more accurate determination of B can be obtained from Eq. (5-92) with the use of measured values of attenuation. From handbooks or manufacturer's specifications, the value of attenuation for 100 ft of RG 58A/U at 5×10^8 Hz is found to be 14 dB or 1.61 radians per 100 ft. For 1,360 ft of this line, $\alpha\ell = 21.9$ radians and from Eq. (5-92), $B = 7.6 \times 10^{-8}$ sec.* The particular normalized curve to be used can be found by evaluating $T_R' = T_R/B = 0.0065$. This is sufficiently small so that the curve for $T_R' = 0$ in Fig. 5-20 can be used with the time scale given by $t = 7.6 \times 10^{-8}\ t'$ sec. A comparison between this calculated output pulse and the applied linear rise time is shown in Fig. 6-21 (lower time scale). It can be seen that the attenuation and distortion are so large that if the pulse of 7 ns base width (Fig. 6-20) were applied, the output would bear little resemblance to the input, thus showing the great improvement obtained with superconductors.

Of course, it could be argued that the superconducting line is operated at low temperature and RG 58A/U would also possess improved response at low temperatures. However, even though the dc resistivity can be lowered by several orders of magnitude, anomalous skin effect becomes important at these high frequencies and greatly limits the improvement to be expected. Let us consider the improvement brought on by lowering the temperature to a point just before the onset of anomalous skin effect. From Eq. (4-120), this point is expected for a frequency of 5×10^8 Hz at $\rho_a = (1.67 \times {}^{-36}f)^{1/3} = 0.094 \times 10^{-8}$ ohm-meter or a temperature of about 60°K. We need to determine the new value of B at this temperature. The skin-effect resistance R_{sk} varies as $\rho^{1/2}$ (Eq. 5-76) but B varies as R_{sk}, so that B varies directly with ρ. Thus the ratio of B at room temperature (298°K) to that at any other temperature must vary directly as the ratio of the resistivities at the respective temperature. Therefore

$$\frac{B(60°)}{B(298°)} = \frac{\rho(60°)}{\rho(298°)} = 5.5 \times 10^{-2}$$

Using the previous value for B at room temperature gives $B(60°K) = 4.2 \times 10^{-9}$, $T_R' = 0.12$ and $t = 4.2 \times 10^{-9}\ t'$ sec. We still must use the normalized curve of $T_R' = 0$ of Fig. 5-20, but the time scale is greatly changed. This is shown by the upper time scale on Fig. 6-21 and indicates significant improvement of more than an order of magnitude in the rise time of the output. However, the distortion and attenuation are still quite severe compared to the superconducting line.

If the temperature were lowered further, anomalous and eventually extreme anomalous effects would set in with the surface resistance following

---

*α and therefore B vary slightly with frequency, which indicates that the line does not strictly follow the $\omega^{1/2}$ law, but the error is small. The rise time portion of the response is determined by the high frequencies: The upper band-pass frequency at 3 dB point is 7×10^8 Hz for a 0.5 ns rise time, so that 5×10^8 was chosen to be representative of the desired response.

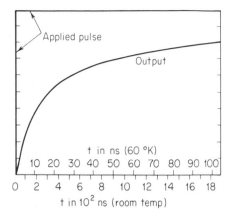

Fig. 6-21. Theoretical comparison of applied and output pulse (neglecting delay) of RG 58A/U, 1,360 feet long at two temperatures; applied pulse rise time is 0.5 ns.

approximately the curve of Fig. 4-11. Decreasing the temperature can only decrease the surface resistance by a factor of about 2 over that at 60°K, an almost insignificant improvement. The analysis and normalized curves of Sec. 5.13 are no longer applicable when anomalous effects are present, so that obtaining a calculated response curve becomes another, more complex problem. It should be understood that for pulses with much larger rise times, anomalous effects will occur at lower temperatures, and so more improvement than that indicated in Fig. 6-21 is possible. Such cases can be treated exactly as above to get the calculated response curve at any temperature.

It can be concluded from the above analyses that for very fast rise time pulses over a substantial length of line, superconducting lines offer a distinct advantage over ordinary lines, as they usually give negligible distortion and attenuation (see [21] for a practical application).

6.8 POWER TRANSMISSION

In Sec. 6.5 we considered transmission lines which were constructed from "soft" (type-I) superconductors and which carried very small amounts of power. It is apparent that if such lines were made to carry large currents, the magnetic field created by this current could become large enough to exceed the critical field, thereby switching the material out of the super-conducting state. Thus, for high-power transmission, it is desirable to use superconductors with very high critical fields, i.e., hard superconductors. Materials such as Nb_3Sn, NbZr, and others have become available in recent years with critical fields in the hundred-thousand gauss (or oersted) range. In addition to the widespread use of such materials in the construction of high-field electromagnets, it may be possible to use such superconductors to transmit large amounts of power over very great distances. Garwin and Matisoo [5] propose a dc transmission line to transmit 10^{11} watts over a distance of 1,000 kilometers (620 miles). Such a line would have no electrical losses within the superconductor at steady state, but it is necessary to tap some power to provide refrigeration of the Nb_3Sn at $4°K$.

It is necessary to use dc rather than ac transmission because of the large amounts of eddy current and other losses associated with the latter. This seems rather surprising at first, since we have seen in previous sections that the normal conduction current, which gives rise to the losses, is usually negligible. However, this is only true for type-I or for low current densities in type-II superconductors. When the current density becomes sufficiently large, bulk flux penetration occurs, i.e., large penetration depth, and the material is said to be in the mixed state, a state of mixed superconducting and normal materials. The current can flow over the entire cross-sectional area, and the transition from a superconducting to a normal state occurs over a very broad transition region, as in Fig. 6-5. Such a mixed state can only occur in type-II superconductors, but it is necessary to use type-II because of the required high critical field.

Type-I superconductors have a very sharp transition from super to normal conduction, and also have a penetration depth which is nearly independent of the magnetic field. However, these critical fields are much too low to be of practical value in power lines. Another disadvantage of type-I materials is that the current must flow entirely on the surface in a depth λ as a result of the Meissner effect. Thus, the current-carrying capacity will vary directly with the circumference or diameter of a circular wire. In type-II materials, the current density can be uniform, so that the current carrying capacity will vary as the square of the diameter, thus requiring a much smaller wire.

For large currents passing through type-II materials in the mixed state, the lines are designed to have a sufficient amount of super electrons available to carry dc without losses. However, if ac is applied, the surface impedance analysis in Sec. 6.5 is no longer applicable. The fields are so high that the hysteresis (Fig. 6-5) gives a significant power loss within the material. These losses can be calculated with the aid of the critical state model. For example, consider a line using Nb_3Sn. At temperatures well below its critical temperature of $18°K$, this material will remain superconducting in a field of

10^5 gauss while carrying 2×10^5 A/cm^2. For a circular wire of 5-cm^2 area carrying $J_c = 10^5$ A/cm^2, the equation for $I_0 = \pi r^2 J_c$ in Sec. 6.2(h) gives the power loss as

$$2 \times 10^{-9} I_0^2 = 500 \text{ joules/cycle cm}$$

for each conductor. At ordinary power frequency of 60 Hz, this gives 3×10^4 W/cm, which is rather large, to say the least. Thus, one solution is to use dc rather than ac; another course of action might be to reduce the magnetic field by decreasing the total current or increasing the conductor size. However, both of these are economically very undesirable since there are other factors to be considered, such as heat loss (which determines the amount of refrigeration and insulation required), wire cost, and others. These exclude the use of ac for superconducting lines, so that dc transmission with the attendant converter, refrigeration, protection, and maintenance facilities and numerous other provisions will be required. The final choice of superconducting versus ordinary lines must be made in view of the tradeoffs of numerous factors and must arrive at some reasonable compromise to satisfy the specific application.

PROBLEMS

6-1. A superconducting strip line with a large W/S ratio is to be constructed of aluminum with $W = 0.010$ in. and is to be operated well below the critical temperature. Find the maximum current capability which permits superconducting operation.

Answer: 2 amperes.

6-2. If the strip line in the above problem is made of lead, what improvement is to be expected?

Answer: 16.2 amperes maximum.

6-3. Using Eq. (6-88), derive γ for a line with small losses by expanding Eq. (6-86) into its real and imaginary parts.

6-4. Given: a very thick superconducting strip of width $W = 32$ K Å separated by 8 K Å from a superconducting ground plane of the same material. The penetration depth is $\lambda = 2$ K Å and the dielectric has $\epsilon_r = 4$, $\mu_r = 1$. The strip is operated such that the losses are negligible. Determine the characteristic impedance of such a line.

Answer: 70 ohms.

6-5. In the previous problem, suppose that both the strip and ground plane are 2 K Å thick. Find the new characteristic impedance.

Answer: 73.5 ohms.

REFERENCES

1. Allen, R. J., and N. S. Nahman: Analysis and Performance of Super-conductive Coaxial Transmission Lines, *Proc. IEEE,* vol. 52, no. 10, p. 1147, October 1964.

2. "American Institute of Physics Handbook," 2nd ed., pp. 9-42, McGraw-Hill Book Company, New York, 1963.

3. Bardeen, J.: Review of the Present Status of the Theory of Super-conductivity, *IBM J. Res. Develop.,* vol. 6, no. 1, p. 3, January 1962.

4. Bean, C. P., *et al.*: "A Research Investigation of the Factors That Affect the Superconducting Properties of Materials," Tech. Rept. AFML-TR-65-431, for the Air Force Materials Laboratory, Wright Patterson Air Force Base, Ohio, 1966.

5. Garwin, R. L. and J. Matisoo: Superconducting Lines for the Trans-mission of Large Amounts of Electrical Power over Great Distances, *Proc. IEEE,* vol. 55, no. 4, p. 538, 1967.

6. Khalatnikov, I. M., and A. A. Abrikosov: The Modern Theory of Super-conductivity, *Advan. Phys.,* vol. 8, no. 29, p. 66, 1959.

7. London, F., and H. London: The Electromagnetic Equations of the Supraconductor, *Proc. Roy. Soc. (London),* vol. A149, p. 71, 1935.

8. Lynton, E. A.: "Superconductivity," pp. 37, 63, Methuen and Co. Ltd., London, 1962.

9. Maxwell, E., P. M. Marcus, and J. C. Slater: Surface Impedance of Normal and Superconductors at 24,000 Megacycles Per Second, *Phys. Rev.,* vol. 76, no. 9, p. 1332, November 1949.

10. Meyers, N. H.: Inductance in Thin-Film Superconducting Structures, *Proc. IRE,* vol. 49, no. 11, p. 1640, 1961.

11. Pippard, A. B.: The Surface Impedance of Superconductors and Normal Metals at High Frequencies: I. Resistance of Superconducting Tin and Mercury at 1200 Mc./sec., *Proc. Roy. Soc. (London),* vol. A191, p. 370, 1947.

12. *Ibid.,* III. The Relation Between Impedance and Superconducting Penetration Depth, p. 399.

13. *Ibid.,* IV. Impedance at 9400 Mc./sec. of Single Crystals of Normal and Superconducting Tin, vol. A203, p. 98, 1950.

14. *Ibid.,* V. Analysis of Experimental Results for Superconducting Tin, p. 195.
15. *Ibid.,* An Experimental and Theoretical Study of the Relation Between Magnetic Field and Current in a Superconductor, vol. A216, p. 547, 1953.
16. *Ibid.,* "Advances in Electronics and Electron Physics," vol. 6, p. 1, Academic Press, New York, 1954.
17. Reuter, G. E. H., and E. H. Sondheimer: The Theory of the Anomalous Skin Effect in Metals, *Proc. Roy. Soc. (London),* vol. A195, p. 336, 1948.
18. Shizume, P. K., and E. Vaher: Superconducting Coaxial Delay Line, *IRE Conv. Record,* vol. 10, pt. 3, p. 95, March 1962.
19. Shoenberg, D.: "Superconductivity," Cambridge University Press, New York, 1952. QCL611.S36 1952
20. Swihart, J. C.: Field Solution for a Thin-Film Superconducting Strip Transmission Line, *J. Appl. Phys.,* vol. 32, no. 3, p. 461, March 1961.
21. Rathbun, D. K., and H. J. Jensen: Nuclear Test Instrumentation with Miniature Superconducting Cables, *IEEE Spectrum,* September, 1968, p. 91.
22. NEWHOUSE, APPLICATIONS OF SUPERCONDUCTIVITY.
23. TAYLOR, SUPER CONDUCTIVITY
24. RATHBUN — JENSEN

7 COUPLED TRANSMISSION LINES AND DIRECTIONAL COUPLERS

7.1 INTRODUCTION

A coupled transmission-line system consists of at least two ordinary transmission lines which have some form of coupling between them such that a voltage and/or current wave in one line will induce a voltage and/or current wave in the other line.

Coupled transmission lines are frequently encountered through design, where they can be exploited in some useful way, or through unavoidable circumstances where the coupling is not desired but cannot easily be eliminated. In the former case, coupling between lines is used in devices such as directional couplers and transmission-line transformers, while with the latter, undesirable coupling in the form of cross talk or interference occurs in adjacent telephone lines, in electronic equipment where high-frequency open-wire lines run parallel for some distance, and in many other places.* It is not possible to cover all the various aspects of coupled lines; instead, we shall concentrate on the general theory of coupled transmission lines, particularly the fundamental concepts with extensions to include an analysis of directional couplers. It is desirable first to derive differential equations which will describe the general case of coupled lines. We then wish to show that for a particular case of ideal open parallel wires, the line parameters are such that the equations reduce to a simple form regardless of whether the lines are symmetrical or unsymmetrical, and that they give equal propagation velocity for all possible traveling waves.

We will then consider the equations which describe the general case of two coupled transmission lines where two different propagation constants and hence two waves, traveling at different velocities, are possible. After analyzing the characteristics of directional couplers, this general analysis of

*For a practical example of unavoidable coupling effects in a high-speed magnetic memory array, see [5, p. 340].

268

coupled lines will be obtained in terms of the "sum" and "difference" mode. Such an analysis can be applied to the class of transmission-line transformers described in [4] for which this approach is applicable and instructive.

The most general case of coupled lines would be a multiconductor system with coupling between all the lines. However, the basic concepts can be illustrated, and numerous useful devices obtained, with just two coupled lines, so that we shall concentrate primarily on this case.

7.2 PARAMETERS AND DIFFERENTIAL EQUATIONS OF TWO COUPLED LINES

In order to analyze coupled lines, it is necessary to derive the differential equations which describe the system. Since many types of configurations can be used, it is desirable to choose a configuration which is sufficiently general so that the fundamental conclusions to be obtained are applicable for all configurations. We shall consider open wire transmission lines above a ground plane (see Fig. 7-1). In general, the wires can have any cross-sectional shape, but each line is assumed to be uniform and parallel to any other nearby line. There are a number of forms in which one can write the equations for voltages and currents on the lines, depending on the definition of the inductive and capacitive parameters. The usual method is to define the parameters so that the equations are symmetrical, in which case

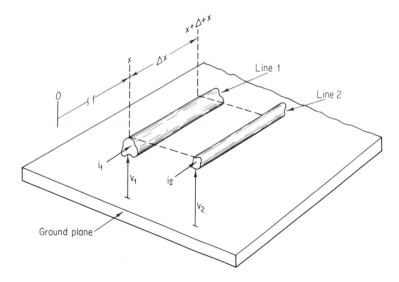

Fig. 7-1. Two general, coupled, open-wire transmission lines.

the equations for lossless lines using the polarities of Fig. 7-1 are

$$\frac{\partial v_1}{\partial x} = -L_{11}\frac{\partial i_1}{\partial t} - M_{12}\frac{\partial i_2}{\partial t} \tag{7-1}$$

$$\frac{\partial i_1}{\partial x} = -C_{11}\frac{\partial v_1}{\partial t} + C_{12}\frac{\partial v_2}{\partial t} \tag{7-2}$$

$$\frac{\partial v_2}{\partial x} = -L_{22}\frac{\partial i_2}{\partial t} - M_{21}\frac{\partial i_1}{\partial t} \tag{7-3}$$

$$\frac{\partial i_2}{\partial x} = -C_{22}\frac{\partial v_2}{\partial t} + C_{21}\frac{\partial v_1}{\partial t} \tag{7-4}$$

In these equations, the line parameters are obtained as indicated in Figs. 7-2 and 7-3(a). L_{11} is the inductance per unit length of line 1 with line 2 open: M_{12} is the mutual inductive coupling per unit length between line 1 and 2, that is, voltage across line 2 with line 2 open and a current applied to line 1. C_{11} is the capacitance per unit length of line 1 with line 2 shortened to ground or, in other words, it is the capacitance of line 1 to line 2 and the ground plane. $C_{12} = C_{21}$ is obtained by again grounding line 2 and applying a voltage to line 1, but measuring the current in line 2, that is, the current in ground wire from line 2 to the ground plane. L_{22} and C_{22} are similarly defined for line 2. In the general case, $C_{11} \neq C_{22}, L_{11} \neq L_{22}$, but for an isotropic, linear medium, e.g., air, $L_{12} = L_{21}$ and $C_{12} = C_{21}$ for

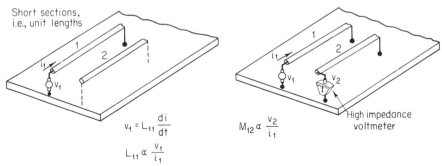

Short sections, i.e., unit lengths

$$v_1 = L_{11}\frac{di}{dt}$$

$$L_{11} \propto \frac{v_1}{i_1}$$

$$M_{12} \propto \frac{v_2}{i_1}$$

High impedance voltmeter

(a) Self-inductance of Line 1 in presence of Line 2: (b) Mutual inductance: Line 1 shorted,
     1 short-circuited,   2 open-circuited   Line 2 shorted one end and open
                                                     on other

Fig. 7-2. Inductance parameters of coupled lines.

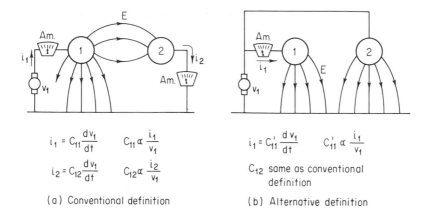

$$i_1 = C_{11}\frac{dv_1}{dt} \qquad C_{11} \propto \frac{i_1}{v_1}$$

$$i_2 = C_{12}\frac{dv_1}{dt} \qquad C_{12} \propto \frac{i_2}{v_1}$$

(a) Conventional definition

$$i_1 = C'_{11}\frac{dv_1}{dt} \qquad C'_{11} \propto \frac{i_1}{v_1}$$

C_{12} same as conventional definition

(b) Alternative definition

Fig. 7-3. Capacitance parameters of coupled lines.

all cases. In the case of practical interest here, the lines are assumed to be identical, so that $C_{11} = C_{22}$ and $L_{11} = L_{22}$.

It is possible to define the capacitance values differently from those above and to obtain a different set of equations. In the above case, C_{11} is obtained by grounding line 2 and measuring the capacitance of line 1 to line 2 and the ground plane. Referring to Fig. 7-3(b), suppose that C_{11} is obtained not by grounding line 2 but rather by applying the same voltage to lines 1 and 2 and measuring the current to line 1. This results in a different value of C_{11} than that obtained in Fig. 7-3(a). $C_{12} = C_{21}$ is obtained by grounding line 2, applying a voltage on line 1, and measuring the current in line 2, that is, in the ground wire connecting line 2 to the ground plane (same value as used previously). The inductance parameters are the same as previously used, so the new set of equations becomes

$$\frac{\partial v_1}{\partial x} = -L_{11}\frac{\partial i_1}{\partial t} - M_{12}\frac{\partial i_2}{\partial t} \qquad (7\text{-}5)$$

$$\frac{\partial i_1}{\partial x} = -C'_{11}\frac{\partial v_1}{\partial t} - C_{12}\frac{\partial}{\partial t}(v_1 - v_2) \qquad (7\text{-}6)$$

$$\frac{\partial v_2}{\partial x} = -L_{22}\frac{\partial i_2}{\partial t} - M_{21}\frac{\partial i_1}{\partial t} \qquad (7\text{-}7)$$

$$\frac{\partial i_2}{\partial x} = -C'_{22}\frac{\partial v_2}{\partial t} - C_{21}\frac{\partial}{\partial t}(v_2 - v_1) \qquad (7\text{-}8)$$

The various expressions which one obtains from the solution of Eqs. (7-1) through (7-4) will differ somewhat from those obtained from the solution of Eqs. (7-5) through (7-8) because of the slightly different definition of parameters. However, the ultimate conclusions will obviously be independent of the particular definition of parameters used, so that either set is adequate, provided that the definitions are consistent. We shall use the former for all subsequent analyses.

In all of the above equations, all the parameters are positive values. The reader is cautioned that in order to make Eqs. (7-1) through (7-4) symmetrical, some authors will choose the polarities such that a negative value must be assigned to the capacitance C_{12}. This is very undesirable, since negative capacitance is unrealistic: It is much more desirable to choose positive parameters and derive the equations with the proper algebraic signs to take care of polarities.

The solution to Eqs. (7-1) through (7-4) for the conventional definition of capacitance is most easily obtained by assuming sinusoidal excitation and letting*

$$Z_{11} = j\omega L_{11} \qquad Z_{12} = j\omega M_{12} \qquad Z_{21} = j\omega M_{21}$$

$$Z_{22} = j\omega L_{22} \qquad Y_{12} = j\omega C_{12} \qquad Y_{21} = j\omega C_{21}$$

$$Y_{11} = j\omega C_{11} \qquad Y_{22} = j\omega C_{22}$$

These equations then become

$$\frac{dv_1}{dx} = -Z_{11}\, i_1 - Z_{12}\, i_2 \tag{7-9}$$

$$\frac{di_1}{dx} = -Y_{11}\, v_1 + Y_{12}\, v_2 \tag{7-10}$$

$$\frac{dv_2}{dx} = -Z_{22}\, i_2 - Z_{21}\, i_1 \tag{7-11}$$

$$\frac{di_2}{dx} = -Y_{22}\, v_2 + Y_{21}\, v_1 \tag{7-12}$$

*It is obvious that for lines with losses, these parameters would be $Z_{11} = j\omega L_{11} + R_{11}$, $Y_{11} = j\omega C_{11} + G_{11}$, etc.

These can be solved by first taking the derivative of each with respect to x

$$\frac{d^2v_1}{dx^2} = -Z_{11}\frac{di_1}{dx} - Z_{12}\frac{di_2}{dx} \qquad (7\text{-}13)$$

$$\frac{d^2i_1}{dx^2} = -Y_{11}\frac{dv_1}{dx} + Y_{12}\frac{dv_2}{dx} \qquad (7\text{-}14)$$

$$\frac{d^2v_2}{dx^2} = -Z_{22}\frac{di_2}{dx} - Z_{21}\frac{di_1}{dx} \qquad (7\text{-}15)$$

$$\frac{d^2i_2}{dx^2} = -Y_{22}\frac{dv_2}{dx} + Y_{21}\frac{dv_1}{dx} \qquad (7\text{-}16)$$

If Eqs. (7-10) and (7-12) are substituted into Eqs. (7-13) and (7-15), the result (after collecting terms and simplifying) is

$$\frac{d^2v_1}{dx^2} = J_1 v_1 + K_1 v_2 \qquad (7\text{-}17)$$

$$\frac{d^2v_2}{dx^2} = J_2 v_2 + K_2 v_1 \qquad (7\text{-}18)$$

where

$$J_1 = Z_{11}Y_{11} - Z_{12}Y_{21} \qquad (7\text{-}19)$$

$$J_2 = Z_{22}Y_{22} - Z_{21}Y_{12} \qquad (7\text{-}20)$$

$$K_1 = Z_{12}Y_{22} - Z_{11}Y_{12} \qquad (7\text{-}21)$$

$$K_2 = Z_{21}Y_{11} - Z_{22}Y_{21} \qquad (7\text{-}22)$$

In a similar manner, the equations for current can be found as

$$\frac{d^2i_1}{dx^2} = J_3 i_1 + K_3 i_2 \qquad (7\text{-}23)$$

$$\frac{d^2 i_2}{dx^2} = J_4 i_2 + K_4 i_1 \tag{7-24}$$

where

$$J_3 = Z_{11} Y_{11} - Z_{21} Y_{12} \tag{7-25}$$

$$J_4 = Z_{22} Y_{22} - Z_{12} Y_{21} \tag{7-26}$$

$$K_3 = Z_{12} Y_{11} - Z_{22} Y_{12} \tag{7-27}$$

$$K_4 = Z_{21} Y_{22} - Z_{11} Y_{21} \tag{7-28}$$

Equations (7-17) through (7-28) represent the most general case of two coupled, ideal lines. If the lines are located in a linear isotropic medium, then

$$Z_{12} = Z_{21} \qquad Y_{12} = Y_{21} \tag{7-29}$$

and it is apparent that

$$J_3 = J_1 \qquad J_4 = J_2 \tag{7-30}$$

$$K_3 = K_2 \qquad K_4 = K_1 \tag{7-31}$$

Thus the current equations for such a medium become

$$\frac{d^2 i_1}{dx^2} = J_1 i_1 + K_2 i_2 \tag{7-32}$$

$$\frac{d^2 i_2}{dx^2} = J_2 i_2 + K_1 i_1 \tag{7-33}$$

while the voltage equations remain as in Eqs. (7-17) and (7-18). Equations (7-17), (7-18), (7-23), and (7-24) represent the general case in any medium, while Eqs. (7-17), (7-18), (7-32), and (7-33) represent the general case in a linear, isotropic medium. However, even for the general case of unsymmetrical lines, if the medium is linear and isotropic, then $J_1 = J_2$ and $K_1 = K_2 = 0$. The former can be shown in several possible ways. For instance, we know that in linear isotropic media, $Z_{12} = Z_{21}$ and $Y_{12} = Y_{21}$, so that it is only necessary to show that the first two terms in Eqs. (7-19) and (7-20) are equal, i.e., to show that

$$Z_{11} Y_{11} = Z_{22} Y_{22} \qquad (7\text{-}34)$$

assuming a TEM mode. Using the impedance parameters with no losses, the above can be written as

$$\omega^2 L_{11} C_{11} \overset{?}{=} \omega^2 C_{22} L_{22} \qquad (7\text{-}35)$$

The terms on the left- and right-hand side are proportional to the phase velocity of lines 1 and 2, respectively, in the presence of the other line (see Eq. (7-131)). Since a TEM was assumed to be present, these velocities must be equal and independent of geometry; thus

$$L_{11} C_{11} = L_{22} C_{22} \qquad (7\text{-}36)$$

and

$$J_1 = J_2 \qquad (7\text{-}37)$$

In order to show that $K = 0$, it is necessary to show from Eq. (7-21) that

$$K = Z_{12} Y_{22} - Z_{11} Y_{12} = 0 \qquad (7\text{-}38)$$

Using the coupling coefficients (see Sec. 8.7)

$$k_L = \frac{M_{12}}{\sqrt{L_{11} L_{22}}} \qquad k_C = \frac{C_{12}}{\sqrt{C_{11} C_{22}}} \qquad (7\text{-}39)$$

and assuming no losses, Eq. (7-38) becomes

$$k_L C_{22} \sqrt{L_{11} L_{22}} - k_C L_{11} \sqrt{C_{11} C_{22}} \overset{?}{=} 0 \qquad (7\text{-}40)$$

After simplification, this becomes

$$k_L \left(\frac{L_{22} C_{22}}{L_{11} C_{11}} \right)^{1/2} - k_C \overset{?}{=} 0 \qquad (7\text{-}41)$$

Since the velocities of propagation must be equal for the two lines (Eq. (7-36)), then K is zero only if $k_L = k_C$. But for a TEM (assumed), the latter must be true and therefore

$$K = 0 \qquad (7\text{-}42)$$

This is true in general whenever $k_L = k_C$.* Thus, for unsymmetrical (or symmetrical) lines in linear, isotropic media, the differential equations are

$$\frac{d^2 v_1}{dx^2} = J v_1 \qquad (7\text{-}43)$$

$$\frac{d^2 v_2}{dx^2} = J v_2 \qquad (7\text{-}44)$$

$$\frac{d^2 i_1}{dx^2} = J i_1 \qquad (7\text{-}45)$$

$$\frac{d^2 i_2}{dx^2} = J i_2 \qquad (7\text{-}46)$$

where J is given by either Eq. (7-19) or Eq. (7-20). It is apparent that by making the appropriate substitutions, this constant becomes

$$J = Z_{11} Y_{11} (1 - k^2) = Z_{22} Y_{22} (1 - k^2) \qquad (7\text{-}47)$$

It is obvious from the general form of the wave equation of (2-9) that in the above equations, the propagation constant is $\gamma^2 = J$.

For the more general case of lines which are *not* straight lines above a ground plane, but are rather coiled in some fashion, or if the dielectric between the wires and ground plane is different from that between the two wires themselves, then $k_L \neq k_C$ and K need not be zero; thus, the more general equations become

$$\frac{d^2 v_1}{dx^2} = J v_1 + K v_2 \qquad (7\text{-}48)$$

---

*See Sec. 7.5 and 8.7 for further discussion of the significance of this and coupling coefficients.

$$\frac{d^2v_2}{dx^2} = Jv_2 + Kv_1 \tag{7-49}$$

$$\frac{d^2i_2}{dx^2} = Ji_1 + Ki_2 \tag{7-50}$$

$$\frac{d^2i_2}{dx^2} = Ji_2 + Ki_1 \tag{7-51}$$

It is instructive to derive the propagation constant for this more general case. This can be done in the usual manner by assuming general solutions to v_1 and v_2 of the form

$$v_1 = Ae^{-\gamma x} + Be^{\gamma x} \qquad v_2 = Ce^{-\gamma x} + De^{\gamma x} \tag{7-52}$$

where γ is the propagation constant. Substituting this into Eqs. (7-48) and (7-49)

$$\gamma^2 v_1 = Jv_1 + Kv_2 \tag{7-53}$$

$$\gamma^2 v_2 = Kv_1 + Jv_2 \tag{7-54}$$

Solving for v_2 in terms of v_1

$$v_2 = \frac{K}{\gamma^2 - J} v_1 \tag{7-55}$$

Substituting this into Eq. (7-53) and simplifying

$$\gamma^4 - 2J\gamma^2 + J^2 - K^2 = 0 \tag{7-56}$$

or

$$G^2 - 2JG + (J^2 - K^2) = 0 \tag{7-57}$$

where

$$G = \gamma^2 \tag{7-58}$$

The solution to this, by means of the quadratic formula, is

$$G = \gamma^2 = J \pm K \qquad (7\text{-}59)$$

Since, for an ideal line, γ is proportional to the velocity of propagation, it is apparent that two different velocities are possible for this general coupled line. In Sec. 7.5 we will derive this expression in a different manner, and it will be seen that the (+) sign in Eq. (7-59) corresponds to the "sum" mode, while the (-) sign represents the "difference" mode.

7.3 DIRECTIONAL COUPLING FOR DISCRETE ELEMENTS AND SHORT LINES

Devices which make use of two coupled transmission lines and which are commonly known as directional couplers find a number of uses, including directional coupling of power and separation of the incident (direct) and reflected waves occurring on an ordinary transmission line. For the latter, an indirect means of separating incident and reflected waves is to measure the voltage standing wave ratio (Sec. 2.8) by means of a calibrated detector which can be moved along the line. However, a much more direct measurement can be made by using a directional coupler to separate the incident and reflected waves and hence measure them individually. Such devices have found widespread use with transmission lines, and even greater use in the field of microwaves where probing into wave guides is often difficult.*

Directional coupling can take many forms, ranging from very simple to very complex structures. However, the fundamental principles remain the same in all cases and rely on the fact that an applied wave traveling in one direction on a given transmission line will induce essentially two waves on a second line coupled to the first line. By proper choice of the coupling effects, waves traveling on the second line can be made to cancel in one direction and add in the opposite direction; hence the name directional coupling.

There are numerous ways in which the coupling effects can be obtained, but they generally fall into two categories, namely, coupling over short sections and coupling over long sections of transmission lines. The terms long and short are used with reference to the wavelength of the applied sinusoidal

*For a discussion of waveguide directional couplers, see [8]; for a brief history and bibliography of directional couplers, see [1].

voltage. For the former, the coupling section can be considered to be a lumped circuit element, whereas for the latter, the coupling section must be analyzed in terms of the coupled differential equations derived in Sec. 7.2. We will consider the former, i.e., coupling over short lines, in this section, and will defer the latter until Sec. 7.4.

There are, in fact, two cases of particular interest in this category, one merely being a limiting case of long coupled lines where the two lines are inductively and capacitively coupled but the length of coupling is very small. Thus, L_{11}, M_{12}, C_{11}, and C_{12} of Sec. 7.2 can be considered to be discrete elements coupling two transmission lines at a single point. Another case would be that for which there exists only inductive* coupling at a point between the two lines. This could be obtained, for instance, with a small transformer. Clearly, the two cases are different and must be analyzed separately.

The simplest case is that of two lines which are connected through coupling elements which are sensitive only to current and not to voltage, i.e., inductive coupling such as that shown in Fig. 7-4. Any number of such coupling elements can be placed between the lines with a different response obtained for the different cases. We will see that with two or more coupling elements, the system always exhibits directional coupling and the bandwidth over which infinite directivity (ideally) exists will be found to increase with an increasing number of such coupling elements. We shall now explore such coupling in greater detail.

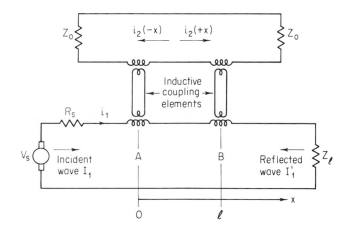

Fig. 7-4. Two lines coupled by two inductive elements.

*Another case would obviously be capacitive coupling alone, but this is not considered here.

(a) *Two-element inductively coupled directional coupler.** The case to be considered is shown in Fig. 7-4. In order to simplify the analysis and illustrate the fundamental features, it will be assumed that the lines are all lossless and have the same phase velocity, that line 2 is perfectly matched while line 1 need not be, that the coupling coefficient is very small so as not to cause any additional reflections, and that the coupling coefficients have no losses or phase shift, i.e., ideal.

The current in the driven line 1 at any point x is given by the general form of Eq. (2-13a) and is (assuming sinusoidal excitation)

$$i_x = I_1 e^{-j\beta x} + I_1' e^{j\beta x} \tag{7-60}$$

I_1 and I_1' are the amplitudes of the incident and reflected waves traveling in the $+x$ and $-x$ directions, respectively, with x being equal to 0 at the location of the first coupling element as shown. If the second coupling element is located at the distance $x = \ell$, then the amplitude and phase (with respect to $x = 0$) of the currents in line 1 at the locations of these two coupling elements are

$$i_1 \Big|_{x=0} = I_1 + I_1' = I_A \tag{7-61}$$

$$i_1 \Big|_{x=\ell} = I_1 e^{-j\beta\ell} + I_1' e^{j\beta\ell} = I_B \tag{7-62}$$

The coupling element is characterized by a coupling coefficient k, which relates the amount of current induced in line 2 per unit current flowing in line 1

$$i_2(x) = k i_1(x) \tag{7-63}$$

for any given value of x. Each coupling element will individually induce two waves in line 2, one traveling toward the right and the other toward the left. The two waves traveling in the $+x$ direction toward the right will add to give

$$i_2(+x) = k i_1(x=0) e^{-j\beta\ell} + k i_1(x=\ell) e^{-j\beta(x-\ell)} \tag{7-64}$$

$$= k I_A e^{-j\beta\ell} + k I_B e^{j\beta(x-\ell)}$$

*This analysis is similar to that given by Mumford [6].

The second exponential term is written in terms of $x - \ell$, since the current induced in line 2 at coupling element B must be out of phase with that at A by an amount $x = \ell$. The $x - \ell$ term then simply represents the inherent delay. Substituting Eqs. (7-61) and (7-62) into Eq. (7-64) gives

$$i_2(+x) = k e^{-j\beta x}[2I_1 + I_1'(1 + e^{j2\beta\ell})]$$

$$= 2k[I_1 e^{j\beta x} + I_1' e^{-j\beta(x-\ell)} \cos\beta\ell] \tag{7-65}$$

At the end of the line, $x = \ell$, and Eq. (7-65) yields

$$i_2(+x)\bigg|_{x=\ell} = 2k[I_1 e^{j\beta\ell} + I_1' \cos\beta\ell] \tag{7-66}$$

In a similar manner, the current in line 2 traveling in the $-x$ direction toward the left is

$$i_2(-x) = kI_A e^{j\beta x} + kI_B e^{j\beta(x-\ell)} \tag{7-67}$$

$$= 2k[I_1' e^{j\beta x} + I_1 e^{j\beta(x-\ell)} \cos\beta\ell] \tag{7-68}$$

At the left end of line 2, $x = 0$, and the above becomes

$$i_2(-x)\bigg|_{x=0} = 2k[I_1' + I_1 e^{-j\beta\ell} \cos\beta\ell] \tag{7-69}$$

Equations (7-65) and (7-68) are similar to the usual exponential forms for traveling waves except that the amplitude of each is determined by the line parameters as given by the terms in brackets in each equation. It can be seen that if the distance ℓ between couplers is changed, the amplitude of these two oppositely traveling waves, and hence the coupling, will vary. For instance, the current in line 2 flowing in the $+x$ direction is composed of the initial incident wave I_1 applied to line 1 multiplied by the coupling coefficient, plus the reflected wave I_1' in line 1 multiplied by a factor which includes the distance between couplers.

The amplitudes of these two oppositely traveling waves are given by the real parts of Eqs. (7-65) and (7-68) or Eqs. (7-66) and (7-69)

$$I_2(+x) = 2k[I_1 + I_1' \cos\beta\ell] \tag{7-70}$$

$$I_2(-x) = 2k[I_1' + I_1 \cos\beta\ell] \tag{7-71}$$

These two amplitudes are plotted in Fig. 7-5(a) and (b). It is easily seen from Fig. 7-5(b) that if the couplers are separated by a distance $\ell = \lambda/4$ (λ is the wavelength of the applied excitation), then the current and therefore voltage appearing at the left side of line 2 will be proportional only to the reflected wave $2kI_1'$ in the primary line, while the current flowing to the right will be proportional only to the incident wave $2kI_1$ (Fig. 7-5(a)). Thus, the reflection coefficient of the primary line can easily be obtained by taking the ratio of these two currents, provided that line 2 is properly terminated, i.e.,

$$\rho_1 = \frac{i_2(-x)}{i_2(+x)} \quad \text{at} \quad \ell = \frac{\lambda}{4} \tag{7-72}$$

If the spacing between the couplers is not exactly one quarter wavelength, the reflection coefficient cannot be obtained from Eq. (7-72). If line 1 is properly terminated so that $I_1' = 0$, Fig. 7-5(b) shows that a current proportional to

$$kI_1(1 + e^{-j2\beta\ell})$$

will still flow to the left side of line 2 while a current of kI_1 flows to the right side. The ratio of these two currents is commonly known as the directivity of a coupler. The directivity is expressed in terms of the power ratio in dB. Since the power is proportional to the current squared times terminating impedance, and since line 2 is assumed to properly matched,

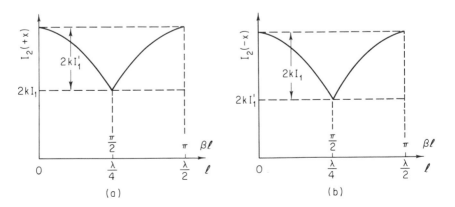

Fig. 7-5. Amplitude of currents flowing in the secondary line of a two-element directional coupler. (a) Current in $+x$ direction; (b) current in $-x$ direction.

the directivity for any wavelength is easily seen to be (assuming line 1 properly terminated)

$$dB = 10 \log \left| \frac{i_2(-x)}{i_2(+x)} \right|^2 = 20 \log \left| \frac{1 + e^{-j2\beta\ell}}{2} \right| \qquad (7\text{-}73)$$

It is obvious from Eq. (7-73) that the directivity for this case can never be infinite and in fact will always be negative (or zero), since the current flowing toward the right will exceed or equal the current flowing toward the left.

It is apparent that this two-element coupler is not an especially good device since the directivity is not good and the wavelength must be properly adjusted to enable one to directly obtain the reflection coefficient of line 1. Nevertheless, it illustrates the basic principles and serves as a foundation for more complex couplers.

(b) Multielement inductively coupled directional coupler. The properties of the above two-element coupler can be greatly improved by the incorporation of several coupling elements, all equally spaced. This will provide a larger frequency range of usefulness, as will be evidenced by a less-sensitive frequency dependence of the reflection coefficient.

Let us now consider the multielement coupler shown in Fig. 7-6. It is assumed that there are $n + 1$ coupling elements all a distance ℓ apart which are similar to those of Fig. 7-4. The value of the coupling coefficient k is

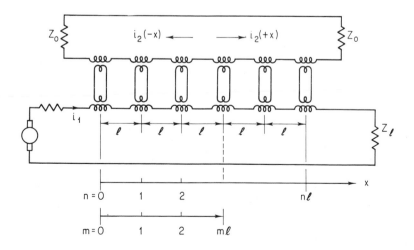

Fig. 7-6. Two lines coupled by many inductive elements.

not necessarily identical for each element, and we shall see that a judicious choice can result in an improved bandwidth.

In order to avoid confusion, we will arbitrarily label the coupling elements starting with 0: The first physical element is called the 0th order ($n = 0$), since it is located at $x = 0$; the second physical element is called the 1st ($n = 1$) order, since it is located at a distance $x = 1\ell$, etc. If there are, for example, seven physical elements, then $n = 6$. If this is not done, factors of ± 1 must be included in the expressions which are unnecessary. We will consider an arbitrary element, call it the mth element, located at a distance $x_m = m\ell$ from the first element, as shown. The current in line 1 at this mth element is given by an equation similar to Eq. (7-60), namely

$$i_1\bigg|_{x = m\ell} = I_1 e^{-j\beta m\ell} + I_1' e^{j\beta m\ell} \tag{7-74}$$

where I_1 and I_1' are the incident and reflected waves on line 1, respectively. This represents the total amplitude and phase of current at any one element along line 1.

Proceeding as in the previous example and referring all waves to the input at $x = 0$, the current in line 2, traveling in the $+x$ direction, is given by the summation of all currents from elements m, where m is summed from 0 to n

$$i_2(+x) = \sum_{m=0}^{n} k_m i_1 (x = m\ell) e^{-j\beta(x - m\ell)} \tag{7-75}$$

$$= \sum_{m=0}^{n} k_m \left(I_1 e^{-j\beta m\ell} + I_1' e^{j\beta m\ell} \right) e^{-j\beta(x - m\ell)} \tag{7-76}$$

$$= e^{-j\beta x} \sum_{m=0}^{n} \left[k_m \left(I_1 + I_1' e^{j2\beta m\ell} \right) \right] \tag{7-77}$$

where k_m is the coupling coefficient at the mth element.

In a similar manner, the current in line 2 traveling in the $-x$ direction is

$$i_2(-x) = \sum_{m=0}^{n} k_m i_1 (x = m\ell) e^{j\beta(x - m\ell)} \tag{7-78}$$

$$= e^{j\beta x} \sum_{m=0}^{n} \left[k_m \left(I_1 e^{-j2\beta m\ell} + I_1' \right) \right] \tag{7-79}$$

It can be seen that the current expressions of Eqs. (7-77) and (7-79) are somewhat complex and in general cannot be specified in a closed-form analytical expression. However, by properly choosing the coupling coefficients, some very simple expressions can be obtained. Let us investigate the current in line 2 traveling in the $+x$ direction and obtain an expression at the end of the line, namely, at $x = n$. From Eq. (7-77), this would be

$$i_2(+x) = e^{-j\beta n\ell}\left[I_1 \sum_{m=0}^{n} k_m + I_1' \sum_{m=0}^{n} k_m e^{j2\beta m\ell}\right] \qquad (7\text{-}80)$$

Writing out the first and last few terms of the above, we get, after some simplification

$$i_2(+x) = e^{-j\beta n\ell} I_1 (k_0 + k_1 + \cdots k_{n-1} + k_n) + I_1'\left[\left(k_0 e^{-j\beta n\ell} + k_n e^{j\beta n\ell}\right)\right.$$
$$\left. + \left(k_1 e^{-j\beta\ell(n-2)} + k_{n-1} e^{j\beta\ell(n-2)}\right) + \left(k_2 e^{-j\beta\ell(n-4)} + \text{etc.}\right)\right]$$

$$(7\text{-}81)$$

The first part of this involving summation of k_m is relatively simple, and it is the remaining expression which requires simplification. It is easily seen by inspection that there is a certain symmetry which is suggestive of the symmetry appearing in the binomial theorem. If we specify that

$$k_0 = k_n \qquad k_1 = k_{n-1} \qquad k_2 = k_{n-2} \qquad \text{etc.} \qquad (7\text{-}82)$$

then terms in parentheses are all of the form $\cos[\beta\ell(n - 2m)]$. In order to get the final closed-form solution, it is necessary to recognize that

$$(2\cos\theta)^n = (e^{j\theta} + e^{-j\theta})^n = e^{jn\theta} + e^{-jn\theta}$$

$$\frac{n}{1!}(e^{j\theta(n-2)} + e^{-j\theta(n-2)}) + \frac{n(n-1)}{2} \quad \text{etc.} \qquad (7\text{-}83)$$

where $\theta = \beta\ell$. If we make the coupling coefficients equal to the coefficients of this binomial expansion,* namely

$$k_m = k_0 \frac{n!}{m!(n-m)!} \qquad (7\text{-}84)$$

*From [6, p. 162].

then the second part of Eq. (7-81) reduces to a very simple form, and current in the load at $x = n$ from Eq. (7-77) is simply

$$i_2(+x) = e^{-j\beta n} \, 2^n k_0 I_1 + k_0 (2 \cos \beta \ell)^n I_1' \qquad (7\text{-}85)$$

where use has been made of the identity

$$\sum_{m=0}^{n} k_m = k_0 \sum_{m=0}^{n} \frac{n!}{m!(n-m)!} = 2^n k_0 \qquad (7\text{-}86)$$

to simplify the first part of Eq. (7-81).

The current in line 2 flowing in the $-x$ direction at the left-hand load of Fig. 7-6 can be found from Eq. (7-79). If we let $x = 0$ and proceed in a manner similar to that above, we will find that

$$i_2(-x) = k_0 I_1 \, e^{-jn\beta\ell} (2 \cos \beta \ell)^n + 2^n k_0 I_1' \qquad (7\text{-}87)$$

where the coupling coefficients are still given by Eq. (7-84). This equation is similar to Eq. (7-85) except that the roles of I_1 and I_1' as well as the role of the phase-shift factor are reversed.

The amplitudes of the current in line 2 in the right- and left-hand load are given by the real parts of Eqs. (7-85) and (7-87) and are, respectively

$$I_2(+x) = 2^n k_0 \left(I_1 + I_1' \cos^n \beta \ell \right) \qquad (7\text{-}88)$$

$$I_2(-x) = 2^n k_0 \left(I_1' + I_1 \cos^n \beta \ell \right) \qquad (7\text{-}89)$$

It is obvious that when there are only 2 coupling elements, that is, $n = 1$, then these equations become identical to Eqs. (7-66) and (7-69), as expected. Equations (7-88) and (7-89) are plotted in Fig. 7-7 for several values of n. The curve $n = 1$ is identical to that of Fig. 7-5 and requires a well-tuned arrangement in order for the reflection coefficient to be found. It is apparent that as the number of coupling elements is increased, the circuit becomes more broadband, permitting a much more poorly-tuned arrangement. It should be kept in mind that as the number of coupling elements is increased, more energy will be taken out of line 1 and may eventually become large enough to invalidate the above analysis.

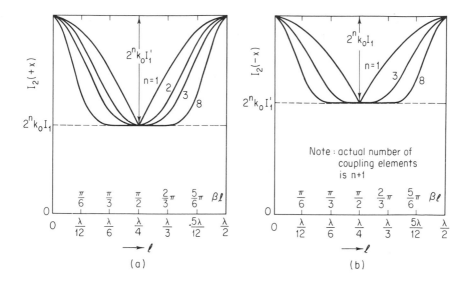

Fig. 7-7. Amplitude of current in line 2 vs. distance between the coupling elements for several values of n: current flowing in (a) $+x$ direction and (b) $-x$ direction.

Equation (7-72) can be used, as before, to obtain the reflection coefficient, and a direct measurement can be made for a substantial range of frequencies if nine coupling elements, that is, $n = 8$, are used, permitting approximately a 3/1 frequency variation.

The above two examples illustrate the basic properties of directional couplers, namely, that the addition and subtraction of various waves traveling in opposite directions on the secondary line can have interesting properties by the nature in which these various waves add to give the composite wave. Such devices find applications as attenuators and in the measurement of reflection coefficients, for example.

When a more complicated coupling element is used, the analysis is more involved, but the essential idea remains the same, namely, interesting properties result from the way in which various oppositely traveling waves add together.

(c) Inductive and capacitive coupling. When both capacitive and inductive coupling are present, the analysis and results are somewhat different, although directional properties are still obtained. Consider the case of a long transmission line, excited by a sinusoidal voltage and coupled to a very short line segment, as shown in Fig. 7-8. The general case, for which the two lines have different parameters but are otherwise uniform and lossless,

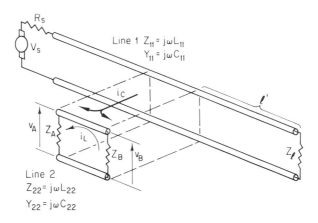

Fig. 7-8. Two lines coupled over a short distance.

is assumed. The lines are coupled such that

$$M_{12} = k\sqrt{L_{11}L_{22}} \qquad C_{12} = k\sqrt{C_{11}C_{22}} \qquad (7\text{-}90)$$

where the capacitive and inductive coupling coefficients are assumed equal.* We wish to examine the coupling properties of these two lines, and, in particular, to find the voltage v_B across load Z_B on line 2 when line 1 is terminated in its characteristic impedance. A wave traveling from source to load will induce a capacitive current and inductive voltage given by

$$i_C = C_{12}\frac{dv_1}{dt} \qquad v_L = M_{12}\frac{di_1}{dt} \qquad (7\text{-}91)$$

where the expression for i_C is valid only for small coupling k.[†] It is apparent that the capacitance coupling attempts to produce voltages v_A and v_B of the same polarity (as shown) across Z_A and Z_B. On the other hand, the inductive coupling attempts to make v_A more positive, but gives a negative contribution to v_B. The extent to which these two induced components add and/or subtract will obviously depend on the terminating resistors. The

---

*See Secs. 7.2 and 8.7 for further discussion of coupling coefficients.

   †The difficulty with large values of k occurs because of the fact that voltage on line 2 for large k cannot be neglected, which results in a requirement for additional terms in i_C (see Eq. (7-12)) and these terms become difficult to define in lumped-circuit terms.

voltage v_B can easily be found by subtracting the individual contribution from the capacitive and inductive components

$$v_B = Z_B \left(i_C \frac{Z_A}{Z_A + Z_B} - \frac{v_L}{Z_A + Z_B} \right) \tag{7-92}$$

Substitution of Eq. (7-91) and simplifying yields

$$v_B = \frac{Z_B}{Z_A + Z_B} \left(Z_A C_{12} \frac{dv_1}{dt} - M_{12} \frac{di_1}{dt} \right) \tag{7-93}$$

Observe that this voltage will be zero when the term in parentheses is zero or

$$Z_A \Big|_{v_B = 0} = \frac{M_{12}}{C_{12}} \frac{di_1}{dv_1} \tag{7-94}$$

Substituting Eq. (7-90) and recognizing that

$$\frac{di_1}{dv_1} = \sqrt{\frac{C_{11}}{L_{11}}} \tag{7-95}$$

reduces Eq. (7-94) to

$$Z_A \Big|_{v_B = 0} = \frac{k \sqrt{L_{11} L_{22}}}{k \sqrt{C_{11} C_{22}}} \sqrt{\frac{C_{11}}{L_{11}}} = \sqrt{\frac{L_{22}}{C_{22}}} \tag{7-96}$$

This is just the characteristic impedance of line 2 (in the presence of line 1 as described in Sec. 7.2). This result is of fundamental importance since it indicates that if line 2 is terminated at the left end in its characteristic impedance of Z_A, there is no voltage across Z_B, provided that k is small and that line 1 is properly terminated. Obviously, there must be a finite voltage across Z_A, so that such coupled lines are inherently directional, i.e., a wave traveling from left to right in line 1 induces a wave in line 2 which travels from right to left. It is further interesting to note that Z_B does not enter into the above specifications; thus, it can be any value and v_B will still be zero, provided that Z_A equals the line characteristic impedance. The reason that the directional coupling is independent of Z_B is easily seen from Eq. (7-92). The inductively and capacitively induced current in line 2 depends on Z_B in the same manner, so that changing Z_B changes these currents by the same amount. Thus, they are always equal in magnitude but opposite in sign and they give perfect directional coupling for all Z_B. It is to be further noticed that the directional coupling properties for this small coupled section are independent of the applied frequency. Thus, any frequency or sum of frequencies, and therefore any pulses, will still display the directional characteristics, provided that the original assumptions are not violated.

The question naturally arises as to what the directional properties of these coupled lines are when line 1 is not properly terminated. When this happens, there will be reflection which will result in a standing wave pattern on line 1. When Z_ℓ does not equal the line characteristic impedance, the input impedance seen at any point is a function of Z_ℓ and the length of line at that point. In order to see what effect this has on the directional properties, let us return to Eq. (7-94), which gives the fundamental relationship required for perfect directional coupling. From this, we find that $v_B = 0$ for

$$Z_A \frac{dv_1}{dt} = \frac{M_{12}}{C_{12}} \qquad \text{for small } k \qquad (7\text{-}97a)$$

where v_1 and i_1 are the values at the location of the coupled section. Since this ratio dv_1/di_1 merely represents the effective line impedance at that point, then it is only necessary that the product of this impedance multiplied by Z_A be equal to the ratio given by Eq. (7-97). Z_A and v_1/i_1 may individually contain real and/or imaginary parts, but the product must be a real number. This is a rather profound conclusion, namely, that for any given frequency the lines exhibit infinite directivity for any value of Z_ℓ, provided that the product of Z_A and the impedance of line 1, looking toward the load from the point of location of the coupler, equals M_{12}/C_{12}. A corollary of this, and a rather far-reaching conclusion, is that in Fig. 7-8, if $\ell' = 0$, then $dv_1/di_1 = Z_\ell$ and the lines exhibit infinite directivity for all Z_A and Z_ℓ which satisfy

$$Z_A Z_\ell = \frac{M_{12}}{C_{12}} \qquad \text{for small } k \qquad (7\text{-}97b)$$

The above equation is seen to be independent of the symmetry of the lines;

thus, for unsymmetrical lines, substitution of Eq. (7-90) for M_{12} and C_{12} yields

$$Z_A Z_\ell = \frac{\sqrt{L_{11} L_{22}}}{\sqrt{C_{11} C_{22}}} = Z_{011} Z_{022} \qquad \text{for small } k \qquad (7\text{-}98)$$

where $Z_{011} = \sqrt{L_{11}/C_{11}}$ and is the characteristic impedance of line 1 in the presence of line 2, and similarly for Z_{022}. Any amount of mismatch is tolerable on line 1 and infinite directivity can still be obtained. For the more general case of large k, the requirement for infinite directivity will be considered in the next section, where it will be seen that it is identical to Eq. (7-98), a rather interesting result.

7.4 DIRECTIONAL COUPLING FOR ANY LENGTH OF COUPLED SECTION ON OPEN WIRES

In Sec. 7.3, we considered the directional coupling properties of a section of line which was small compared to the wavelength of the applied frequency so that the coupling could be considered a lumped-circuit phenomenon. This approach provided some physical insight and showed the mechanism of directional coupling, but is of course just a special case of two lines coupled over a long length comparable to or longer than the applied wavelength. We now wish to consider this more general case of long coupled sections, where it will be found that for two coupled unsymmetrical lines with any value of coupling and a single frequency excitation, the condition for infinite directivity depends only on the coupling k and is independent of the applied frequency.

The problem we wish to consider consists of two lossless transmission line systems which are coupled over any given length, as shown in Fig. 7-9. The transmission lines can be of finite length with terminating resistors as shown, or the terminating resistors can be lengths of connecting transmission lines with the proper impedance, shown in dotted lines. For generality, we shall assume that the lines are unsymmetrical over their coupled length but that the coupling coefficients $k_L = k_C = k$ and can have any value, large or small, with $k \leq 1$ by definition. Thus, for sinusoidal excitation, the coupled transmission line equations (7-43) and (7-44) are

$$\frac{d^2 v_1}{dx^2} = Z_{11} Y_{11} (1 - k^2) v_1 \qquad (7\text{-}99)$$

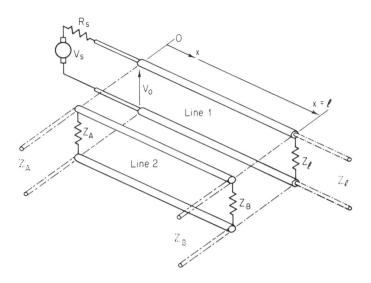

Fig. 7-9. Two transmission lines coupled over a long distance.

$$\frac{d^2 v_2}{dx^2} = Z_{11} Y_{11} (1 - k^2) v_2 \tag{7-100}$$

where the parameters are as defined in Sec. 7.2. Equations identical in form to those above exist for the currents i_1 and i_2 on the lines, i.e., Eqs. (7-45) and (7-46).

The general solution to these equations can be obtained by assuming that both positive and negative traveling waves exist on each line. In fact, from the results of Sec. 7.3 for the coupling of a short section of line, it is to be expected that both positive and negative traveling waves are present, since directional coupling is obtained only by these two oppositely traveling waves adding in one direction to give a net energy flow, and subtracting in the other direction to give no energy flow. Thus, we will assume that

$$v_1 = V_1 e^{-\gamma x} + V_1' e^{\gamma x} \tag{7-101}$$

$$v_2 = V_2 e^{-\gamma x} + V_2' e^{\gamma x} \tag{7-102}$$

$$i_1 = I_1 e^{-\gamma x} + I_1' e^{\gamma x} \tag{7-103}$$

$$i_2 = I_2 e^{-\gamma x} + I_2' e^{\gamma x} \tag{7-104}$$

The primed constants represent the amplitudes of the negatively traveling waves while unprimed constants are the amplitudes of the positively traveling waves. If these expressions are substituted into their respective differential equations, it is found that

$$\gamma = \sqrt{Z_{11}Y_{11}\,(1-k^2)} \;=\; \sqrt{Z_{22}Y_{22}\,(1-k^2)} \tag{7-105}$$

Since current and voltage of a given traveling wave are always related by the line parameters, it is possible to get the current expressions of Eqs. (7-103) and (7-104) in terms of voltage. This can be done by substituting Eqs. (7-101), (7-102), and (7-103) into Eq. (7-10) to get

$$-\gamma l_1 e^{\gamma x} + \gamma l_1' e^{\gamma x} = -Y_{11}\!\left(V_1 e^{-\gamma x} + V_1' e^{\gamma x}\right) + Y_{12}\!\left(V_2 e^{-\gamma x} + V_2' e^{\gamma x}\right) \tag{7-106}$$

Equating like terms in $e^{-\gamma x}$ and $e^{\gamma x}$ separately and using

$$G_{11} = \frac{Y_{11}}{\gamma} = \imath C_{11} \qquad G_{22} = \frac{Y_{22}}{\gamma} = \imath C_{22} \qquad G_{12} = \frac{Y_{12}}{\gamma} = \imath C_{12} \tag{7-107}$$

results in

$$I_1 = \imath(C_{11}V_1 - C_{12}V_2) = G_{11}V_1 - G_{12}V_2 \tag{7-108}$$

$$I_1' = -\imath\!\left(C_{11}V_1' - C_{12}V_2'\right) = -\left(G_{11}V_1' - G_{12}V_2'\right) \tag{7-109}$$

where

$$\imath = (\mu\epsilon)^{-\frac{1}{2}} \tag{7-110}$$

is the phase velocity in the bulk dielectric. Proceeding in a similar manner from Eq. (7-12)

$$I_2 = G_{22}V_2 - G_{12}V_1 \tag{7-111}$$

$$I_2' = -\left(G_{22}V_2' - G_{12}V_1'\right) \tag{7-112}$$

The set of equations which describe this system is therefore

$$v_1 = V_1 e^{-\gamma x} + V_1' e^{\gamma x} \tag{7-113}$$

$$v_2 = V_2 e^{-\gamma x} + V_2' e^{\gamma x} \tag{7-114}$$

$$i_1 = (G_{11}V_1 - G_{12}V_2)e^{-\gamma x} - \left(G_{11}V_1' - G_{12}V_2'\right)e^{\gamma x} \tag{7-115}$$

$$i_2 = (G_{22}V_2 - G_{12}V_1)e^{-\gamma x} - \left(G_{22}V_2' - G_{12}V_1'\right)e^{\gamma x} \tag{7-116}$$

These four equations require boundary conditions in order to evaluate the unknown constants, the most general conditions being as follows. Referring to Fig. 7-9, at the end of line 1 at $x = \ell$, the total voltage must equal total current times the load impedance. On line 2 at $x = 0$ and $x = \ell$, the total voltage must be the total current times the load impedances at those points respectively. At $x = 0$, the voltage across line 1 is the source voltage minus the iR drop in the source impedance R_s. These boundary conditions can be summarized as

$$v_1\Big|_{x=\ell} = Z_\ell i_1\Big|_{x=\ell} \tag{7-117}$$

$$v_2\Big|_{x=\ell} = Z_B i_2\Big|_{x=\ell} \tag{7-118}$$

$$v_2\Big|_{x=0} = -Z_A i_2\Big|_{x=0} \tag{7-119}$$

$$v_1\Big|_{x=0} = V_s - R_s i_1\Big|_{x=0} \tag{7-120}$$

It is possible to simplify the analysis by assuming infinite directivity, which makes Eq. (7-118) equal to 0. However, this assumption will not be made; instead, the result will be derived as follows. If Eqs. (7-113) through (7-116) are substituted into the above boundary conditions, then Eqs. (7-117) through (7-120) become, after rearrangement of terms

$$e^{-\gamma \ell}(1 - Z_\ell G_{11})V_1 + e^{\gamma \ell}(1 + Z_\ell G_{11})V_1' + e^{-\gamma \ell}Z_\ell G_{12}V_2 - e^{-\gamma \ell}Z_\ell G_{12}V_2' = 0$$

$$e^{-\gamma \ell}Z_B G_{12}V_1 - e^{\gamma \ell}Z_B G_{12}V_1' + e^{-\gamma \ell}(1 - Z_B G_{22})V_2 + e^{\gamma \ell}(1 + Z_B G_{22})V_2' = 0$$

$$-Z_A G_{12}V_1 + Z_A G_{12}V_1' + (1 + Z_A G_{22})V_2 + (1 - Z_A G_{22})V_2' = 0$$

$$(1 + R_s G_{11})V_1 + (1 - R_s G_{11})V_1' - R_s G_{12}V_2 + R_s G_{12}V_2' = V_s$$

where R_s = source impedance. In order to express more simply and solve the above set of simultaneous equations, it is convenient to express these in terms of matrix notation

$$
\begin{bmatrix}
a_1 & b_1 & c_1 & d_1 \\
a_2 & b_2 & c_2 & d_2 \\
a_3 & b_3 & c_3 & d_3 \\
a_4 & b_4 & c_4 & d_4
\end{bmatrix}
\times
\begin{bmatrix}
V_1 \\
V_1' \\
V_2 \\
V_2'
\end{bmatrix}
=
\begin{bmatrix}
0 \\
0 \\
0 \\
V_s
\end{bmatrix}
\tag{7-121}
$$

where

$$
a_1 = e^{-\gamma \ell}(1 - Z_\ell G_{11}), \quad a_2 = e^{-\gamma \ell} Z_B G_{12}, \quad b_1 = e^{\gamma \ell}(1 + Z_\ell G_{11})
$$

etc., respectively. The four unknown constants can be evaluated by the use of determinants. By expanding in terms of minors to make use of the zero terms, it is easily shown that

$$
V_1 = -V_s \frac{\begin{vmatrix} b_1 & c_1 & d_1 \\ b_2 & c_2 & d_2 \\ b_3 & c_3 & d_3 \end{vmatrix}}{\Delta}
\qquad
V_1' = V_s \frac{\begin{vmatrix} a_1 & c_1 & d_1 \\ a_2 & c_2 & d_2 \\ a_3 & c_3 & d_3 \end{vmatrix}}{\Delta}
\tag{7-122a}
$$

$$
V_2 = -V_s \frac{\begin{vmatrix} a_1 & b_1 & d_1 \\ a_2 & b_2 & d_2 \\ a_3 & b_3 & d_3 \end{vmatrix}}{\Delta}
\qquad
V_2' = V_s \frac{\begin{vmatrix} a_1 & b_1 & c_1 \\ a_2 & b_2 & c_2 \\ a_3 & b_3 & c_3 \end{vmatrix}}{\Delta}
\tag{7-122b}
$$

where Δ is the determinant of the coefficients a, b, c, and d in Eq. (7-121). It is apparent that the solutions for these constants will be very unwieldy. Fortunately, many significant conclusions and insights can be obtained without the complete solution. For instance, the condition for infinite directivity can be found by noting from Eq. (7-114) that in order to have $v_2 = 0$ at the end of the line, it is necessary to have

$$
V_2 e^{-\gamma \ell} + V_2' e^{\gamma \ell} = 0 \quad \text{or} \quad \frac{V_2'}{V_2} = -e^{-2\gamma \ell}
$$

From Eq. (7-122b), this requires that

$$\frac{-V_2'}{V_2} = e^{-2\gamma\ell} = \frac{\begin{vmatrix} a_1 & b_1 & c_1 \\ a_2 & b_2 & c_2 \\ a_3 & b_3 & c_3 \end{vmatrix}}{\begin{vmatrix} a_1 & b_1 & d_1 \\ a_2 & b_2 & d_2 \\ a_3 & b_3 & d_3 \end{vmatrix}} \tag{7-123}$$

After substitution of appropriate expressions and much simplification, the result is

$$Z_A Z_\ell = \frac{1}{G_{11} G_{22} (1 - k^2)} = Z_{011} Z_{022} = \frac{M_{12}}{C_{12}} \tag{7-124}$$

where

$$Z_{011} = \sqrt{L_{11}/C_{11}} \qquad Z_{022} = \sqrt{L_{22}/C_{22}}$$

and they represent the characteristic impedance of line 1 in the presence of line 2 and that of line 2 in the presence of line 1, respectively. This is the necessary and sufficient condition for infinite directivity for the most general case of unsymmetrical lines and any value of k, R_s, and Z_B* and is identical to Eq. (7-98), which was derived for small k. This condition is also independent of frequency[†] and is therefore applicable to pulses, an important result. It is further interesting to note that infinite directivity is obtained not only independent of R_s and Z_B, but also even though there are negatively traveling waves on line 1, that is, V_1' is not zero, which is obvious since any value of Z_ℓ can be used so long as Eq. (7-124) is satisfied. It is possible to match line 1 to make $V_1' = 0$; the value of Z_ℓ required to achieve this can be obtained by setting $V_1' = 0$ in Eq. (7-122a). After much algebraic manipulation, the result is

$$Z_\ell \bigg|_{V_1'=0} = \frac{Z_{011}}{\sqrt{1 - k^2}} = Z_{01} \tag{7-125}$$

---

*These results, although derived in a somewhat different manner, are in agreement with Cristal [2], although the notation is different.

†Knechtli [3] derives a condition for infinite directivity which depends on the applied frequency, thus contradicting these results. He sets up the problem correctly, but after a painstaking effort and thorough checking, the author concludes that Knechtli apparently made an algebraic error.

where Z_{01} is the characteristic impedance of line 1 in the absence of line 2. Since Eq. (7-124) must still be satisfied, it is necessary that

$$Z_A\Big|_{V_1'=0} = Z_{022}\sqrt{1-k^2} = Z_{02}(1-k^2) \tag{7-126}$$

where Z_{02} is the impedance of line 2 in the absence of line 1. Note that from Eq. (7-125) the termination required to prevent reflections on line 1 is not the line characteristic impedance Z_{011} but rather must be the impedance of line 1 in the absence of line 2.

It should be realized that even though the condition for infinite directivity, that is, $v_2 = 0$ at $x = \ell$, is independent of frequency, the voltage amplitude on line 2 at $x = 0$ (across Z_A) will vary with frequency. So for the general case, either with or without $V_1' = 0$, pulses observed at Z_A will be distorted even though infinite directivity is still obtained.

It should be noted that even though the directivity of the line is independent of R_s, the amplitude of V_1, V_1', V_2, and V_2' will change with R_s, as is to be expected since the power delivered to the line from the generator depends on the source impedance. If the lines are assumed to be symmetrical so that $Z_{011} = Z_{022}$ and if the source impedance is assumed equal to zero, then the expressions for the voltage amplitudes are greatly simplified* and can be found from Eq. (7-121), assuming infinite directivity

$$\frac{V_1}{V_s} = \frac{1 + Z_n(1-k^2)}{1 + Z_n(1-k^2) - e^{-2\gamma\ell}\left[1 - Z_n(1-k^2)\right]} \tag{7-127}$$

$$\frac{V_1'}{V_s} = \frac{1 - Z_n(1-k^2)}{1 - Z_n(1-k^2) - e^{2\gamma\ell}\left[1 + Z_n(1-k^2)\right]} \tag{7-128}$$

$$\frac{V_2}{V_s} = \frac{k}{e^{-2\gamma\ell}\left[Z_n(1-k^2) - 1\right] + 1 + Z_n(1-k^2)} \tag{7-129}$$

$$\frac{V_2'}{V_s} = \frac{k}{1 - Z_n(1-k^2) - e^{2\gamma\ell}\left[1 + Z_n(1-k^2)\right]} \tag{7-130}$$

*This special case of $R_s = 0$ with the additional constraint that $Z_\ell = Z_{011} = Z_A$ is the case considered by Oliver [7], although he does not specify these limitations.

where $Z_n = Z_\ell/Z_{011}\sqrt{1-k^2}$. If it is further assumed that $Z_\ell = Z_{011} = Z_A$, then infinite directivity is obtained with a standing wave pattern on lines 1 and 2; the voltages on the lines can be obtained by substituting Eqs. (7-127) through (7-130) into Eqs. (7-101) and (7-102). Assuming no losses so that $\gamma = j\beta$, the results, after much simplification, are*

$$\frac{v_1}{V_s} = \frac{\sqrt{1-k^2}\,\cos\beta(\ell-x) + j\sin\beta(\ell-x)}{\sqrt{1-k^2}\,\cos\beta\ell + j\sin\beta\ell}$$

$$\frac{v_2}{V_s} = \frac{jk\,\sin\beta(\ell-x)}{\sqrt{1-k^2}\,\cos\beta\ell + j\sin\beta\ell}$$

These are valid for any k provided that $R_s = 0$ and $Z_\ell = Z_{011} = Z_A$. Some interesting conclusions for this case are as follows: the output on line 1 at $x = \ell$ is always $90°$ out of phase with the output on line 2 at $x = 0$. This can be seen by evaluating these two voltages as

$$\frac{v_1(x=\ell)}{v_2(x=0)} = \frac{k}{\sqrt{1-k^2}}\,\frac{e^{\gamma\ell}-e^{-\gamma\ell}}{2} = \frac{jk}{\sqrt{1-k^2}}\,\sin\gamma\ell$$

Thus, the ratio of the amplitudes will vary with the length ℓ of the coupled section as well as with k, but the two voltages will always be in quadrature.

Figure 7-10 shows the variation of the amplitude of the voltage along lines 1 and 2 for a fixed length of coupled section and medium coupling. For the particular case shown in Fig. 7-10(a), $\ell = \lambda_w/2$ and no voltage appears on the output of line 2 (at $x = 0$). If $\ell = \lambda_w/4$, maximum output on line 2 at $x = 0$ is obtained as in Fig. 7-10(b). Since we have assumed ideal lines, it is necessary that the power on line 1 plus that on line 2 be equal to the applied power or, since we assumed that $Z_A = Z_\ell$

$$|v_1|^2\Big|_{x=\ell} + |v_2|^2\Big|_{x=0} = V_s^2$$

This is seen actually to be the case, that is, $0.866^2 + 0.25^2 = 1$. The variation of the amplitude of the voltage on line 2 at $x = 0$ as a function of

*These are equivalent to those of Oliver [7].

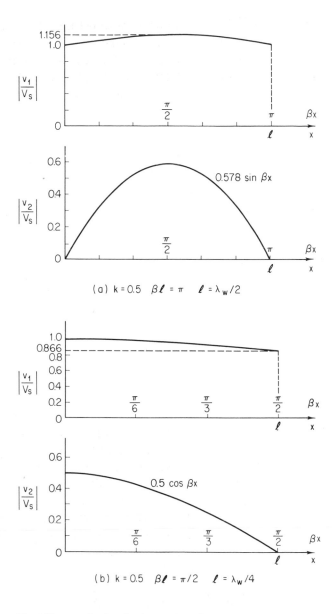

Fig. 7-10. Amplitude of voltages along lines 1 and 2 as a function of distance for two coupled lines.

the length of coupled section, for a fixed applied frequency, is shown in Fig. 7-11. This shows that maximum voltage and hence maximum power is delivered to line 2 when ℓ is an odd multiple of a quarter wavelength and

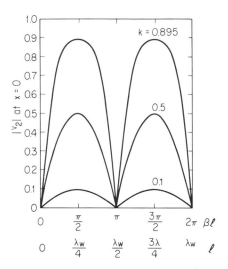

Fig. 7-11. Amplitude of v_2 $(x = 0)$ for various coupling coefficients k, vs. length of coupled section.

the maximum voltage always has a value of kV_s. No power is delivered to line 2, that is, to Z_A, when ℓ is an even multiple of a quarter wavelength, for example, a half wavelength.

For small value of coupling coefficient, k, Fig. 7-11 shows that for maximum power transfer to line 2, the frequency must be reasonably well adjusted to satisfy $\beta\ell = \pi/2$; in other words, the bandwidth of the device is not very broad. The bandwidth can be increased slightly by increasing the coupling coefficient as shown in Fig. 7-11, although the improvement is not obvious unless the curves are normalized. The phase angle of the various voltages relative to the input V_s will vary with both ℓ (or frequency) and k. We have already pointed out that v_2 at $x = 0$ is always in quadrature with v_1 at $x = \ell$. It can also be shown that if the coupled section is adjusted in length or the applied frequency adjusted such that maximum coupling occurs, for example, if $\beta\ell = \pi/2$ or integer multiples, then v_2 at $x = 0$ is always in phase with the applied voltage V_s. This example has been given for a special case; for more general values of R_s, Z_ℓ and Z_A, it is necessary to solve Eqs. (7-122a), (7-122b), and (7-123) with the coefficients given by Eq. (7-121) and the expression becomes very unwieldy.

For all the cases considered above, it is apparent that as long as the dielectric is uniform throughout all space, lines 1 and 2 must have the same velocity, assuming a TEM mode. The velocity can be obtained with the aid of Eq. (7-105) and can easily be shown to be for symmetrical or unsymmetrical lines

$$\varkappa = [L_{11} C_{11} (1 - k^2)]^{-\frac{1}{2}} = [L_{22} C_{22} (1 - k^2)]^{-\frac{1}{2}} \qquad (7\text{-}131)$$

For an applied sine wave, the output across Z_A in Fig. 7-9 will be sinusoidal in time although there may be a phase shift; however, for an applied pulse, the voltage across Z_A will not duplicate the input pulse. In fact, if a step function is applied to line 1, the output across Z_A will be a pulse with a base width equal to twice the delay time of line 2, that is, $2T_0 = 2\ell/\varkappa$, where $\varkappa$ is the phase velocity. If the applied voltage is a pulse of width greater than $2T_0$, then a positive pulse is generated on line 1 during the rise time, and a negative pulse during the fall time, both of width $2T_0$.

The reason why a pulse is obtained on line 2 for a step function applied pulse is that as the applied wavefront travels from $x = 0$ to ℓ there is a time changing flux and electric field coupled to line 2. Thus, energy is being stored in line 2 during this interval while at the same time some of the energy is being discharged into Z_A. When the applied wavefront passes point $x = \ell$, no more energy is being coupled to line 2 and it must complete its discharge in an additional time period T_0; thus, the line is charged and partially discharged for a time T_0 while the wavefront passes from $x = 0$ to ℓ and fully discharged in time T_0 thereafter, giving a pulse width of $2T_0$.

It should be noted that the rise time of the pulse on line 2 is unaffected by line losses and will be the same as the rise time of the applied pulse, but since the remainder of the energy must travel on the lines, the amplitude (droop) and fall time will be affected.*

Directional coupling effects can and often do occur in integrated circuits work where various signal lines run parallel to other lines for short distances. These distances are usually much shorter than the pulse width, so that the above effects can occur. However, in integrated circuits, the transmission lines are usually not properly terminated and the loading effects (transistors) are nonlinear; thus the problem becomes more complex. Nevertheless, directional coupling will occur where lines are close together and it is important to understand the fundamental mechanisms and possible consequences.

---

*The same effect occurs with the discharging of a line as in Sec. 5.5—if the line of Fig. 5-4 is lossy, the pulse rise time is determined only by the switching mechanism and is uneffected by line characteristics. However, the pulse droop and fall time behavior will be affected by the losses.

7.5 SOLUTION TO COUPLED EQUATIONS IN TERMS OF SUM AND DIFFERENCE MODES

In Sec. 7.2, the differential equations for coupled lines were derived in terms of the individual voltages on the lines. In Sec. 7.4 these equations were used to examine the characteristics of a directional coupler. This approach is sufficient to give all the necessary information concerning the behavior of such a coupler, and, in general, will always be sufficient. However, it is sometimes more convenient to analyze certain cases in terms of the so-called "sum" and "difference" modes,* which allows easier visualization of the physical case. Such situations do, in fact, occur, as discussed in [4]. The advantage of using the sum and difference mode analyses is that it allows us to determine the impedances and propagation constants for these two modes directly. Since these quantities are, in general, different for these two modes, it provides physical insight, especially since these are equivalent to the primary and secondary modes discussed in [4]. We shall concentrate on deriving the basic equations as well as general expressions for impedance and propagation velocity.

In order to obtain the equations of coupled lines in the proper form, we can start with Eqs. (7-17) and (7-18) for the general case. If we assume the lines to be symmetrical, so that $J_1 = J_2$ and $K_1 = K_2$, then the equations to be solved are

$$\frac{d^2 v_1}{dx^2} = Jv_1 + Kv_2 \tag{7-132}$$

$$\frac{d^2 v_2}{dx^2} = Jv_2 + Kv_1 \tag{7-133}$$

To obtain a general solution; assume

$$v_1(x, t) = \left(V_1 e^{-\gamma_1 x} + V_1' e^{\gamma_1 x}\right) e^{j\omega t} \tag{7-134}$$

$$v_2(x, t) = \left(V_2 e^{-\gamma_2 x} + V_2' e^{\gamma_2 x}\right) e^{j\omega t} \tag{7-135}$$

V_1' and V_2' are the amplitudes of the negatively traveling waves in line 1 and 2, respectively. The propagation constants must be equal for both lines 1

---

*The sum and difference modes are sometimes referred to as the even and odd modes, respectively.

and 2, since they are assumed to be in the same medium. Thus

$$\gamma_1 = \gamma_2 = \gamma \tag{7-136}$$

Substituting Eqs. (7-134), (7-135) and (7-136), and leaving out the explicit time-dependence terms, Eqs. (7-132) and (7-133) become

$$\gamma^2 v_1 = J v_1 + K v_2 \tag{7-137}$$

$$\gamma^2 v_2 = K v_1 + J v_2 \tag{7-138}$$

where J and K are given by Eqs. (7-19) and (7-21), respectively. Adding Eqs. (7-137) and (7-138) and collecting terms yields

$$\gamma^2 (v_1 + v_2) = (J + K)(v_1 + v_2) \tag{7-139}$$

This represents the so-called "sum" mode of propagation, with the propagation constant given by

$$\gamma_s^2 = J + K \tag{7-140}$$

$$= K_{11} Y_{11} - Z_{12} Y_{21} + Z_{12} Y_{22} - Z_{11} Y_{12} \tag{7-141}$$

or for ideal lines

$$\gamma_s = j\omega [(L_{11} + M_{12})(C_{11} - C_{12})]^{1/2} = j\beta_s \tag{7-142}$$

Proceeding in a similar manner and subtracting Eq. (7-138) from Eq. (7-137) yields

$$\gamma^2 (v_1 - v_2) = (J - K)(v_1 - v_2) \tag{7-143}$$

This is commonly known as the "difference" mode, with the propagation constant given by

$$\gamma_d^2 = J - K \tag{7-144}$$

For an ideal line, this becomes

$$\gamma_d = j\omega \left[(L_{11} - M_{12})(C_{11} + C_{12}) \right]^{1/2} = j\beta_d \qquad (7\text{-}145)$$

The general solution is therefore

$$v_s = v_1 + v_2 = V_s e^{-\gamma_s x} + V'_s e^{\gamma_s x} \qquad (7\text{-}146)$$

where $V_s = V_1 + V_2$ and $V'_s = V'_1 + V'_2$ and they represent the positively and negatively traveling waves, respectively, for the sum mode. For the difference mode, the general solution is

$$v_d = v_1 - v_2 = V_d e^{-\gamma_d x} + V'_d e^{\gamma_d x} \qquad (7\text{-}147)$$

where $V_d = V_1 - V_2$ and $V'_d = V'_1 - V'_2$ and they represent the positively and negatively traveling waves, respectively, for the difference mode. The currents can be obtained in a similar manner

$$i_s = i_1 + i_2 = \frac{1}{Z_s} \left(V_s e^{-\gamma_s x} - V'_s e^{\gamma_s x} \right) \qquad (7\text{-}148)$$

$$i_d = i_1 - i_2 = \frac{1}{Z_d} \left(V_d e^{-\gamma_d x} - V'_d e^{\gamma_d x} \right) \qquad (7\text{-}149)$$

where Z_s and Z_d are the characteristic impedances (to be derived later).

The physical interpretations of the sum and difference modes are very simple and will be seen to be quite logical, since these interpretations are usually used naturally without one's awareness. Consider the two very long, ideal wires above a ground plane, as in Fig. 7-12, and assume that two traveling waves v_1 and v_2 are simultaneously launched down the lines. If these two voltages have different amplitudes, then it is obvious that there will exist a potential difference $v_1 - v_2$ between the two wires. In general, the electric and magnetic field pattern will be highly distorted, but, nevertheless, a plane wave will exist with the magnetic field orthogonal to E. If $v_2 = -v_1$, the line will be balanced with a symmetrical field pattern as shown in Fig. 7-12(b). Thus, as v_1 and v_2 propagate between line 1 and the ground plane, respectively, it is also apparent that a plane wave propagates

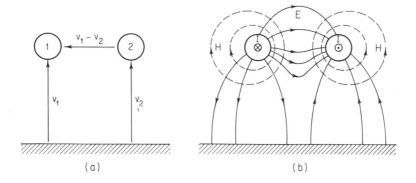

Fig. 7-12. Parallel wires above a ground plane. (a) Sum and difference voltages; (b) approximate field pattern for balanced excitation.

between the wires. It is easily seen that as the spacing and, therefore, the impedance is changed between the wires themselves or between the wires and the ground plane, the field pattern will change with a resulting change in the current distribution within the wires. The difference mode therefore represents the wave propagating between the wires, while the sum mode represents the waves traveling between each wire and the ground plane.

It is apparent from the previous discussion and physical reasoning that if the lines are symmetrical and located in the same medium, then the sum and difference modes must propagate at the same velocity. This can be obtained from Eqs. (7-142) and (7-145) by letting

$$Z_{11} = Z_{22} \quad \text{and} \quad Y_{11} = Y_{22} \tag{7-150}$$

Thus

$$Z_{12} = k_L \sqrt{Z_{11} Z_{22}} = k_L Z_{11} = k_L Z_{22} \tag{7-151}$$

$$Y_{12} = k_C \sqrt{Y_{11} Y_{22}} = k_C Y_{11} = k_C Y_{22} \tag{7-152}$$

Thus the propagation constants become

$$\gamma_s = \gamma_d = [Z_{11} Y_{11} (1 + k_L)(1 - k_C)]^{1/2} \tag{7-153}$$

which is identical to Eq. (7-47), since k_L must equal k_C for this case. Since phase velocity is proportional to the reciprocal of the propagation constant, then the velocities of propagation become

$$v_s = \frac{j\omega}{\gamma_s} = \left[Z_{11} Y_{11}(1 - k^2)\right]^{-\frac{1}{2}} = \frac{j\omega}{\gamma_s} = v_d \qquad (7\text{-}154)$$

Thus, the two modes for this case propagate at the same velocity as anticipated. However, in general, this is not necessarily true and, in fact, the different propagating velocities lead to advantages as well as difficulties in practical devices, as detailed in [4].

The characteristic impedance for the sum and difference modes can easily be obtained with the aid of Eqs. (7-9) and (7-11); substitution of Eqs. (7-134) and (7-135), using only positively traveling waves, yields

$$\frac{dv_1}{dx} = -Z_{11} i_1 - Z_{12} i_2 = -\gamma V_1 e^{-\gamma x} = -\gamma v_1$$

$$\frac{dv_2}{dx} = -Z_{12} i_1 - Z_{22} i_2 = -\gamma V_2 e^{-\gamma x} = -\gamma v_2$$

Adding these two for the sum mode and using Eq. (7-150) yields

$$\gamma_s(v_1 + v_2) = (Z_{11} + Z_{12})(i_1 + i_2)$$

Thus

$$Z_s = \frac{v_1 + v_2}{i_1 + i_2} = \frac{Z_{11} + Z_{12}}{\gamma_s} \qquad (7\text{-}155)$$

Subtracting the two equations for the difference mode yields

$$\gamma_d(v_1 - v_2) = -(Z_{12} - Z_{11})(i_1 - i_2)$$

Thus

$$Z_d = \frac{v_1 - v_2}{i_1 - i_2} = \frac{Z_{11} - Z_{12}}{\gamma_d} \qquad (7\text{-}156)$$

These impedance expressions can be simplified by assuming symmetrical lines and making use of Eqs. (7-150) through (7-153)

$$Z_s = \frac{Z_{11}(1 + k_L)}{\gamma_s} = \left[\frac{Z_{11}}{Y_{11}} \frac{(1 + k_L)}{(1 - k_C)} \right]^{1/2} \tag{7-157}$$

$$Z_d = \frac{Z_{11}(1 - k_L)}{\gamma_d} = \left[\frac{Z_{11}}{Y_{11}} \frac{(1 - k_L)}{(1 - k_C)} \right]^{1/2} \tag{7-158}$$

For the special case when $k_L = k_C$, it is apparent that since these coupling coefficients are always less than unity, then Z_s must be larger than Z_d. It should be pointed out that Z_d, because of the way it is obtained mathematically, is really only one-half of the actual characteristic impedance which would exist between line 1 and 2 if Z_s were infinitely large. This factor of one-half comes about because of the way the inductance is defined.

Additional discussions on evaluation of coupling coefficients are given in Sec. 8.7.

PROBLEMS

7-1. Prove that for lossless lines, the velocity of propagation on two coupled lines is $\nu = \left[L_{11} C_{11} (1 - k^2) \right]^{-1/2}$.

7-2. In the above problem, show that ν is independent of k, so that C_{11} must vary as $1/(1 - k^2)$; see Prob. 8-8.

7-3. Given: a word-organized memory array of drive conductors consisting of strip lines as in Fig. 8-4 between which are interposed figure 8 type sense loops as illustrated below (see [5] for practical example).

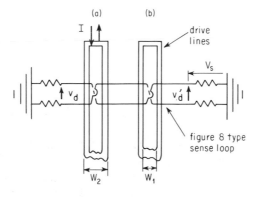

Fig. 7P-1.

Pulses are applied to the drive line and a voltage is measured across the two sense-line conductors.

(a) Show that in order to minimize the difference noise v_d on the sense line, it is desirable to have $W_1 = W_2$ and the sense line midway between the drive conductors.

(b) Show that a balanced drive and balance sensing arrangement give minimum difference noise on sense loop.

7-4. In the above problem, show that if an unbalanced, i.e., single-ended, drive is used on the drive lines and if the rise time is longer than the electrical length of the drive lines, then the amplitude of the single-ended or sum mode voltage v_s on the sense line will be directly proportional to the distance of the sense line from the generator end of the drive line. (Hint: assume that a ground plane is far away.) Also show that as the width to separation ratio increases, v_s decreases.

7-5. In Prob. 7-3, suppose that the sense loop is displaced off center, closer to one or the other conductor of the drive line. If the same situation exists for both lines (a) and (b), show that when drive line (a) is excited, the voltage v_d (on left end) will be larger or smaller than v'_d (on the right end) depending on the direction of displacement relative to the sense line geometry and the voltage polarity on the drive-line conductors. Also show that just the opposite effect is obtained when drive line (b) is excited, e.g., if exciting (a) causes $v_d > v'_d$, then exciting (b) will cause $v_d < v'_d$.

7-6. Prove that for two coupled lines with a very large sum-mode impedance (Sec. 7.5), the difference mode impedance is only one-half the characteristic impedance between lines 1 and 2.

REFERENCES

1. Caswell, W. E., and R. F. Schwartz: The Directional Coupler—1966, *IEEE Trans. Microwave Theory Tech.*, vol. MTT 15, no. 2, p. 120, February 1967.

2. Cristal, E. G.: Coupled Transmission Line Directional Couplers with Coupled Lines of Unequal Characteristic Impedances, *ibid.*, vol. MTT 14, no. 7, p. 337, July 1966.

3. Knechtli, R. C.: Further Analysis of Transmission Line Directional Couplers, *Proc. IRE*, p. 867, July 1955.

4. Matick, R. E.: Transmission Line Pulse Transformers—Theory and Applications, *Proc. IEEE*, vol. 56, no. 1, p. 47, January 1968.

5. Matick, R. E., P. Pleshko, C. Sie, and L. M. Terman: A High Speed Read Only Store Using Thick Magnetic Films, *IBM J. Res. Develop.*, vol. 10, no. 4, p. 333, July 1966.
6. Mumford, W. W.: Directional Couplers, *Proc. IRE,* vol. 35, p. 160, February 1947.
7. Oliver, B. M.: Directional Electromagnetic Couplers, *ibid.*, p. 1688, November 1954.
8. Levy, R., Directional Couplers, in L. Young (ed.), "Advances in Microwaves," vol. 1, p. 115, Academic Press, New York, 1966.

8 TRANSMISSION-LINE PARAMETERS

8.1 INTRODUCTION

In Sec. 1.2, we very briefly consider the parameters required to construct the equivalent incremental or lumped circuit of a transmission line. We considered the fundamental laws which govern the inductive and capacitive elements but did not evaluate these parameters for any given geometries. In this chapter, we shall evaluate the line parameters for various cases of practical interest and present normalized curves versus the geometrical ratios which are useful as a reference. Coupling coefficients as well as a method for measuring parameters of simple or complex geometries will also be discussed. The purpose of this chapter is to bring together a number of important and useful pieces of information into a coherent from with insights and some consequences of certain fundamental facts which bear on the day-to-day application of transmission lines and wave propagation. We shall discuss some important concepts which can be summarized as follows.

For a typical set of conductors of any cross-sectional geometry, excited at dc or low frequency, a true TEM mode cannot exist except for special cases because of the internal or surface impedance of the conductors (due to penetration of H). This increased inductance slows down a traveling wave because of the reduced phase velocity. For a round wire with uniform current distribution, this internal inductance is a constant, independent of wire radius or any other dimensions. This internal inductance also increases the impedance of the line:* however, for most cases of practical interest, this inductance is small compared to the external (circuit) inductance, so that its effect on phase velocity and Z_0 is often negligibly small. However, for cases of low Z_0, this surface impedance must be considered.

---

*This is also true for certain cases at very high frequency, e.g., a strip line at high frequency will have current crowding near the edges, causing an increase in Z_0.

As the frequency increases, the penetration depth decreases, and the mode becomes more nearly a true TEM. As the frequency is still further increased, the skin depth becomes so small that the current does indeed flow entirely on the surface of the conductor. A true TEM would exist except for the fact that the series resistance, increased by the decreased penetration depth, gives a small component of E in the direction of propagation. However, this is usually small in comparison to the transverse E component, so that the wave is very nearly a TEM.

For a transmission line with any cross-sectional geometry, but dimensions much smaller than the applied wavelength,* a TEM field configuration exists at any frequency, provided that the losses and conductor internal impedance are negligible—the latter is equivalent to assuming that the current flows entirely on the surface of the conductors, while the former is equivalent to assuming very large conductivity. When a TEM mode exists or is assumed to exist, there are several consequences of this. First, L and C are not independent parameters. If C is known, L is automatically known and will vary with the reciprocal of the capacitance, that is, L is proportional to $1/C$ with appropriate multiplying constants.† The coupling coefficients for L and C will thus be equal (see Sec. 8.7). Another consequence is that the velocity of propagation is independent of geometry and equals the velocity of light in the dielectric, that is, $\nu = (LC)^{-\frac{1}{2}}$. Thus, in order to calculate the characteristic impedance of a line to a TEM wave, it is only necessary to calculate either C or the external inductance.

The skin depth, series resistance, and internal self-inductance of the conductors are all intimately related. In Secs. 4.7 and 6.3, the penetration depth and surface impedance were derived for plane conductors with an assumed TEM wave traveling into or along the conductor surface, and it was found that the surface resistance and inductive reactance were equal in magnitude for all cases. The surface inductance is just the internal self-inductance that one would calculate from simple flux linkages and ac theory. An important consequence of this is that when the series resistance is negligible, the self (surface) inductance of the conductors is also negligible. Likewise, when the resistance is not negligible, the surface inductance is of the same order of magnitude and may need to be included in some cases, depending on the ratio of internal to external inductance. We will consider a typical coaxial cable as an example.

---

*If the dimensions become comparable to, or smaller than the applied wavelength, higher-order modes and even radiation effects may become important—see Sec. 8.8.

†This is proved and discussed in greater detail in Appendix 8A.

In Chap. 1, and throughout the book, in fact, it has been shown that the inductance and capacitance per unit length give rise to the characteristic impedance Z_0, phase shift per unit length β, and phase velocity ν or delay time T_0 for a lossless line. Thus these are two of the more important parameters to consider. Assuming that certain fundamental laws are known, we shall calculate L and C for some simple geometries, such as coaxial cables and strip lines with large conductor width-to-separation ratio. For more complex geometries, more sophisticated techniques are necessary such as Schwartz-Cristoffel transformations. We will not consider these techniques in detail but will rather present results for the more common lines, such as the general strip line, strip in a rectangular box, and helical line.

8.2 COAXIAL LINE

We will consider a unit length of an assumed infinitely long line, so that the fields are everywhere uniform with no fringing, and will calculate the capacitance per unit length. Consider the line in Fig. 8-1(a) with inner radius r_1 and outer radius r_2 and a voltage V_0 between the two conductors. There will be (+) and (−) charges on the surfaces of the two conductors as

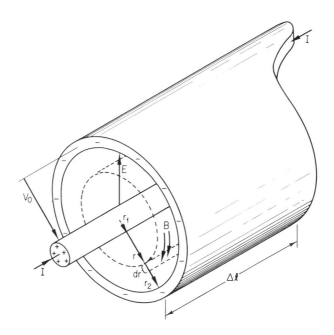

Fig. 8-1(a). Voltage and current applied to a coaxial cable.

shown. The total capacitance is just total charge divided by voltage, so that the capacitance per unit length is just charge per unit length divided by voltage

$$C = \frac{Q/\Delta\ell}{V_0} = \frac{\rho}{V_0} \tag{8-1}$$

where ρ is charge per unit length. In order to evaluate ρ, we can make use of Gauss' law, namely, the surface integral of the normal components of electric flux density $\mathbf{D}$ (displacement) over any closed surface equals the total charge enclosed

$$\oiint \mathbf{D} \cdot \mathbf{ds} = Q \tag{8-2}$$

Since $\mathbf{D} = \epsilon \mathbf{E}$ and since $\mathbf{E}$ is everywhere in the radial direction between the conductors, we can use a unit surface as a cylinder of radius r or circumference $2\pi r$ and unit length $\Delta\ell$ as shown in dotted lines. Thus, Gauss' law gives

$$\epsilon \mathbf{E} \, 2\pi r \, \Delta\ell = \rho\Delta\ell = Q \quad \text{so} \quad E = \frac{\rho}{2\pi r\epsilon} \tag{8-3}$$

But the line integral of $\mathbf{E}$ from r_1 to r_2 must be the total applied voltage

$$\int_{r_1}^{r_2} \mathbf{E} \cdot \mathbf{dl} = V_0 = \int_{r_1}^{r_2} \frac{\rho}{2\pi\epsilon r} \, dr = \frac{\rho}{2\pi\epsilon} \ln \frac{r_2}{r_1} \tag{8-4}$$

Thus

$$C = \frac{\rho}{V_0} = \frac{2\pi\epsilon}{\ln(r_2/r_1)} \tag{8-5}$$

or

$$C = \frac{55.6\,\epsilon_r}{\ln(r_2/r_1)} \,\mu\mu\text{F/m} \tag{8-6}$$

where ϵ_r is relative permittivity.

The inductance can be found as follows: Assume that a current is applied to the coaxial line, as shown in Fig. 8-1(a). If we assume that the current is dc or very low frequency, then the current will be uniformly distributed over the conductor cross-sectional areas. There will then be two separate components of inductance, one from the flux linkages occurring in the region between the conductors and the other as a result of partial flux linkages within the conductors. The latter component will disappear at sufficiently high frequencies where skin effect causes the current to flow only on the outer surface of the inner conductor and the inner surface of the outer (larger) conductor.

If a TEM mode is assumed, then the internal inductance is zero and only the external inductance, associated with the magnetic field between the conductors, exists. This external inductance is given simply by the reciprocal of the capacitance with ϵ replaced by $1/\mu$ (see Appendix 8A). However, we will calculate this inductance from first principles.

The flux density at any point will be in a circumferential direction and can be related to the current by means of Ampere's circuital law. For the region between the conductors, using a circle of circumference $2\pi r$

$$\oint \mathbf{H} \cdot \mathbf{dl} = \oint \frac{\mathbf{B}}{\mu} \cdot \mathbf{dr} = I \quad \text{or} \quad B = \frac{\mu I}{2\pi r} \tag{8-7}$$

The inductance per unit length is simply

$$L = \frac{\phi/\Delta \ell}{I} \tag{8-8}$$

where ϕ is the total flux between the conductors given by

$$d\phi = \mathbf{B} \cdot \mathbf{dA} \quad \text{or} \quad \phi = \int \mathbf{B} \cdot \mathbf{dr} \, \Delta \ell \tag{8-9}$$

where the incremental area $dr \, \Delta \ell$ is as shown in dotted lines. Thus the flux per unit length is

$$\frac{\phi}{\Delta \ell} = \int_{r_1}^{r_2} \mathbf{B} \cdot \mathbf{dr} = \frac{\mu I}{2\pi} \int_{r_1}^{r_2} \frac{dr}{r} = \frac{\mu I}{2\pi} \ln \frac{r_2}{r_1} \tag{8-10}$$

The inductance between the conductors is thus

$$L = \frac{\mu}{2\pi} \ln \frac{r_2}{r_1} \text{ h/m or } 0.2\mu_r \ln \frac{r_2}{r_1} \mu\text{h/m} \qquad (8\text{-}11)$$

where μ_r is the relative permeability.

If we do not assume a TEM mode but rather assume that dc or a sufficiently low frequency is applied, then the current will be uniformly distributed over the cross-sectional surface of the conductors, which gives rise to the internal inductance of the conductor as derived in Sec. 4.2. This inductance can often be neglected. As an example, consider a typical cable such as RG 58A/U, which has the following parameters:

$$2r_1 = 0.0375 \text{ inch} = 9.5 \times 10^{-4} \text{ meters},$$

$$\epsilon_r = 2.26 \text{ (polyethylene)}, \quad C = 97 \ \mu\mu\text{F/meter}$$

(from manufacturer's specifications)

At sufficiently low frequencies where the current is uniformly distributed over the inner conductor, the total inductance is $L = L_{in} + L_{ex}$. If we neglect the internal inductance of the outer sheath of the cable, then using Eqs. (4-6) and (8-11), the above becomes

$$L = \frac{\mu}{4\pi}\left(\frac{1}{2} + 2\ln\frac{r_2}{r_1}\right) \quad \text{h/m} \qquad (8\text{-}12)$$

From the specifications for RG 58A/U and from Eq. (8-6), it is seen that the effective radius ratio is

$$\ln \frac{r_2}{r_1} = \frac{55.6\,\epsilon_r}{C} = 1.3$$

Thus

$$L_{in} = \frac{0.5\mu}{4\pi} \qquad L_{ex} = \frac{2.6\mu}{4\pi} \quad \text{h/m}$$

Thus the internal component is only 16 percent of the total inductance. From Fig. 4-7(a), at a frequency around 20×10^3 Hz, the inner wire radius approximately equals the skin depth. From Fig. 4-4, it is seen that up to about this frequency the internal inductance is essentially constant but as the frequency increases, the internal inductance decreases, and at sufficiently high frequencies, it is usually negligible compared to the external inductance.

For most cases of interest, the characteristic impedance neglecting the internal wire inductance, is

$$Z_0 = 60 \left(\frac{\mu_r}{\epsilon_r} \right)^{1/2} \ln \frac{r_2}{r_1} \quad \text{ohms} \qquad (8\text{-}13)$$

Fig. 8-1(b) gives a normalized curve of this impedance for typical values of r_2/r_1.

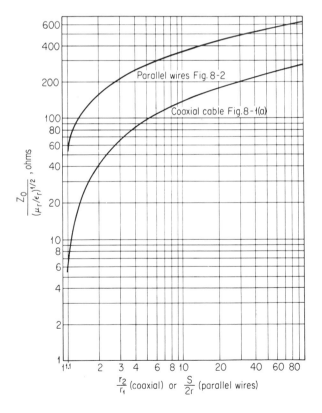

Fig. 8-1(b). Characteristic impedance for a coaxial cable and open, two-wire line.

8.3 OPEN, TWO-WIRE LINE

Rather than deriving both L and C for a two-wire line, such as in Fig. 8-2, it is expedient to evaluate only one parameter and make use of the reciprocity theorem (Appendix 8A) for other.

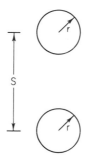

Fig. 8-2. Open, two-wire line.

For the general case when the wires are close together, proximity effects will cause a redistribution of charge and current on the wire surfaces. The capacitance can be found in a relatively simple manner and is easily shown to be*

$$C = \frac{\pi \epsilon}{\ln \left[\frac{S + (S^2 - 4r^2)^{1/2}}{2r} \right]} \qquad \frac{\text{farads}}{\text{meter}} \qquad (8\text{-}14)$$

From the reciprocity theorem for a TEM mode, the external circuit inductance is given by Eq. (8-75)

$$L = \frac{\epsilon \mu}{C} = \frac{\mu}{\pi} \ln \left[\frac{S + (S^2 - 4r^2)^{1/2}}{2r} \right] \qquad \frac{\text{henries}}{\text{meter}} \qquad (8\text{-}15)$$

The characteristic impedance is easily seen to be

$$Z_0 = 120 \left(\frac{\mu_r}{\epsilon_r} \right)^{1/2} \ln \left[\frac{S + (S^2 - 4r^2)^{1/2}}{2r} \right] \qquad \text{ohms} \qquad (8\text{-}16)$$

*[14, p. 77] or [12, p. 55].

and is plotted in Fig. 8-1(b). It is apparent from symmetry conditions that the same wire of radius r, located a distance $S/2$ above a ground plane, will have an impedance of one-half that given by Eq. (8-16).

8.4 STRIP LINE

(a) Uniform field case. Referring to Fig. 8-3, when W/S is very large, the calculations are greatly simplified and the results can easily be remembered. The case will now be considered; the more general case will be deferred until later.

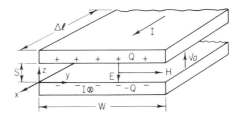

Fig. 8-3. Strip line with large width-to-separation ratio.

Consider the case of parallel plates of equal width and infinite length with an applied voltage and current as shown: if W/S is sufficiently large, the E and H fields external to the structure are negligible. It is assumed there is no variation of any quantity in the x direction. Between the conductors, E is everywhere normal to and H is parallel to the conductor surface. It is then obvious from Gauss' law (Eq. (8-2)) that $E = Q/(\epsilon W \Delta \ell)$, but

$$V_0 = \int_0^S \mathbf{E} \cdot d\mathbf{z} = ES$$

so that

$$V_0 = \frac{SQ/\Delta \ell}{W \epsilon}$$

Thus the capacitance per unit length is

$$C = \frac{Q/\Delta\ell}{V_0} = \frac{\epsilon W}{S}$$

$$= 8.85 \, \epsilon_r \frac{W}{S} \quad \mu\mu F/m \tag{8-17}$$

The inductance is easily obtained with the aid of Ampere's circuital law

$$\int_0^W \mathbf{H} \cdot d\mathbf{y} = \frac{B}{\mu} W = I \quad *$$

so that

$$B = \frac{\mu I}{W} \tag{8-18}$$

or, in other words, B equals the surface current density.

The total flux between the conductors is just

$$\phi = \Delta\ell \int_0^S B \, dz = BS \, \Delta\ell \tag{8-19}$$

so that the inductance per unit length is from Eqs. (8-18) and (8-19)

$$L = \frac{\phi/\Delta\ell}{I} = \mu \frac{S}{W} = 4\pi \times 10^{-7} \mu_r \frac{S}{W} \quad \frac{\text{henries}}{\text{meter}} \tag{8-20}$$

The characteristic impedance for the line is easily found

$$Z_0 = \sqrt{\frac{L}{C}} = \frac{S}{W}\left(\frac{\mu}{\epsilon}\right)^{1/2} = 377 \frac{S}{W}\left(\frac{\mu_r}{\epsilon_r}\right)^{1/2} \quad \text{ohms} \tag{8-21}$$

The range of accuracy of this simplified case will be investigated later.

(b) *General strip line.* The determination of the capacitance and impedance of two strip conductors of unequal widths W_1 and W_2 and any

---

*The integral need be taken only from $y = 0$ to W since H $= 0$ outside the conductors.

separation S, as in Fig. 8-4, is far more complex, requiring the use of conformal mapping to transform the actual geometry into one which can be more readily solved. We will not attempt to derive the desired quantities but will instead present only the results which are of greater concern. Extensive literature is available for more details.

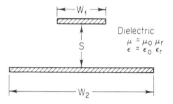

Fig. 8-4. General strip line.

For this general case, C and Z_0 can be determined exactly, with no approximations other than that the conductors be very thin with respect to other dimensions. However, explicit formulas for C and Z_0 in terms of W_1, W_2, and S cannot be obtained; a two-step process is instead necessary. First assume a value of the parameter k, the modulus of elliptic integrals. With the aid of elliptic integral tables, the values of W and S corresponding to the k, as well as C and Z_0, can be determined. If W and S are not the desired values, a new value of k must be assumed, etc. By plotting a curve of W/S versus k, the required value of k is easily obtained and therefore C and Z_0 are also. We will now consider several specific examples.

First consider the case of parallel plates of equal width $W_1 = W_2$. The exact value of capacitance per unit length is derived by Palmer [18] to be

$$C = \epsilon \frac{K'}{K} = \epsilon_0 \epsilon_r \frac{K'}{K} \quad \frac{\text{farads}}{\text{meter}} \tag{8-22}$$

where K and K' are the complete elliptic integral* of the first kind of modulus k and k', respectively. For an assumed value of k, C is easily found from tables. The value of W/S corresponding to this k is found from

$$\frac{W}{S} = K'E(\phi, k') - E'F(\phi, k') \tag{8-23}$$

*See Appendix 8B for a complete list of functions used.

$E(\phi, k')$ = elliptic integral of second kind (incomplete) of modulus k' for which the value of ϕ is found from

$$\sin^2 \phi = \frac{K' - E'}{(1 - k^2) K'} \qquad (8\text{-}24)$$

By assuming various values of k, a curve of k versus W/S can be obtained as in Fig. 8-5. Thus, for a given W/S, the corresponding value of k is first obtained and then C can be obtained from Eq. (8-22). These steps have been carried through and are presented in Fig. 8-6. The value of Z_0 is easily determined from

$$Z_0 = \frac{\sqrt{\mu\epsilon}}{C} = \frac{3.33 \times 10^{-9} \sqrt{\mu_r \epsilon_r}}{C} \quad \text{ohms} \qquad (8\text{-}25)$$

as described in Appendix 8A. The curves are plotted for W/S only up to 2. For larger values, accurate evaluation of elliptic integrals near $\phi = 90°$ are required which become tedious, so that the approximate formula

$$C = \frac{W}{S} \epsilon \left[1 + \frac{S}{\pi W} \left(1 + \log 2\pi \frac{W}{S} \right) \right] \quad \frac{\text{farads}}{\text{meter}} \qquad (8\text{-}26)$$

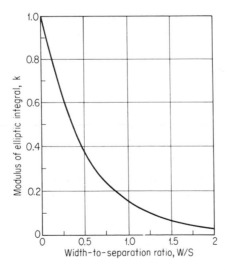

Fig. 8-5. Modulus k vs. W/S for a general strip line with $W_1 = W_2$. (*After Palmer.*)

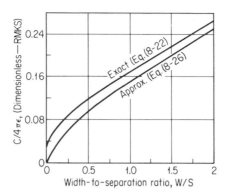

Fig. 8-6. Capacitance for a general strip line with $W_1 = W_2$. (*After Palmer.*)

can be used with reasonable results. For comparison, this function is plotted in Fig. 8-6 along with the exact value. At $W/S = 2$, the error is only 4.35 percent and decreases as W/S increases. It is easily seen that if $W_2 = $ infinity, the capacitance of the structure must be twice that of the previous structure with $W_1 = W_2$ but with separation $2S$. This is obviously true since the mid-plane of the latter must be an equipotential surface and is equivalent to a very thin ground plane. The phase velocity must, of course, be identical for the two structures since it is independent of geometry.

When both conductors are of finite, but different widths, the determination of C for a specified geometry becomes a more involved trial and error process since there are two geometric ratios W_1/S and W_2/S leading to two independent variables k and θ which are required along with the aid of various tables. Black and Higgins have considered this problem in detail. The essential points are as follows.

Assume a value of k (modulus of elliptic integrals) and θ (a numerical constant appearing in the transformation). Using these assumed parameters, it is necessary to calculate two constants c_1 and c_2 (which merely represent particular points on the transformed geometry) from

$$c_1 c_2 = \frac{\theta^2 E' - K'}{E' - k^2 \theta^2 K'}$$

$$c_1 + c_2 = \theta \frac{k^2 K' + K' - 2E'}{E' - k^2 \theta^2 K'}$$

Two further constants v_1 and v_2 are required; these are found from

$$c_1 = \text{sn}(K + jv_1) \qquad c_2 = \text{sn}(-K + jv_2)$$

where K is the complete elliptic integral. Having obtained these, the geometrical ratios W_1/S and W_2/S corresponding to the assumed values of k and θ are found from

$$\frac{W_2}{S} = \frac{2K'}{j\pi} \left[\text{zn}(K + jv_1) + \frac{j\pi v_1}{2KK'} + \frac{\text{cn}(K + jv_1)\,\text{dn}(K + jv_1)}{\theta + \text{sn}(K + jv_1)} \right] \qquad (8\text{-}27)$$

$$\frac{W_1}{S} = \frac{2K'}{j\pi} \left[\text{zn}(-K + jv_2) + \frac{j\pi v_2}{2KK'} + \frac{\text{cn}(-K + jv_2)\,\text{dn}(-K + jv_2)}{\theta + \text{sn}(-K + jv_2)} \right] \qquad (8\text{-}28)$$

The value of capacitance corresponding to the assumed values of k and θ is given by Eq. (8-22). This capacitance as a function of the geometric ratios is given in Fig. 8-7 for various cases; Z_0 can easily be obtained from Eq. (8-25), or more precisely

$$Z_0 = \frac{0.333 \times 10^{-8} (\mu_r \epsilon_r)^{\frac{1}{2}}}{C} \quad \text{ohms}$$

where C is in farads/meter with dielectric constant included. If the capacitance is taken as that of air, then the above becomes

$$Z_0 = \frac{0.333 \times 10^{-8}}{C_{\text{air}}} \left(\frac{\mu_r}{\epsilon_r} \right)^{\frac{1}{2}} \quad \text{ohms}$$

The characteristic impedance for some typical cases obtained from Fig. 8-7 and the above equation are shown in Fig. 8-8 for $W_2/W_1 = 1$ and 4. Also shown is the approximate curve obtained from Eq. (8-21), which assumed uniform fields. It can be seen that just as would be expected, Eq. (8-21) is very inaccurate for small W_1/S but becomes a reasonable approximation for large W_1/S, that is, about 10 or larger. It should be noted that the impedance is actually smaller than that given by the uniform field approximation.

It is apparent that all the cases considered thus far have been assumed to have a uniform dielectric material throughout all space. In actual practice,

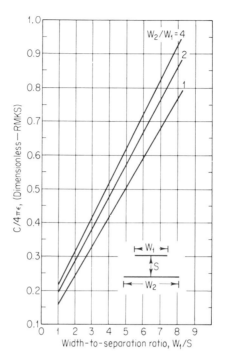

Fig. 8-7. Capacitance vs. geometrical ratios for a general strip line. (*After Black and Higgins.*)

the strip line will more often be made as in Fig. 8-9. It is apparent from Fig. 8-8 for sufficiently large W/S (approximately 10 or larger) that the fringe field is rather small; thus the fact that the dielectric does not fill all space is of little concern for such cases. However, for much smaller W/S, a significant amount of the fringe field will be in air and will cause a small increase in Z_0 and decrease in delay time. An exact analysis for the general case when the strip conductor has a finite thickness is not available. If the strip is assumed to be a very thin perfect conductor carrying a TEM mode, the problem is solvable by using conformal transformations as shown by Wheeler [27]. The complete solution for this case of mixed dielectric does not yield an explicit formula but rather approximations are necessary which give adequate results. Figure 8-9 gives the characteristic impedance for this case for several values of relative dielectric constant found with the use of Wheeler's analysis. Seckelmann [23] has performed several measurements which are in good agreement with the theoretical calculations.

For the more general case of a strip of width W and finite thickness t, an exact analysis is not possible, but approximate values can be obtained

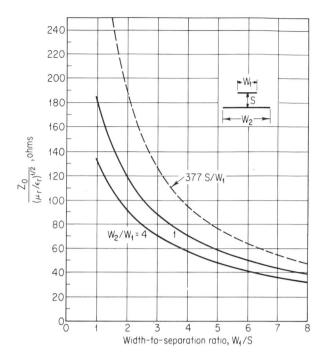

Fig. 8-8. Characteristic impedance of a general strip line.

over a limited range of parameters. One such simple technique requires measurement of the delay time of the actual line to determine the equivalent or "effective" dielectric constant. Then the rectangular strip is represented by an equivalent circular wire above a ground plane and the equation for Z_0 of the latter can be used with appropriate equations for diameter in terms of W and t and for the effective dielectric constant (see [13]). This technique is not general, being limited mainly to small W/S, and is quite approximate, thus requiring care in its application.

8.5 SHIELDED STRIP LINE

(a) Shielded strip line with open ends: for the geometry shown in Fig. 8-10, the exact expression with an infinitely thin center conductor $t = 0$ is*

$$Z_0 = 30\pi \frac{K}{K'} \tag{8-29}$$

*See [16] or [5].

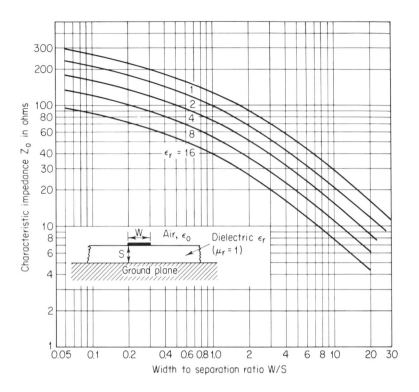

Fig. 8-9. Characteristic impedance of a thin, perfect conductor of width W separated from an infinite ground plane by an infinite dielectric sheet of thickness S.

where K and K' are complete elliptic integrals of the first kind of modulus

$$k = \operatorname{sech} \frac{\pi W}{2S'} \qquad k' = \tanh \frac{\pi W}{2S'} \tag{8-30}$$

If the center conductor has a finite thickness t, then the characteristic impedance is

$$Z_0 = \frac{94.15}{\sqrt{\epsilon_r}} \left(\frac{W/S'}{1 - t/S'} + \frac{C_f}{8.85 \epsilon_r} \right)^{-1} \text{ ohms} \tag{8-31}$$

where C_f is the fringe capacitance in $\mu\mu\text{F/m}$. If W is sufficiently large such that the fringe fields on the right and left side do not interact, then C_f can be

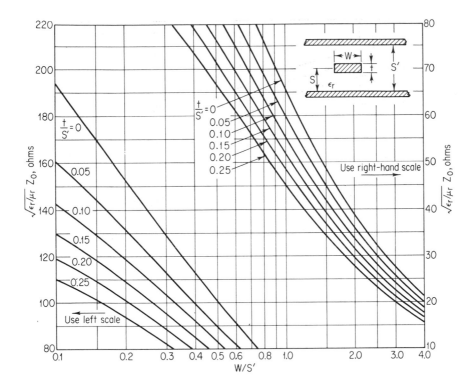

Fig. 8-10. Z_0 for a partially shielded strip line with finite thickness strip. (*After Cohn* [4].)

taken as twice that for a single edge [25]. Thus

$$C_f = \frac{8.85\,\epsilon_r}{\pi}\left[\frac{2}{1-t/S'}\ln\left(\frac{1}{1-t/S'}+1\right)-\left(\frac{1}{1-t/S'}-1\right)\ln\left(\frac{1}{[1-t/S']^2}-1\right)\right]$$

$$\mu\mu F/m \quad (8\text{-}32)$$

As W decreases, the error increases; however, the above equations are good for $W/(S'-t) \geq 0.35$ with a maximum error of 1.2 percent.

Curves of Z_0 as a function of W/S' for various thickness ratios t/S' are given in Fig. 8-10.

The shielded strip line (Fig. 8-10) can be used to decrease the Z_0 of the unshielded strip (Fig. 8-4). For a given spacing between strip and ground plane, the addition of a second ground plane an equal distance from the strip as the first ground plane will cut Z_0 in half only for rather large ratios of W_1/S, e.g., those greater than 10. For smaller ratios, the second ground plane is not nearly so effective in reducing Z_0 and its effect becomes smaller as W_1/S becomes smaller. This is shown in Fig. 8-11 where Z_0 for the two cases is compared assuming a strip of zero thickness, located the same distance from the ground plane(s) in both geometries.

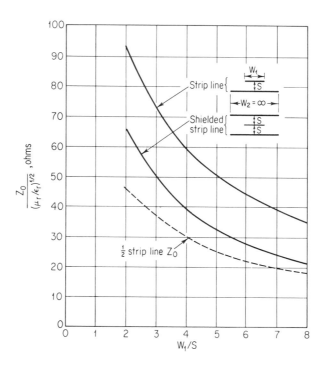

Fig. 8-11. Comparison of Z_0 for strip line and shielded strip line for equivalent geometries and $W_2 = \infty$.

(b) Completely shielded strip line: The method of conformal mapping can be used once again to obtain the parameters of a completely enclosed, infinitely thin strip as in Fig. 8-12. Primozich *et al.* [20] employed two conformal transformations with the essential steps as follows. The impedance is given by

$$Z_0 = \frac{1}{4}\left(\frac{\mu_r}{\epsilon_r}\right)^{1/2} \frac{K}{K'} \tag{8-33}$$

In order to find K and K', it is necessary to relate the modulus k to the physical parameters of the line. As before, this can be done only

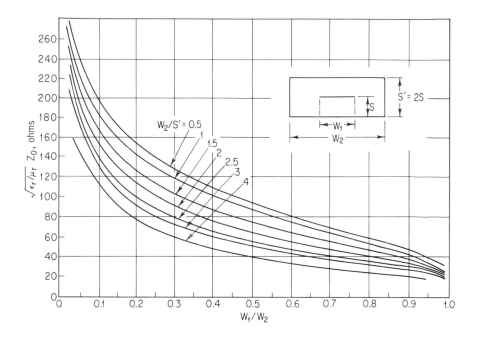

Fig. 8-12. Z_0 for a completely shielded, thin strip. (*After Primozich et al.*)

indirectly. A number of constants will be used here which enter because of the various transformations required. Their meaning and significance will not be discussed, but it is necessary to carry out the following steps to arrive at the final solution.

From a specified geometrical ratio W_2/S', it is possible to find the modulus k_0 of elliptic integrals K_0 and K_0' from

$$\frac{K_0}{K_0'} = \frac{W_2}{S'} \tag{8-34}$$

with the aid of tables. Once k_0 is determined, K_0 can be found separately. From a specified geometrical ratio W_1/W_2 and the value of K_0 above, it is possible to find a constant Z_2 (not impedance) from

$$Z_2 = \left(1 - \frac{W_1}{W_2}\right) K_0$$

Once k_0 and Z_2 are known, the modulus k of the necessary elliptic integrals is found from

$$k^2 = (k_0')^2 \frac{sn(k_0, Z_2)}{dn(k_0, Z_2)} \tag{8-35}$$

With this modulus, K and K' are found from tables and the impedance is given by Eq. (8-33). Curves of impedance for various ratios of the physical dimensions are given in Fig. 8-12.

The impedance of this completely shielded strip differs by only a small amount from that of the partially shielded strip of Fig. 8-10, that is, ends removed, for all the values of W_2/S' in Fig. 8-12 when W_1/S' is less than 0.4.

8.6 HELICAL LINES

In pulse work, it is often necessary to employ delay lines as timing, temporary storage, and numerous other devices. Often the required delay times are quite large, making ordinary transmission lines impractical. It is possible to increase the delay line of, say, a coaxial cable by constructing the inner conductor of a helical winding as shown in Fig. 8-13, rather than straight. This will usually require a larger outer-diameter cable but the delay per unit of axial length of cable can be substantially increased.

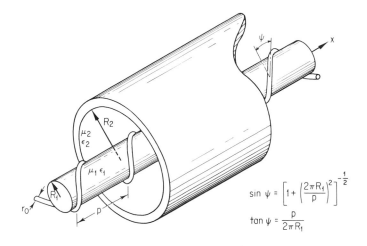

$$\sin \psi = \left[1 + \left(\frac{2\pi R_1}{p}\right)^2\right]^{-\frac{1}{2}}$$

$$\tan \psi = \frac{p}{2\pi R_1}$$

Fig. 8-13. Helical transmission line.

The determination of the inductance and capacitance of such a line is an involved mathematical problem and has been solved in great detail by Primozich for the case where μ_1 and μ_2 as well as ϵ_1 and ϵ_2 differ. It is assumed that the line is carrying a TEM mode, so that the internal conductor inductance is negligible, i.e., only the external (circuit) inductance requires evaluation. It should be noted that there are two values of L and C (per unit length) of interest, namely, those in the helical direction with distance measured along the helix or wire, and L_x and C_x measured along the axial direction of the cylinder. We are only interested in the axial parameters; these parameters are as follows.*

$$\frac{1}{C_x} = \frac{1}{2\pi\epsilon_2} \left\{ \ln\frac{R_2}{R_1} + 2\sum_{m=1}^{\infty} \frac{\left[1 - \dfrac{I_m(bR_1)K_m(bR_2)}{I_m(bR_2)K_m(bR_1)}\right] I_m(bR_1)K_m(bR_2)\cos\left(\dfrac{br_0}{\cos\psi}\right)}{\dfrac{\epsilon_1}{\epsilon_2} + \left(\dfrac{\epsilon_1}{\epsilon_2} - 1\right)\left[1 - \dfrac{I_m'(bR_1)K_m(bR_2)}{I_m(bR_2)K_m'(bR_1)}\right](bR_1)I_m(bR_1)K_m'(bR_1)} \right\} \left(\frac{\text{farads}}{\text{meter}}\right)^{-1}$$

(8-36)

where $b = 2\pi m/p$, and m is an integer. (Other functions are defined in Appendix 8B).

If the pitch p is small, the above becomes

$$\frac{1}{C_x} = \frac{1}{2\pi\epsilon_2} \left\{ \ln\frac{R_2}{R_1} - \frac{2\epsilon_2}{\epsilon_1 + \epsilon_2}\sin\psi \ln\left[2\sin\left(\frac{\pi r_0}{p\cos\psi}\right)\right] \right\} \qquad (8\text{-}37)$$

For large pitch p the capacitance is

---

*In these equations, different parts of the dielectric are allowed to have different values of ϵ or μ so that L and C are not necessarily reciprocal relations. Reciprocity is valid only when the dielectric is isotropic and linear throughout the entire region.

$$\frac{1}{C_x} = \frac{1}{2\pi\epsilon_2} \left\{ \ln\frac{R_2}{R_1} + \frac{2\epsilon_2}{\epsilon_1 + \epsilon_2} \sum_{m=1}^{\infty} \frac{\left[1 - \left(\dfrac{R_1}{R_2}\right)^{2m}\right] \cos\dfrac{mr_0}{R_1}}{\left[1 - \left(\dfrac{\epsilon_1 - \epsilon_2}{\epsilon_1 + \epsilon_2}\right)\left(\dfrac{R_1}{R_2}\right)^{2m}\right] m} \right\} \quad (8\text{-}38a)$$

If $\epsilon_1 = \epsilon_2$, the above simplifies to

$$\frac{1}{C_x} = \frac{1}{2\pi\epsilon} \left\{ \ln\frac{R_2}{R_1} + \sum_{m=1}^{\infty} \frac{\left[1 - (R_1/R_2)^{2m}\right] \cos(mr_0/R_1)}{m} \right\} \quad (8\text{-}38b)$$

Evaluation of capacitance is still rather complex, but if $r_0/R_1 \ll 1$, a simple closed-form expression results, namely (from Eq. (8-38b))

$$C_x = \frac{2\pi\epsilon}{\ln[(R_2/R_1) - (R_1/R_2)]} \quad \frac{\text{farads}}{\text{meter}} \quad (8\text{-}38c)$$

In a similar manner, the external inductance, neglecting the internal conductor inductance, is

$$L_x = 2\pi\left(\frac{R_1}{p}\right)^2 \left\{ \frac{\beta}{2}\left[1 - \left(\frac{R_1}{R_2}\right)^2\right] + \mu_2 \tan^2\psi \ln\left(\frac{R_2}{R_1}\right) \right.$$

$$\left. -2\sum_{m=1}^{\infty} \frac{1}{\alpha}\left[1 - \frac{I'_m(bR_1) K'_m(bR_2)}{I'_m(bR_2) K'_m(bR_1)}\right] I'_m(bR_1) K'_m(bR_1) \cos\left(\frac{br_0}{\cos\psi}\right) \right\} \quad \frac{\text{henries}}{\text{meter}}$$

$$(8\text{-}39)$$

where

$$\alpha = bR_1 \left[\frac{I'_m(bR_1) K_m(bR_1)}{\mu_2} - \frac{K'_m(bR_1) I_m(bR_1)}{\mu_1} \right.$$

$$\left. + \frac{I'_m(bR_1) K'_m(bR_2) I_m(bR_1)}{I'_m(bR_2)} \left(\frac{1}{\mu_1} - \frac{1}{\mu_2} \right) \right] \tag{8-40}$$

and

$$\beta = \frac{1}{\left[\frac{1}{\mu_1} - \left(\frac{R_1}{R_2} \right) \left(\frac{1}{\mu_1} - \frac{1}{\mu_2} \right) \right]} \tag{8-41}$$

If the pitch p is sufficiently small, then

$$L_x = 2\pi\mu_2 \left(\frac{R_1}{p} \right)^2 \left\{ \frac{\mu_1/\mu_2}{\left(2 \left[1 - \left(\frac{R_1}{R_2} \right)^2 \left(1 - \frac{\mu_1}{\mu_2} \right) \right] \right)} \left[1 - \left(\frac{R_1}{R_2} \right)^2 \right] \right.$$

$$\left. + \tan^2\psi \left[\ln \frac{R_2}{R_1} - \frac{2\mu_1}{\mu_1 + \mu_2} \frac{1}{\sin\psi} \ln \left(2 \sin \left[\frac{\pi r_0}{p \cos\psi} \right] \right) \right] \right\} \quad \frac{\text{henries}}{\text{meter}}$$

$$\tag{8-42}$$

If the pitch p is sufficiently large, then

$$L_x = \frac{\mu_2}{\pi} \left\{ \frac{1}{2} \ln \frac{R_2}{R_1} + \right.$$

$$+ \frac{\mu_1}{\mu_1 + \mu_2} \sum_{m=1}^{\infty} \frac{1 - \left(\frac{R_1}{R_2}\right)^{2m}}{\left[1 + \left(\frac{\mu_1 - \mu_2}{\mu_1 + \mu_2}\right)\left(\frac{R_1}{R_2}\right)^{2\overline{m}}\right]} \left[\frac{\cos\left(\frac{m r_0}{R_1}\right)}{m}\right] \right\} \qquad (8\text{-}43a)$$

This is very similar in form to Eq. (8-38a), and if $\mu_1 = \mu_2$, the above becomes

$$L_x = \frac{\mu}{2\pi} \left\{ \ln \frac{R_2}{R_1} + \sum_{m=1}^{\infty} \frac{\left[1 - \left(\frac{R_1}{R_2}\right)^{2m}\right]\cos\left(\frac{m r_0}{R_1}\right)}{m} \right\} \quad \frac{\text{henries}}{\text{meter}} \quad (8\text{-}43b)$$

This is obviously identical to Eq. (8-38b) for reciprocal capacitance except for a constant. If Eq. (8-43b) is divided by Eq. (8-38b), the result is $L_x C_x = \mu\epsilon$, in agreement with Eq. (8-75), thus demonstrating the reciprocity between L_x and C_x when the dielectric is uniform throughout the entire region. If $r_0/R_1 \ll 1$, then Eq. (8-43b) gives a closed-form expression identical in form to Eq. (8-38c), namely

$$L_x = \frac{\mu}{2\pi} \ln\left[\frac{R_2}{R_1} - \frac{R_1}{R_2}\right] \quad \frac{\text{henries}}{\text{meter}} \qquad (8\text{-}43c)$$

The previous equations give the characteristic impedance for the general case

$$Z_0 = \frac{R_1}{p}\left(\frac{\mu_2}{\epsilon_2}\right)^{1/2} (F_1 F_2)^{1/2} \quad \text{ohms} \qquad (8\text{-}44)$$

and the phase velocity in the axial direction

$$\mathcal{Q}_x = \frac{\tan\psi}{(\mu_2 \epsilon_2)^{1/2}}\left(\frac{F_1}{F_2}\right)^{1/2} \qquad (8\text{-}45)$$

where

$$F_1 = \left\{ \ln \frac{R_2}{R_1} + 2 \sum_{m=1}^{\infty} \frac{\left[1 - \dfrac{I_m(bR_1) K_m(bR_2)}{I_m(bR_2) K_m(bR_1)} \right] I_m(bR_1) K_m(bR_1) \cos\left(\dfrac{br_0}{\cos\psi}\right)}{\dfrac{\epsilon_1}{\epsilon_2} + \left(\dfrac{\epsilon_1}{\epsilon_2} - 1\right) \left[1 - \dfrac{I'_m(bR_1) K_m(bR_2)}{I_m(bR_2) K'_m(bR_1)} \right] bR_1 I_m(bR_1) K'(bR_1)} \right\}$$

(8-46)

$$F_2 = \frac{1}{\mu_2} \left\{ \frac{\beta}{2} \left[1 - \left(\frac{R_1}{R_2}\right)^2 \right] + \mu_2 \tan^2\psi \ln \frac{R_2}{R_1} \right.$$

$$\left. - 2 \sum_{m=1}^{\infty} \frac{1}{\alpha} \left[1 - \frac{I'_m(bR_1) K'_m(bR_2)}{I'_m(bR_2) K'_m(bR_1)} \right] I'_m(bR_1) K'_m(bR_1) \cos\left(\frac{br_0}{\cos\psi}\right) \right\}$$

(8-47)

For the special case when $\epsilon_1 = \epsilon_2$, $\mu_1 = \mu_2$ and $r_0 \ll R_1$, the impedance is simply

$$Z_0 = \frac{(\mu\epsilon)^{1/2}}{C} = \frac{1}{2\pi}\left(\frac{\mu}{\epsilon}\right)^{1/2} \ln\left[\frac{R_2}{R_1} - \frac{R_1}{R_2}\right] \quad \text{ohms}$$

Since such lines are usually employed to increase the delay time, a simple way to achieve this is to wind the helix on a ferrite core of radius R_1 and insert this structure into a hollow cylinder of radius R_2. Of course, this will tend to increase Z_0 also, but compensation for this would require filling the region with a material of high dielectric constant, which becomes difficult in practice. Curves for impedance and phase velocity for various geometrical ratios are shown in Fig. 8-14 and 8-15 for a case with the entire line assumed to have a dielectric constant ϵ_2^* and with very large μ_1/μ_2. As would be expected, the phase velocity in the axial direction decreases or delay time increases rapidly as the turns become more closely spaced. The

*Even with ϵ_1 different from ϵ_2, there is relatively little effect on the line capacitance and hence on Z_0 and v_x, because the E field is concentrated primarily within the region between R_1 and R_2.

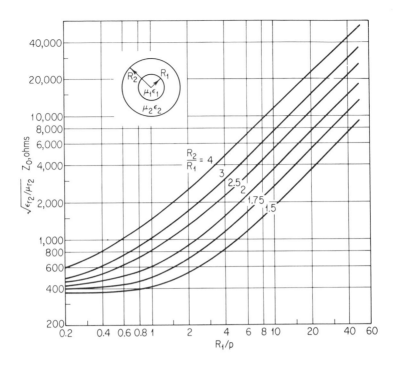

Fig. 8-14. Normalized impedance of a helical line with μ_1/μ_2 very large, $\epsilon_1 \approx \epsilon_2$, and $r_0/R_1 = 0.01$.

impedance of the line also increases and a compromise between these two is often necessary.

8.7 COUPLING COEFFICIENTS

When two transmission lines are coupled, as in Chap. 7, the capacitive and inductive coupling coefficients are defined as

$$k_C = \frac{C_{12}}{(C_{11} C_{22})^{1/2}} \qquad k_L = \frac{M_{12}}{(L_{11} L_{22})^{1/2}} \tag{8-48}$$

and must always be less than or equal to unity. The double subscripted parameters are defined in Sec. 7.2. For two open wires in a uniform, isotropic, linear medium supporting a TEM mode, k_L and k_C are always equal for any cross-sectional geometry (this follows from the reciprocity

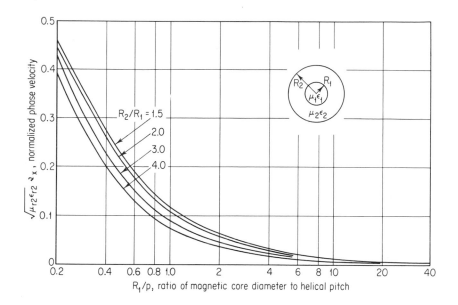

Fig. 8-15. Axial phase velocity of uniform helical line with magnetic inner core of $\mu_1 \gg \mu_2$, $\epsilon_1 \approx \epsilon_2$, and $r_0/R_1 = 0.01$.

theorem of Appendix 8A). In other words, this is true when and only when E and H are orthogonal so that the equipotential lines for capacitance are identical to the magnetic flux lines for inductance. It is apparent that for cases when the internal impedance of a conductor is not negligible, a TEM mode will not exist, since the H field will be different, and an axial E field will exist with the result that the coupling coefficients are no longer equal for dc or ac. For many cases, this difference is negligible, while for others it is important.

The coupling coefficient for a pair of parallel wire lines,* as in Fig. 8-16(a), assuming a TEM mode is easily obtained by way of inductance calculations as follows: It is necessary to calculate both the self and mutual inductance. For the latter, the flux linking line 2 because of the current in line 1 (opposite currents in top and bottom conductor) can be obtained from the Biot-Savart law (Eq. (1-3)). If the wire radii are assumed to be small compared to S and H, this flux is

$$\phi_x = 2 \frac{\mu I}{2\pi} \int_0^S \frac{\sin \theta}{\rho} dy \tag{8-49}$$

*Cohn [6] and Shelton [24] consider coupled, shielded strip lines.

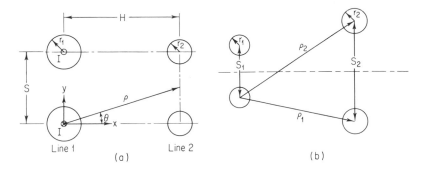

Fig. 8-16. Coupled, two-wire lines. (a) Equal separation; (b) unequal separation but symmetircal.

where $\sin\theta = y/(x^2 + y^2)^{1/2}$, $x = H$, and $\rho = (x^2 + y^2)^{1/2}$. Substituting and simplifying

$$\frac{\phi_x}{I} = \frac{\mu}{\pi} \int_0^S \frac{y}{x^2 + y^2}\, dy = \frac{\mu}{\pi}\left[\ln(x^2 + S^2)^{1/2} - \ln x\right] \tag{8-50}$$

But mutual inductance is given by $M = L_{21} = \phi_{21}/I_1$, which is the above expression. Thus, since $x = H$, the mutual inductance is

$$M = \frac{\mu}{\pi}\ln\frac{(H^2 + S^2)^{1/2}}{H} \tag{8-51}$$

Using the same assumption of thin wires, the self-inductance is easily obtained as was done above or from Eq. (8-15)

$$L_{11} = L_1 = \frac{\mu}{\pi}\ln\frac{S}{r_1} \tag{8-52}$$

$$L_{22} = L_2 = \frac{\mu}{\pi}\ln\frac{S}{r_2} \tag{8-53}$$

Thus, the coupling coefficient for wires with diameters much smaller than any spacing and supporting a TEM mode is

$$k_L = k_C = k = \frac{\ln\left[1 + (S/H)^2\right]^{1/2}}{[\ln(S/r_1)\,\ln(S/r_2)]^{1/2}} \tag{8-54}$$

It can be shown in a similar manner that for symmetrical lines with $S_1 \neq S_2$, as in Fig. 8-16(b), the coupling coefficient is

$$k = \frac{\ln(\rho_2/\rho_1)}{[\ln(S_1/r_1)\,\ln(S_2/r_2)]^{1/2}} \tag{8-55}$$

provided that the wire diameters are much smaller than the spacings.

It should be apparent that when the wires are of large diameter and close together, there can be a significant amount of partial flux linkages in some cases: if the latter is true, then a TEM mode is not present* and k_L and k_C are not equal. We will not analyze such cases but rather will consider a number of situations where $k_L \neq k_C$ and the resulting consequences. It is apparent from Eq. (7-41), as discussed in Sec. 7.2, that when $k_L \neq k_C$, the constant K given by Eq. (7-38) is not 0. Furthermore, from Eqs. (7-140) and (7-144) it is apparent that when K is not 0, the sum and difference modes will propagate at different velocities. It is shown in [4] of Chap. 7 that when the sum and difference modes or the so-called primary and secondary modes propagate at different velocities, the waveforms of transmission-line transformer devices can be affected in various ways. Thus, an understanding of the coupling coefficients is essential for the study of such devices.

A simple case for which $k_L \neq k_C$ is obtained when there are different dielectric media between conductors 1 and 2, and the conductors and ground plane are as shown in Fig. 8-17(a). For the case shown, it is apparent that the dielectric between the wires slows down the difference-mode wave, while the sum mode is only slightly affected in comparison as long as the conductor-to-ground separation is significantly larger than the conductor-to-conductor separation. This result can be obtained analytically by using Eqs. (7-140) and (7-144) and assuming a lossless line to obtain for the ratio of difference-mode to sum-mode propagation velocity

$$\left(\frac{v_d}{v_s}\right)^2 = \left(\frac{\gamma_s}{\gamma_d}\right)^2 = \frac{J+K}{J-K} = \frac{1 + k_L(1-k_C) - k_C}{1 - k_L(1+k_C) + k_C} \tag{8-56}$$

If k_C is increased with a dielectric between the wires while k_L is constant, then v_d/v_s must decrease. If k_C is held constant while k_L is increased

---

*This results from the fact that the internal conductor impedance must be included.

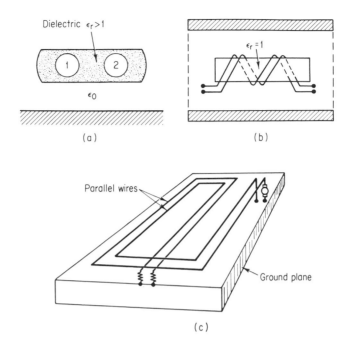

Fig. 8-17. Parallel wire lines for which k_L and k_C are not equal.

with a ferrite between the wires and ground plane, the opposite is true, i.e., the sum mode travels more slowly.

If the pair of parallel wires is coiled, the situation becomes very complex because k_L and k_C both change, not necessarily by the same amount. For a helical coil as in Fig. 8-17(b), if the total length of the parallel pair of wires is constant, i.e., delay time of difference mode is fixed, it is apparent that for a very loosely wound coil, the sum and difference mode have the same delay time (assuming that the dielectric is the same throughout). However, if the coil is tightly wound, then the capacitance coupling between turns is so large that for any reasonable frequency, the helical coil looks more like a solid conductor to the sum mode. The delay time of the sum mode is now much smaller than that of the difference mode—this phenomenon leads to the problems associated with designing baluns, as in [4] of Chap. 7.

The situation is much the same for parallel wires above a ground plane as in Fig. 8-17(c). Any coiling of the wire pair will only cause a decrease in the sum mode delay time, while the difference mode delay remains constant for a closely spaced parallel pair of wires.

8.8 HIGH-FREQUENCY EFFECTS ON CAPACITANCE
AND INDUCTANCE PARAMETERS

In calculating the capacitance of any given line, it is assumed that the conductors are equipotential surfaces. This capacitance arises from free charges that distribute themselves on the surfaces of the conductor, as can be determined from a field plot or a solution to Laplace's equation. This distribution of charges will be independent of frequency, as long as the cross-sectional dimensions of the structure are much smaller than the wavelength of the applied frequency, i.e., as long as there is no propagation in the transverse direction of the line. This results from the fact that the relaxation time for charges to distribute themselves in a conductor due to mutual coulomb repulsion is extremely small, on the order of 10^{-19} sec to decrease to $1/e$ of its initial value* for copper, for instance. This corresponds to frequencies in the x-ray region. Thus, the capacitance as previously calculated is essentially independent of frequency for cases of practical interest.

In determining the inductance of any structure, it is necessary to know the current distribution within the conductors. For very low frequencies, a uniform current distribution can be assumed throughout the conductor cross-sectional area. As the frequency increases, the current will tend to concentrate on the adjacent surfaces of the two conductors. In general, the determination of current distribution is best obtained with the aid of Maxwell's equations and by solving the boundary value problem as was done in Sec. 4.5. As was shown, this not only gives the field and current distribution in the conductor, but also evaluates the internal inductance and resistance, which are the desired parameters.

If the conductor conductivity is large enough that the internal impedance is negligible compared to the geometrical impedance, then a true TEM exists and the inductance (external) can be obtained from the capacitance via the reciprocity relationship of Appendix 8A.

It is important to point out that as the frequency becomes so high that the line cross-sectional dimensions become a significant fraction of the applied wavelength, the simple transmission-line theory is no longer valid. The simple expressions for L and C are no longer applicable, and it is necessary to use the concept of retarded potentials for their evaluation. If this is done, it will be found that for such cases (wavelength approaching line dimensions), open-line structures such as strip lines or parallel wires will have an impedance containing another term in addition to the usual characteristic impedance. The additional term is easily identified as the radiation resistance,

*See [14, p. 399].

i.e., the structure will act as an antenna and radiate a small amount of energy from the open ends. Closed structures such as ideal coaxial lines cannot radiate because of the shield provided by the outer conductor; however, as the cable radii become a significant fraction of the wavelength, the boundary conditions will change and higher-order modes can be excited.* In other words, the structure begins to look more like a waveguide and its propagation characteristics cannot be determined from simple transmission-line theory. It is necessary to employ Maxwell's equations and to solve the specific boundary value problem similar to that done in Sec. 4.5. Such analyses are beyond the scope of this book.

8.9 DETERMINATION OF IMPEDANCE OF COMPLEX GEOMETRIES USING CONDUCTING-PAPER TECHNIQUES

The characteristic impedance of a line of any cross-sectional geometry which carries a TEM mode can be determined by a very simple dc measurement. For example, suppose that it is desirable to determine the impedance of an arbitrary line whose cross-sectional area is as shown in Fig. 8-18. It is apparent that a mathematical analysis would be extremely difficult. However, a relatively accurate measurement could be made by laying out the cross-sectional geometry on a large sheet of conducting paper, e.g., Western Union Teledeltos Paper. The conductors with proper dimensions can be put on the paper with highly conductive silver paint.†

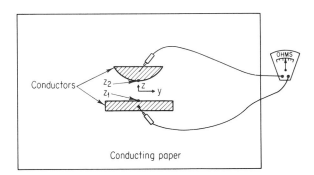

Fig. 8-18. Conducting-paper technique for obtaining Z_0 of complex geometries.

*See [21, p. 364] or [9].

†In some cases where the conductors are large, the paint can have sufficient resistance to give a nonnegligible voltage drop along its surface. In such cases, it is advisable to overlay the paint with a thin piece of copper shim metal (0.005-0.010 in. is often adequate) and to use pressure contact. This will ensure that the conductor contour is an equipotential surface.

In cases where the actual dimensions are very small, the geometry can be scaled up (or down, if large) for convenience and accuracy. An ohm meter is used to measure the resistance between the two electrodes and this resistance is directly proportional to the characteristic impedance of an ideal transmission line of the same cross-sectional geometry.

Proof that this technique does give the characteristic impedance of the two-dimensional geometry follows from the fact that the current density in the conducting paper obeys Laplace's equation and the same boundary conditions as the electric field in the actual line. From Appendix 8A, it follows that the current density will have the same field pattern as the electric field, while the equipotential lines will have the same pattern as the magnetic field in the actual transmission line.

The resistance of the geometry in Fig. 8-18 is easily obtained as voltage divided by current. The voltage is the line integral of the electric field, while the total current is obtained by integrating the normal current density J_n over the conductor surface

$$R_0 = \frac{V}{I} = \frac{-\int_{z_1}^{z_2} E_z \cdot dz}{T \oint J_n \, dp} \tag{8-57a}$$

where dp is the incremental distance along the perimeter of the wire and T is the paper thickness. From Ohm's law, $J = \sigma E$, so that

$$R_0 = \frac{\int E_z \cdot dz}{\sigma T \oint E_n \cdot dp} \tag{8-57b}$$

Now we know that E in the above equation must be identical to that in Fig. 8-18, except for possibly a constant which cancels out in the above equation. It is apparent that Eq. (8-57b) is identical in form to the reciprocal of Eq. (8-70) except for a constant. Thus, it is apparent that by substitution of Eq. (8-57b) in Eq. (8-70), the capacitance of the actual transmission line is

$$C = \frac{\epsilon}{\sigma T R_0} \tag{8-58}$$

where R_0 is the dc resistance (in ohms) of the geometry on conducting paper.

The characteristic impedance of the actual line is given by Eq. (8-76) and substitution of Eq. (8-58) for C yields

$$Z_0 = \frac{(\mu\epsilon)^{\frac{1}{2}}}{C} = \sigma T \left(\frac{\mu}{\epsilon}\right)^{\frac{1}{2}} R_0$$

But $R_{sq} = 1/(\sigma T)$ is the resistance per square of the conducting paper; thus, it follows that

$$Z_0 = \left(\frac{L}{C}\right)^{\frac{1}{2}} = \left(\frac{\mu}{\epsilon}\right)^{\frac{1}{2}} \frac{R_0}{R_{sq}} = 376.7 \left(\frac{\mu_r}{\epsilon_r}\right)^{\frac{1}{2}} \frac{R_0}{R_{sq}} \qquad (8\text{-}59)$$

where R_0 is the measured resistance between the conductors of Fig. 8-18 and R_{sq} is the resistance per square of the conducting paper for the thickness used.

For accurate results, it is apparent that the size of the conducting paper should be sufficiently larger than the conductors to allow the fringe fields to take on their proper shape.

The same experimental setup can be used to plot the field pattern of complex geometries. This can be done by application of a small potential difference between the electrodes and using a sensitive galvameter to trace the equipotential lines: if one probe is held fixed while the other traces out points of the same potential, the field pattern obtained will thus be independent of any variations in power-supply voltage.

APPENDIX 8A

RECIPROCITY BETWEEN L AND C

We wish to show that for any and all transmission-line geometries, the capacitance and external inductance are reciprocal functions of the line cross-sectional dimensions and are related by a constant which will be determined. In other words, we wish to show that a TEM field configuration exists, provided that the internal conductor inductance is negligible. The latter condition can usually (but not always) be achieved by assuming that the frequency is sufficiently high such that the skin depth is very small, or that the conductivity is infinite so that the assumption is valid at low frequency as well. At true dc, the current will flow throughout the conductor cross-sectional area, except in a superconductor, but this is another problem, one which was treated in Sec. 6.3.

Consider two long parallel wires of any cross section with an applied voltage difference V_0 and current I_0, as in Fig. 8-19. It is assumed that the fields are uniform in the axial direction (the problem thus reduces to a two-dimensional field), and that the current flows only in a thin surface shell as shown. The electric charges will always be on the surface as shown and no assumptions are necessary concerning these.* It is well known that in the dielectric region outside the conductors, the electric field is given by the gradient of V

$$E = -\nabla V \qquad (8\text{-}60)$$

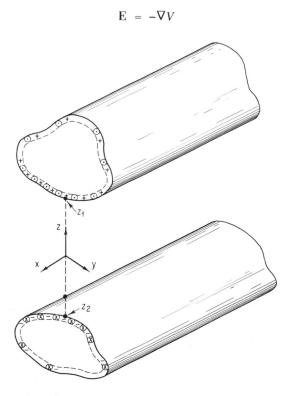

Fig. 8-19. Two long parallel wires of arbitrary cross section.

---

*There is one implicit assumption, namely, that for applied sinusoids, the cross-sectional dimentions are small compared to the wavelength of the highest frequency of interest, so that wave propagation in the cross-sectional plane need not be considered. But this is usually true for practical transmission lines of interest.

If we take the divergence of both sides, the above becomes

$$\nabla \cdot \mathbf{E} = -\nabla \cdot \nabla V = -\nabla^2 V \tag{8-61}$$

But $\nabla \cdot \mathbf{E} = \rho_v/\epsilon$. However, in the dielectric, we assume that there are no charges; thus, $\rho_v = 0$, which gives

$$\nabla^2 V = 0$$

This is the well known Laplace equation. To find V and $\mathbf{E}$ would require solution of the differential equation with arbitrary constants which would be determined by the boundary conditions. Rather than solving this, we will show that the magnetic field $\mathbf{H}$ can be determined by a similar equation and that matching boundary conditions will show that $\mathbf{E}$ and $\mathbf{H}$ are always orthogonal and related to each other in amplitude by a constant factor at each point in the xy plane. From this fact and the definition of L and C, the reciprocity is easily proved.

To do this, it is expedient to employ the concept of magnetic vector potential $\mathbf{A}$. It is well known that

$$\nabla \times \mathbf{A} = \mathbf{B} = \mu \mathbf{H} \tag{8-62}$$

Taking the curl of both sides of the above equation yields

$$\nabla \times \nabla \times \mathbf{A} = -\nabla^2 \mathbf{A} = \mu \nabla \times \mathbf{H} \tag{8-63}$$

It is also known that

$$\nabla \times \mathbf{H} = \mathbf{J} + \frac{\partial \mathbf{D}}{\partial t} \tag{8-64}$$

Since only the fields within the dielectric are of interest, the current density $\mathbf{J}$ is zero. It is desirable and, in fact, necessary to show that the reciprocity is valid for both ac and dc. To do this, it is necessary to assume that in Fig. 8-19 there are only y and z components of $\mathbf{E}$ and $\mathbf{H}$, that the current is entirely on the conductor surface, so that there is no $\mathbf{H}$ internal to the conductors (no internal impedance), and that there are no variations of these quantities with respect to x for the incremental distance $\Delta \ell$ of interest.

If Eq. (8-64) is expressed in its component parts, it gives (using standard vector notation)

$$i\left(\frac{\partial H_z}{\partial y} - \frac{\partial H_y}{\partial z}\right) + j\left(\frac{\partial H_x}{\partial z} - \frac{\partial H_z}{\partial x}\right) + k\left(\frac{\partial H_y}{\partial x} - \frac{\partial H_x}{\partial y}\right)$$

$$= i\frac{\partial D_x}{\partial t} + j\frac{\partial D_y}{\partial t} + k\frac{\partial D_z}{\partial t} \tag{8-65}$$

(i, j, and k are unit vectors along x, y, and z axes). Since it is assumed that $E_x = 0$, then the first term on the right and therefore that also on the left side of Eq. (8-65) must be zero. Also, since $H_x = 0$ and there is no variation of the field quantities in the x direction, the second and third terms (within parentheses) on the left side must also be zero. But D_y and D_z are assumed to change sinusoidally with time. However, if the frequency is not too high, these time derivatives will be negligibly small so that no inconsistency can result. This is equivalent to assuming a lumped-circuit approximation to a distributed network, just as was done in Sec. 4.5 and was shown to be valid for sufficiently low frequencies. Thus, for ac cases with sufficiently low frequency or sufficiently small $\Delta\ell$,* $\nabla \times H = 0$ and the vector potential satisfies Laplace's equation

$$\nabla^2 A = 0 \tag{8-66}$$

Thus, both the scalar and vector potential V and A_x satisfy the same equation for the system. It is now necessary to show that they both also satisfy the same boundary conditions.

It is possible to express the y and z components of H in terms of A from Eq. (8-62)

$$H_y = \frac{1}{\mu}\frac{\partial A_x}{\partial z} \qquad H_z = -\frac{1}{\mu}\frac{\partial A_x}{\partial y} \tag{8-67}$$

The y and z components of E in terms of V are easily obtained from Eq. (8-60)

$$E_y = -\frac{\partial V}{\partial y} \qquad E_z = -\frac{\partial V}{\partial z} \tag{8-68}$$

---

*Since $\Delta\ell$ can always be made smaller, the conclusion is true in general, provided that the original assumptions are valid.

Since it was assumed that the wires of Fig. 9-18 are good conductors, then the divergence equation $\nabla \cdot \mathbf{B} = 0$ indicates that there can be only tangential components of $\mathbf{H}$ at the wire surfaces. Such being the case, Eq. (8-67) shows that there can be no variation of A_x around the periphery of the wires or that the wires must be an equal magnetic potential surface. In a similar manner, the divergence equation $\nabla \cdot \mathbf{D} = \rho_v$ and the high conductivity requires that there be only normal components of $\mathbf{E}$ at the conductor surfaces. From Eqs. (8-68), the wires must be equal electric potential surfaces, as is well known. Thus, since V and A_x satisfy the same equation and boundary conditions, they must have exactly the same functional dependence on y and z, and at most, can differ only by a constant of proportionality. It is therefore apparent from Eqs. (8-67) and (8-68) that

$$\mathbf{E}_y = K\mathbf{H}_z \qquad\qquad \mathbf{E}_z = -K\mathbf{H}_y \qquad\qquad (8\text{-}69)$$

where K is the proportionality constant.

In order to demonstrate that C and L are reciprocal quantities, it is necessary to express C in terms of E and L in terms of H. Capacitance is defined as charge per unit voltage or q/V. Since there will be only surface charge density on the conductors, Gauss' law gives

$$q = \int \epsilon \mathbf{E}_n \cdot ds$$

where $ds = \Delta\ell\, dp$ is the surface element, dp is the elemental distance along the perimeter of the conductor, and $\mathbf{E}_n$ is the field normal to ds. The voltage can be expressed in terms of any path: if we choose the z axis

$$V = -\int_{z_1}^{z_2} \mathbf{E}_z\, dz$$

so that the capacitance per unit length becomes

$$C = \frac{1}{\Delta\ell} \cdot \frac{q}{V} = \frac{\epsilon \oint \mathbf{E}_n\, dp}{-\int_{z_1}^{z_2} \mathbf{E}_z\, dz} \qquad\qquad (8\text{-}70)$$

The inductance is defined as flux per unit current or ϕ/I. The flux can be evaluated by integrating the flux density over any area; thus, if we choose the xz plane as the area

$$\phi = \mu \int_{z_1}^{z_2} H_y \, \Delta\ell \, dz \qquad (8\text{-}71)$$

The current can be evaluated from Ampere's circuital law

$$I = \oint H_t \cdot dp \qquad (8\text{-}72)$$

where H_t is the tangential field at either conductor surface. Thus, the inductance per unit length becomes

$$L = \frac{1}{\Delta\ell} \frac{\phi}{I} = \frac{\mu \int_{z_1}^{z_2} H_y \, dz}{\oint H_t \cdot dp} \qquad (8\text{-}73)$$

Since E and H must be everywhere orthogonal to each other, it is easily seen that substitution of Eq. (8-69) into Eq. (8-70) gives

$$C = \frac{\epsilon K \oint H_t \cdot dp}{K \int_{z_1}^{z_2} H_y \, dz} \qquad (8\text{-}74)$$

The above is seen to be identical to the reciprocal of Eq. (8-73) except for a constant and it is obvious therefore that

$$C = \frac{\mu\epsilon}{L} \qquad (8\text{-}75)$$

The characteristic impedance is obviously

$$Z_0 = \left(\frac{L}{C}\right)^{1/2} = \frac{(\mu\epsilon)^{1/2}}{C} = \frac{L}{(\mu\epsilon)^{1/2}} \qquad (8\text{-}76)$$

This reciprocity could have been shown by a different technique involving the use of complex variables. The general technique (excluding details) is as follows. For static electric fields, $\mathbf{E} = -\nabla V$, since $\mathbf{E}$ is a conservative field satisfying $\oint \mathbf{E} \cdot \mathbf{dl} = 0$. From Ampere's law, $\oint \mathbf{H} \cdot \mathbf{dl} = I$, and H is, in general, not a conservative field. However, in the dielectric where no conduction current is present, $\oint \mathbf{H} \cdot \mathbf{dl} = 0$; thus, it is possible to represent $\mathbf{H}$ as the gradient of some scalar magnetic potential U, that is, $\mathbf{H} = -\nabla U$. Both $U(x, y)$ and $V(x, y)$ are harmonic functions since they have continuous partial derivatives of the second order and satisfy Laplace's equation. (In the actual geometry of Fig. 8-19, x, y would actually be y, z but we are following standard complex variable notation.) Furthermore, if $U + iV$ is an analytic function of $z = x + iy$, then U and V are conjugate harmonic functions ([3, p. 139]). It can be shown that this is true, so that the equipotential curves of $U = $ constant are everywhere orthogonal to the curves of $V = $ constant. Thus, $U = $ constant must also be the electric flux lines (except for a constant of proportionality) and $V = $ constant must also be the magnetic flux lines except for a constant. From this it follows that L and C are reciprocally related to each other. Note that a function $f(z)$ is analytic at a point if its derivative exists at that point and at every point in some neighborhood of that point; it is analytic in a region if it is analytic at every point of that region or, alternatively, the function $f(z) = U + iV$ is analytic at every point of the two-dimensional region if $U(x, y)$ and $V(x, y)$ together with their partial derivatives of the first order are continuous and single-valued and satisfy the Cauchy-Riemann conditions

$$\frac{\partial U}{\partial x} = \frac{\partial V}{\partial y} \quad \text{and} \quad \frac{\partial U}{\partial y} = -\frac{\partial V}{\partial x}$$

APPENDIX 8B

DEFINITION OF ELLIPTIC, BESSEL, AND JACOBIAN FUNCTIONS USED IN LINE PARAMETER CALCULATIONS

$K \equiv K(k) = $ complete elliptic integral of first kind of modulus k

$K' \equiv K(k') = $ complete integral of first kind of complementary modulus k'

$k' = \sqrt{1 - k^2} = $ complementary modulus

$E \equiv E(k) = $ complete elliptic integral of the second kind

$E' \equiv E(k') = $ complete integral of the second kind of complementary modulus k'

$F(k, \phi)$ = incomplete elliptic integral of first kind of amplitude ϕ and modulus k

$E(k, \phi)$ = incomplete elliptic integral of second kind*

$I_m(bR), K_m(bR)$ = modified Bessel functions of first and second kind, respectively, of order m and argument bR

$I'_m(bR), K'_m(bR)$ = first derivatives with respect to the argument of the above Bessel functions

$cn(w) = cn(k, w)$ ⎫ Jacobian elliptic functions*† (trigonometric type of
$dn(w) = dn(k, w)$ ⎬ modulus k (sometimes omitted); w can be complex,
$sn(w) = sn(k, w)$ ⎭ that is, $w = K + jv$
$zn(w) = zn(k, w)$ = Jacobian zeta function†

Jacobian functions are often required in the evaluation of elliptic integrals. Each elliptic function is the inverse of an elliptic integral and, in some respects, elliptic functions are similar to the trigonometric functions, except that they are doubly periodic, having a real and an imaginary period.

PROBLEMS

8-1. Show that the magnetic field between the conductors of a strip line with a large width-to-separation ratio, e.g., Fig. 8-3, is given approximately by

$$H \text{ (oersted)} = 0.5 \, I \text{ (amperes)} / W \text{ (inches)}$$

8-2. Find the capacitance per unit length and Z_0 of an infinitely long strip line of width $W = 0.5$ mm suspended in air a distance 0.25 mm above an infinite ground plane.

Answer: $C = 35.4 \, \mu\mu F/m$, $Z_0 \approx 94$ ohms.

8-3. Show that for a very long, thin, single-strip conductor of width W carrying a current I, with origin as below, the horizontal magnetic field at any point is given by

$$H_y = \frac{I}{2\pi W} \arctan \left\{ \frac{zW}{z^2 + y^2 - \left(\dfrac{W}{2}\right)^2} \right\} \quad \text{RMKS}$$

*See [7, p. 168], [2], or [15].
†See [8], [10, pp. 92, 98], [22], or [26, p. 51].

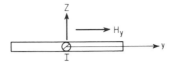

Fig. 8P-1.

Note: For a strip line (two conductors), superimpose the individual fields.

8-4. For the above problem, show that the vertical field at any point is given by

$$H_z = \frac{I}{4\pi W} \left\{ \ln\left[\left(y - \frac{W}{2} \right)^2 + z^2 \right] - \ln\left[\left(y + \frac{W}{2} \right)^2 + z^2 \right] \right\} \quad \text{RMKS}$$

8-5. A strip line for use in a memory array consists of two copper conductors of equal width $W = 0.03$ in., equal thickness 0.001 in., separated 0.005 in. by a dielectric of $\epsilon_r = 4$ (epoxy glass) which fills all space, and a line length of 20 in. A narrow pulse of 5 ns rise time and amplitude V_0 is applied at one end; find the approximate amplitude on the other end assuming perfect termination.

   Answer: $0.98\ V_0$, that is, from Figs. 5-19 and 4-7(a) $\delta \gg$ thickness, so that series resistance is dc value at all frequencies (which is 0.9 ohms ($25°$C)). $\sqrt{L/C} = 26$ ohms (Fig. 8-8), $\alpha\ell = 0.0173$.

8-6. Show that for symmetrical and identical lines in Fig. 8-16(a) with large r/S, when S/H is also large so that the coupling is small, the coupling coefficient is given by

$$k = \frac{1}{2} \frac{S^2}{H^2} \frac{1}{\ln(S/H)} = \frac{60}{Z_{01}} \sqrt{\frac{\mu_r}{\epsilon_r}} \frac{S^2}{H^2}$$

where Z_{01} is the characteristic impedance of either line by itself and a dielectric of μ_r and ϵ_r fills all space.

8-7. In the above problem, show that k can be expressed as

$$k = \frac{60}{Z_{0a}} \frac{S^2}{H^2}$$

where Z_{0a} the characteristic impedance of either line by itself in air, regardless of what the actual dielectric might be. Note: This shows that k does not depend on the dielectric provided that it fills all space.

8-8. Prove that with two identical coupled lines as in Fig. 8-16(a) with small conductor diameters

$$C_{22} = \frac{C_2}{1 - k^2}$$

where C_2 is capacitance of line 2 by itself; prove for the same case that

$$Z_{022} = Z_{02}\sqrt{1 - k^2}$$

where Z_{02} is the impedance of line 2 by itself (in the absence of line 1).

REFERENCES

1. Black, K. G., and T. J. Higgins: Rigorous Determination of the Parameters of Microstrip Transmission Lines, *IRE Trans. Microwave Theory Tech.,* vol. MTT3, p. 93, March 1955.
2. Byrd, P. F., and M. D. Friedman: "Handbook of Elliptic Integrals for Engineers and Physicists," Springer-Verlag, Berlin, 1954.
3. Churchill, R. V.: "Introduction to Complex Variables and Applications," McGraw-Hill Book Company, New York, 1948.
4. Cohn, S.: Characteristic Impedance of the Shielded-Strip Transmission Line, *IRE Trans. Microwave Theory Tech.,* vol. MTT2, p. 52, July 1954.
5. *Ibid.,* Problems in Strip Transmission Lines, vol. 3, p. 119, March 1955.
6. *Ibid.,* Shielded Coupled-strip Transmission Line, p. 29, October 1955.
7. Dwight, H. B.: "Tables of Integrals and Other Mathematical Data," 3rd ed., The Macmillan Company, New York, 1957.
8. Fettis, H. E., and J. C. Caslin: "Ten Place Tables of Jacobian Elliptic Functions," Applied Mathematics Research Laboratory, Wright Patterson Air Force Base, Ohio, September 1965.
9. Gruner, L.: Higher Order Modes in Rectangular Coaxial Waveguides, *IEEE Trans. Microwave Theory Tech.,* p. 483, August 1967.
10. Jahnke, E., and F. Emde: "Tables of Functions," 4th ed., Dover Publications, Inc., New York, 1945.

11. Jahnke, E., F. Emde, and F. Losch: "Tables of Higher Functions," McGraw-Hill Book Company, New York, 1960.
12. Jordan, E. C.: "Electromagnetic Waves and Radiating Systems," Prentice-Hall, Inc., Englewood Cliffs, N. J., 1950, 1955.
13. Kaupp, H. R.: Characteristics of Microstrip Transmission Lines, *IEEE Trans. Electron. Computers,* vol. EC-16, no. 2, p. 185, April 1967.
14. Kraus, J. D.: "Electromagnetics," McGraw-Hill Book Company, New York, 1953.
15. Neville, E. H.: "Jacobian Elliptic Functions," Oxford University Press, London, 1951.
16. Oberhettinger, F., and W. Magnus: "Amvending der Elliptischen Functionen in Physik and Technik," Springer-Verlag, Berlin, 1949.
17. Oliner, A. A.: Proc. Symp. Mod. Advances Microwave Tech., Polytechnic Inst. Brooklyn, p. 379, November 1954.
18. Palmer, H.: The Capacitance of a Parallel-plate Capacitor by the Schwartz-Christoffel Transformation, *AIEE Trans.,* vol. 56, p. 363, March 1937.
19. Primozich, F. G.: "Millimicrosecond Pulse Studies—Engineering Design of Tapered Transmission Lines," doctoral dissertation, Carnegie Inst. of Tech., Pittsburgh, Pa., 1954.
20. Primozich, F. G., E. R. Schatz, and J. B. Woodford: A Tapered Strip Transmission Line for Pulse Transformer Service, *AIEE Trans. Commun. Electron.,* pt. I, p. 158, vol. 74, no. 18, May 1955.
21. Ramo, S., and J. Whinnery: "Fields and Waves in Modern Radio," John Wiley & Sons, Inc., New York, 1958.
22. Schuler, M., and H. Gebelein: "Eight and Nine Place Tables of Elliptical Functions," Springer-Verlag, Berlin, 1955.
23. Seckelmann, R.: On Measurements of Microstrip Properties, *Microwave,* p. 61, January 1968.
24. Shelton, J. P., Jr.: Impedances of Offset Parallel-Coupled Strip Transmission Lines, *IEEE Trans. Microwave Theory Tech.,* vol. MTT-14, no. 1, p. 7, January 1966.
25. Smythe, W. R.: "Static and Dynamic Electricity," McGraw-Hill Book Company, New York, problem 35, p. 105, 1950.
26. Sokolnikoff, I. S., and E. S. Sokolnikoff: "Higher Mathematics for Engineers and Physicists," 2nd ed., McGraw-Hill Book Company, New York, 1941.
27. Wheeler, H. A.: Transmission-Line Properties of Parallel Strips Separated by a Dielectric Sheet, *IEEE Trans. Microwave Theory Tech.,* vol. 13, p. 172, March 1965.

INDEX